Biological and Medical Physics, Biomedical Engineering

For further volumes:
http://www.springer.com/series/3740

The fields of biological and medical physics and biomedical engineering are broad, multidisciplinary and dynamic. They lie at the crossroads of frontier research in physics, biology, chemistry, and medicine. The Biological and Medical Physics, Biomedical Engineering Series is intended to be comprehensive, covering abroad range of topics important to the study of the physical, chemical and biological sciences. Its goal is to provide scientists and engineers with textbooks, monographs, and reference works to address the growing need for information.

Books in the series emphasize established and emergent areas of science including molecular, membrane, and mathematical biophysics; photosynthetic energy harvesting and conversion; information processing; physical principles of genetics; sensory communications; automata networks, neural networks, and cellular automata. Equally important will be coverage of applied aspects of biological and medical physics and biomedical engineering such as molecular electronic components and devices, biosensors, medicine, imaging, physical principles of renewable energy production, advanced prostheses, and environmental control and engineering.

J. Michael Köhler · Brian P. Cahill
Editors

Micro-Segmented Flow

Applications in Chemistry and Biology

Springer

Editors
J. Michael Köhler
Institute of Chemistry and Biotechnology
Technical University Ilmenau
Ilmenau
Germany

Brian P. Cahill
Institute for Bioprocessing
and Analytical Measurement Techniques
Heilbad Heiligenstadt
Germany

ISSN 1618-7210
ISBN 978-3-642-38779-1 ISBN 978-3-642-38780-7 (eBook)
DOI 10.1007/978-3-642-38780-7
Springer Heidelberg New York Dordrecht London

Library of Congress Control Number: 2013950741

Printed on acid-free paper

Springer is part of Springer Science+Business Media (www.springer.com)

Preface

During the last dozen years, droplet-based microfluidics and the technique of micro-segmented flow have been evolving into a key strategy for lab-on-a-chip devices as well as for micro-reaction technology. The unique features and advantages of these technologies with regard to the generation and manipulation of small liquid portions in microsystems have attracted widespread attention from scientists and engineers and promise a large spectrum of new applications. The steep increase of scientific interest in the field corresponds to a quickly rising number of publications and to the increasing importance of the field for numerous scientific conferences. Among them, the CBM workshop on miniaturized techniques in chemical and biological laboratories has dealt with droplet-based methods and micro-segmented flow since 2002. In particular, the sixth workshop—held in Elgersburg/Germany in March 2012—focussed on recent developments in micro-segmented flow. This meeting highlighted the progress of the field over the past few years and reflected a well-developed state in the understanding of droplet-based microfluidics, segment operations, in the development and manufacture of devices and in their applications in chemistry and biotechnology. The focus of the meeting on the state-of-the-art in research and development in the science, technology and application of micro-segmented flow proved an opportune occasion for a summarizing description of the main aspects of *Micro-Segmented Flow* in the form of this book.

The authors and editors of this book understand their writing as a mission for giving a representative overview of the principles and basics of micro-segmented flow as well as a description of the huge number of possibilities for processing micro-fluid segments and their applications in chemistry, material sciences as well as in biomedicine, environmental monitoring, and biotechnology. So, the book is divided into three parts: the first part introduces the fascinating world of droplet and segment manipulation. The described methods range from droplet handling by surface forces and light to electrical switching and chip-integrated systems and to sensing of the presence and content of micro-fluid segments. In the second part, the application of micro-segmented flow in the synthesis and operation of micro and nanoparticles is chosen as a typical example of taking advantage of micro-fluid segments in chemical technology. Beside the large spectrum of applications in the preparation of new and homogeneous materials, the potential of micro-segmented flow for the screening of nanoparticle compositions, shapes, and sizes by combinatorial synthesis is shown by the example of plasmonic nanoparticles and the tuning of their optical properties. Finally in the third part, two important aspects of miniaturized cell cultivation and screenings have been selected for demonstrating the power of micro-segmented flow in biological applications. In both of these chapters, the use of micro-segmented flow for the determination of highly resolved dose/response functions for toxicology, for the characterization of combinatorial effects in two- and three-dimensional concentration spaces and for the application of droplet-based methods and micro segmented flow in the search for new antibiotics are reported.

All authors are active researchers in the field of micro-segmented flow. The chapters follow the concept of connecting a review-like overview of the specific topics with a report on recent examples of the researcher's own research. So, it is expected that the reader will find a very informative survey of the most important aspects and an authentic introduction into the fastly developing and fascinating world of segmented-flow microfluidics.

Ilmenau, April 2013

Contents

Contributors

David William Agar Department of Biochemical and Chemical Engineering, Technical University of Dortmund, Emil-Figge-Straße 66, 044227 Dortmund, Germany, e-mail: agar@bci.tu-dortmund.de

Charles N. Baroud Laboratoire d'Hydrodynamique (LadHyX), Ecole Polytechnique, 91128 Palaiseau cedex, France, e-mail: baroud@ladhyx.polytechnique.fr

Matthias Budden Institute of Chemistry and Biotechnology, Ilmenau University of Technology, PF 10 05 65, 98684 Ilmenau, Germany , e-mail: matthias.budden@tu-ilmenau.de

Brian P. Cahill Institute of Chemistry and Biotechnology, Technical University Ilmenau, Weimarer Str. 32, 98693 Ilmenau, Germany; Institute for Bioprocessing and Analytical Measurement Techniques, Rosenhof, 37308 Heilbad Heiligenstadt, Germany, e-mail: brian-patrick.cahill@tu-ilmenau.de

J. Cao Institute of Chemistry and Biotechnology, Ilmenau University of Technology, PF 10 05 65, 98684 Ilmenau, Germany, e-mail: jialan.cao@tu-ilmenau.de

Lars Dittrich Department Micromechanical Systems, Technische Universität Ilmenau, PF 10 05 65, 98684 Ilmenau, Germany , e-mail: lars.dittrich@tu-ilmenau.de

A. Funfak Institute of Chemistry and Biotechnology, Ilmenau University of Technology, PF 10 05 65, 98684 Ilmenau, Germany

Gunter Gastrock Institute for Bioprocessing and Analytical Measurement Techniques, Rosenhof, 37308 Heilbad Heiligenstadt, Germany

Martin Hoffmann Department Micromechanical Systems, Technische Universität Ilmenau, PF 10 05 65, 98684 Ilmenau, Germany

Andrea Knauer Institute of Chemistry and Biotechnology, Ilmenau University of Technology, PF 10 05 65, 98684 Ilmenau, Germany, e-mail: andrea.knauer@tu-ilmenau.de

J. Michael Köhler Institute of Chemistry and Biotechnology, Ilmenau University of Technology, PF 10 05 65, 98684 Ilmenau, Germany

D. Kürsten Institute of Chemistry and Biotechnology, Ilmenau University of Technology, PF 10 05 65, 98684 Ilmenau, Germany

Karin Martin Leibniz Institute for Natural Product Research and Infection Biology e.V. Hans-Knöll-Institute (HKI), Beutenbergstraße 11a, 07745 Jena, Germany

Thomas Nacke Institute for Bioprocessing and Analytical Measurement Techniques, Rosenhof, 37308 Heilbad Heiligenstadt, Germany

Martin Roth Leibniz Institute for Natural Product Research and Infection Biology e.V. Hans-Knöll-Institute (HKI), Beutenbergstraße 11a, 07745 Jena, Germany

Frederik Scheiff Department of Biochemical and Chemical Engineering, Technical University of Dortmund, Emil-Figge-Straße 66, 044227 Dortmund, Germany, e-mail: frederik.scheiff@bci.uni-dortmund.de

Joerg Schemberg Institute for Bioprocessing and Analytical Measurement Techniques, Rosenhof, 37308 Heilbad Heiligenstadt, Germany

Steffen Schneider Institute of Chemistry and Biotechnology, Ilmenau University of Technology, PF 10 05 65, 98684 Ilmenau, Germany

Miguel Tovar Leibniz Institute for Natural Product Research and Infection Biology e.V. Hans-Knöll-Institute (HKI), Beutenbergstraße 11a, 07745 Jena, Germany

Emerson Zang Leibniz Institute for Natural Product Research and Infection Biology e.V. Hans-Knöll-Institute (HKI), Beutenbergstraße 11a, 07745 Jena, Germany, e-mail: emerson.zang@hki-jena.de

Variables

A	Surface area
A_p	Projected area
A_0	Area between the S11 and S11 = 0 for a loaded sensor
A_l	Area between the S11 and S11 = 0 for an unloaded sensor
C	Capacitance
C_a	Capillary number
C_w	Wall capacitance
C_m	Medium capacitance
C_p	Parasitic capacitance
E	Surface energy
E_0	Initial surface energy
E_∞	Initial surface energy
F	Force
F_B	Buoyant force
F_d	Drag force
F_{DEP}	Dielectrophoretic force
F_{flow}	Force exerted on one segment
F_x	Force in the x direction
F_W	Weight
$F\gamma$	Maximum force due to the change in surface area
G	Geometrical correction parameter
J	Current
J^e	External current density
Q	Electric charge
Q_j	Current source
Q_0	Quality factor for an unloaded sensor
Q_l	Quality factor for a loaded sensor
R	Droplet radius/particle radius
R_a	Droplet radius (case a)
R_b	Droplet radius (case b)

R_m	Medium resistance
R_e	Reynold's number
S_{11}	Reflection coefficient
$SI_{(S11)}$	Reflection index
U	Voltage
U_{ec}	Electrochemical potential for particle charging
U_{RMS}	RMS voltage
V	Volume
V_s	Volume of sphere
V_{cuboid}	Volume of cuboid
W	Work done by the electric field
W_0	Width at half maximum for an unloaded sensor
W_l	Width at half maximum for a loaded sensor
X	Cell constant
Y	Cell constant
Z	Impedance
$\|Z\|$	Impedance modulus
$\|Z\|_\theta$	Impedance modulus at a particular phase value
a	Channel width
b	Distance between electrodes
c	Hole depth
d	Hole diameter
d_w	Thickness of the wall
e	Elementary charge
f	Frequency
f_θ	Frequency at a particular phase value
f	Resonant frequency for an unloaded sensor
f_l	Resonant frequency for a loaded sensor
f_{CM}	Clausius-Mosotti factor
g	Acceleration due to gravity
h	Channel height
h_s	Cuboid height
k_1	Experimentally determined factor
k_2	Experimentally determined factor
l	Length
m	Mass
n	Unit normal
u	Mean velocity of the outer fluid
u_∞	Free-stream velocity
u_+	Mobility of the positive species
u_-	Mobility of the negative species
z	Number of elementary charges
γ	Surface tension
ε	Dielectric constant

ε_0	Permittivity of free space
ε_r	Relative dielectric constant
ε_w	Dielectric constant of the wall
$\varepsilon_{(\infty)}$	Dielectric constant at high frequency
$\varepsilon_{(0)}$	Dielectric constant at low frequency
ε^*	Complex dielectric constant
ε'	Real part of the complex dielectric constant
ε''	Imaginary part of the complex dielectric constant
κ	Curvature
μ	Viscosity
θ	Phase angle
ζ	Ratio of hole diameter to channel height
ρ	Density
ρ_+	Charge density of the positive ion species
ρ_-	Charge density of the negative ion species
σ	Conductivity
τ	Relaxation time
ξ	Drag coefficient
ω	Radial frequency

Chapter 1
Introduction

Brian P. Cahill

1.1 Micro Segmented Flow: A Challenging and Very Promising Strategy of Microfluidics

When scientists and engineers started to realize the idea of the Lab-on-a-Chip, they followed the vision to transfer the power and success of miniaturized systems from solid state electronics into the world of chemistry and molecular biology. The transport and processing of molecules inside highly integrated networks of fluid channels should be controlled in analogy to the transport of electrons through electronic networks and used for powerful analytical procedures, for molecular information management as well as for the synthesis and optimization of new molecules and molecular nanomachines. But during the research on the realization of complex microfluidic systems and experiments guiding homogeneous fluids through microchannels it became more and more clear that this analogy was wrong, that this vision was a delusion.

But, the wrong analogy was only a partial fallacy. The most powerful basic concept behind miniaturization in solid state electronics as well as behind microfluidics is the functional patterning, the hierarchical subdivision of space. It is the same principle which we always observe in living nature and which creates the huge wealth of shapes and structures at the different size scales in the world of organisms. All of the unbelievable plurality of structures and functions in living beings is based on one absolute undispensible concept: fluidic compartmentalization.

The formation of cells is the most fundamental principle of living nature, and liquid compartmentalization is continued in the internal compartmentalization of eucaryotic cells by cell organelles or by the formation of organs and lumens in the development of multicellular organisms. The separation of a small volume from the

B. P. Cahill (✉)
Institute of Chemistry and Biotechnology, Technical University Ilmenau,
Weimarer Str. 32, 98693 Ilmenau, Germany
e-mail: brian-patrick.cahill@tu-ilmenau.de

J. M. Köhler and B. P. Cahill (eds.), *Micro-Segmented Flow*,
Biological and Medical Physics, Biomedical Engineering,
DOI: 10.1007/978-3-642-38780-7_1, © Springer-Verlag Berlin Heidelberg 2014

environment, the subdivision of space into well-defined small units, the partial decoupling of the cell's internal chemistry from the outside conditions and the variation of chemical and biomolecular processes between these compartments have been the essential preconditions for the evolution of life. The generation of droplets and fluid segments in micro fluidic devices was driven by the search of methods for controlled manipulation of small liquid portions, but was not primary motivated by the analogy of liquid compartmentalization in nature. But, the principle of formation, controlled transport and processing of such liquid compartments is an obvious analogy

This book is dedicated to the principle and application potential of micro segmented flow. The recent state of development of this powerful technique is presented in nine chapters by active researchers in this exciting field. In the first section, the principles of generation and manipulation of micro fluid segments are explained. It gives the fundamentals of the fluidic behaviour of micro droplets and microfluidic segments and explains the possibilities for control and reliable manipulation of the liquid compartments. In the second section, the micro continuous-flow synthesis of different types of nanomaterials is shown as a typical example for the use of advantages of the technique in chemistry. These examples show how the specific advantages of transport conditions in segmented fluids can be used in order to improve the conditions for continuous-flow synthesis procedures and for improving the quality of products. In the third part, the particular importance of the technique of micro segmented flow in biotechnical applications is presented demonstrating the progress for miniaturized cell cultivation processes, for cell biology and diagnostics and sequencing as well as for the development of antibiotics and the evaluation of toxic effects in medicine and environment.

There are three main aspects of the use of micro fluid segments in technology:

1. Process homogenization by control of transport processes and realization of highly reproducible local mass and heat transfer conditions ("fluidically determined process homogenization")
2. Subdivision of process volumes in order to generate high numbers of independent process spaces ("fluidically defined separate micro reactors")
3. Interface management by fluidically controlled interaction of liquid compartments ("fluidically designed interface processes").

The first aspect is mainly used in micro reaction technology. It allows the implementation of micro-continuous flow processes with very high homogeneity. These processes are marked by very high rates of mixing and heat transfer as well as by an ultimate narrow distribution of residence times. In addition, the pattern of fluid motion inside micro fluid segments is reproducible. In consequence, it can be expected, that each volume element experiences the same "process history". This quasi-perfect homogenization of all transport and reaction processes in all volume elements means a ultimate step in the quality of chemical engineering in continuous flow processes.

The second aspect concerns the experimental realization of high, but ordered diversity. This aspect is of large interest for combinatorial processes, screenings, variation and investigation of process parameters and for the realization of two- or higher-dimensional concentration spaces. The automated subdivision, the addressing and separate processing of individual fluid segments is, for example, very promising for combinatorial chemistry, for high-throughput diagnostics, for pharmaceutical screenings and for toxicological investigations.

The third aspect relates to the spatial control of interface management. In contrast to suspensions and emulsions, which consist of statistically distributed volume elements, the micro segmented flow realizes well-defined spatial relations between the single liquid compartments, between different types of liquids and between the liquids and the wall. The ordered processing of fluid segments correlates with an ordered transport and processing of interfaces. This is very important for nearly all types of phase-transfer processes and for operations with micro and nanophases. So, the micro segmented flow is, for example, a very promising tool for the synthesis, modification and manipulation of nanoparticles.

The following chapters will introduce us to the fascinating world of micro droplets and fluid segments, will explain the principles of microfluidic functions, describe designs and realization of fundamental devices and give examples for important applications reaching from inorganic chemistry, over organic materials to biological systems.

Part I
Generation, Manipulation and Characterization of Micro Fluid Segments

Chapter 2
Droplet Microfluidics in Two-Dimensional Channels

Charles N. Baroud

Abstract This chapter presents methods for two-dimensional manipulation of droplets in microchannels. These manipulations allow a wide range of operations to be performed, such as arraying drops in two-dimensions, selecting particular drops from an array, or inducing chemical reactions on demand. The use of the two-dimensional format, by removing the influence of the channel side walls, reduces the interactions between droplets and thus simplifies droplet operations, while making them more robust. Finally, the chapter presents further developments on droplet microfluidics without a mean flow of the outer phase.

2.1 Droplets in Linear Channels and on Two-Dimensional Surfaces

The miniaturization of fluid handling tools is a process that has greatly evolved through a large number of independent routes. From pipetting robots or ink-jet printers that can manipulate sub-microliter volumes at high speed, to the formation and transport of liquid segments in straight tubes, several criteria have been considered for determining the optimal approach. Indeed, the robots provide precise and programmable control of a sequence of individual pipetting events and are therefore conceptually simple to program. In contrast, producing a train of liquid segments in a straight tube requires up-front planning, in order to keep track of where the different segments end up, but yields a robust and contamination-free environment for manipulating very large numbers of individual reactors. This tradeoff between the flexibility of programmable machines and the robustness and speed of confined geometries re-appears in more exotic microfluidic tools. In this context again, two independent approaches have also been proposed based on micro-fabricated devices.

C. N. Baroud (✉)
Laboratoire d'Hydrodynamique (LadHyX), Ecole Polytechnique, 91128 Palaiseau cedex, France
e-mail: baroud@ladhyx.polytechnique.fr

J. M. Köhler and B. P. Cahill (eds.), *Micro-Segmented Flow*,
Biological and Medical Physics, Biomedical Engineering,
DOI: 10.1007/978-3-642-38780-7_2, © Springer-Verlag Berlin Heidelberg 2014

The first approach grew out of the micro-electronics community where a vast knowledge was already available for producing electronic components and managing them. This work has lead to the development of so-called "digital microfluidics", where droplets are produced and manipulated on the surface of a solid substrate. These operations take place through surface stresses, applied for example by an electric field [1, 2], differential heating [3, 4], acoustic waves [5], etc. These stresses have been shown to provide basic operations such as drop production, merging, division, or the mixing of the drop contents. In this technique, the position and movement of each drop can be controlled at every moment, so that the user can program the device operation by software. This implies that a generic chip design can provide any number of different functionalities, with the possibility to the re-program it in real time. However, practical implementations of this platform have typically remained limited to a small number of droplets.

In parallel with advances in digital microfluidics, a large body of work has shown that droplets can be generated and transported at high throughput in microfabricated channels [6–8]. These channels can be produced at much lower cost than surface manipulation chips and they typically operate in a passive way, thus displaying excellent robustness and simple operation procedures, in addition to providing a controlled closed environment within the microchannel. Here again, basic tools have been demonstrated for droplet sorting [9, 10], coalescence [11, 10], mixing [12], as described in several recent review articles that describe the state of the art from the applications or fundamental points of view [13–17].

This comparison between the two approaches yields a panorama that shows that each is suitable for a different kind of experiment and that the overlap between the two is very small. The advantages and disadvantages of each method have lead to application areas that are very different for each of the two approaches: digital methods are well suited for experiments that require a high level of control with a low throughput; they have been applied, for example, for long term incubation of biological samples for cell cultures [18]. In contrast, microchannel methods are suited for statistical studies that require little real-time manipulation but a large number of samples, such as the "digital" Polymerase Chain Reaction (PCR), where an initial sample is divided into a large number of subsamples, such that each droplet contains a single DNA strand, before thermocycling (e. g. [19]).

Recent work however has aimed to bridge the gap between the two approaches, namely by developing ways in which microchannel methods can mimic the functionalities of digital microfluidics methods. This includes, for example, the creation of stationary arrays of droplets within microchannels, for long term incubation and observation, or in order to perform successive operations on these drops. The different approaches have generally relied on the ability to microfabricate fine geometric features within the channels, into which drops can enter but where they get trapped. This allows the drops to be held in known locations, against a mean flow, for long term observation. The use of photo-lithography to make these features allows high levels of parallelization and the implementation of concurrent operations in a large number of independent locations.

Below, we will focus on recent extensions of microchannel techniques that have addressed such possibilities. The chapter begins by describing different approaches that have been proposed, which include quasi-two dimensional and true two-dimensional (2D) devices. Further down, we turn our attention to the use of surface energy gradients for true 2D manipulation in devices with no side walls. First the physical concepts of surface energy and energy gradients are introduced, followed by the practical implementation of "rails and anchors". This is followed by some example realizations that show passive and active 2D droplet manipulation. Finally, the chapter ends with the description of very recent methods to completely remove the need for a mean flow of the carrier phase and a discussion of the possibilities that are afforded by such an approach.

2.2 Generating Droplet Arrays in Microchannels

Several approaches have been proposed to array stationary drops in a microfluidic system, in order to observe their contents for extended periods of time. These designs all rely on using the drops' surface tension, which provides a "handle" to push or hold the fluid segment [20–25]. Indeed, this surface tension resists deformations of the interface and therefore requires a minimum force to be able to squeeze through an aperture. As such, moving a drop from a region of low confinement through a region of high confinement requires the interface to deform and the drop will resist moving through this region. The approaches presented for making the 2D arrays all rely on this principle but differ in the details of how the drops are led to the less confined zones, and how they must squeeze to exit from them.

Several practical principles have guided the published designs, depending on the particular application. In all cases however, the leading desire is to produce a two-dimensional array in order to observe droplets for long periods of time. In this respect, placing the drops in a two-dimensional matrix rather than in a straight line increases the number of drops that can be observed in a given area. Several groups have demonstrated the use of a winding linear channel that is patterned with side pockets where droplets can be held stationary, adapting previous designs that were used to trap solid beads [26], as shown in Fig. 2.1a and b. These devices consist of a linear microchannel in which drops are initially formed and flow in series, so that all of the standard droplet microfluidic methods can be applied to the drops. The trapping region then consists of a specific section in which side pockets are added to the main channel. Droplets occupy them individually or in small groups, until they fill the side pockets. Then later droplets flow past until they reach an unoccupied pocket that they fill. The final result is a channel where the individual pockets are filled with drops from the initial train.

More recently, a more parallelized design was developed by connecting the inlet with the outlet through a large number of parallel channels. Those channels have a variable width, as shown in Fig. 2.1c, so that droplets are held in the wider regions when the flow is stopped or reduced [23]. Having a large number of parallel channels

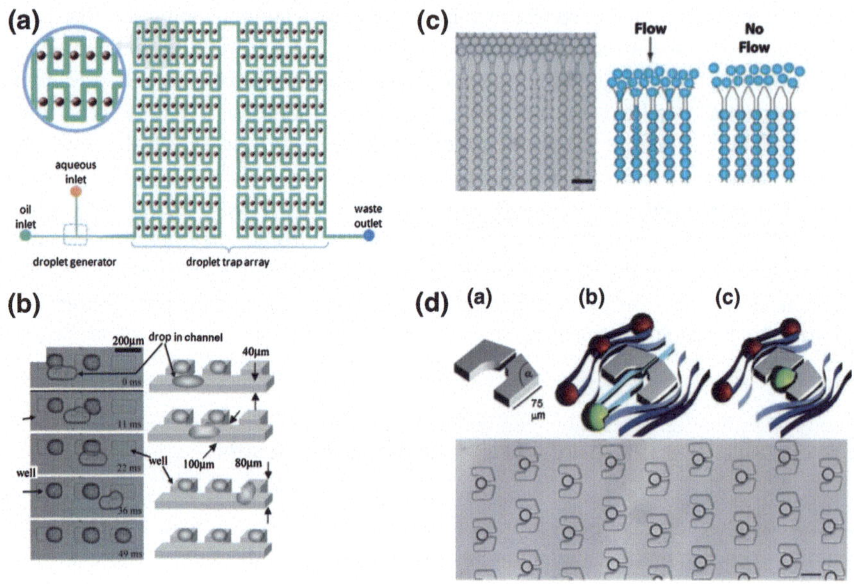

Fig. 2.1 A few trapping devices: **a** from Shi et al. [21]. **b** from Shim et al. [20]. **c** from Schmitz et al. [23]. **d** from Huebner et al. [24]

is favorable to filling them, since once a particular channel fills up its resistance to flow increases, reducing the ability for later drops to enter and leading them to flow into the less occupied channels. Flushing the drops from these channels is however harder for the same reasons: If a channel has been flushed, the carrier phase will preferentially flow through it, leaving the droplets in the full channels unmolested. This device can also be labeled as quasi-2D since droplets interact strongly within each of the individual channels but weakly across channels.

Finally, a truly 2D design was developed by Huebner et al [24], who adapted a design for cell trapping in a microfluidic chamber [27]. In this approach, the side walls are placed very far from the region of interest, such that the droplets are allowed to flow in an open two-dimensional area. This area features micro-fabricated pockets into which a single drop can enter and get trapped. When a drop is trapped it blocks the flow through this region and later drops are directed towards other traps. In this two-dimensional geometry, the behavior of one drop has only a minor influence of others flowing around it and the traps can be filled in a simple and reproducible manner. Moreover, they can be emptied by reversing the flow and the contents can be recovered from the inlet in which they were injected.

These devices all solve the problem of droplet storage in two-dimensional arrays. However they do not offer any method for applying more complex operations, such as guiding the drops into the right spot, selectively removing a single drop, or inducing a droplet fusion for chemical reaction. These operations were treated by later publications [28–30] and will be described below.

2.3 Using Surface Energy Gradients for Droplet Manipulation

The behavior of the devices described above is generally explained by considering the pressure differences in different regions of the microchannels and comparing them with the pressure jump across the drop interface. Indeed, the presence of surface tension introduces a pressure jump (the Laplace pressure jump) across the liquid interface, with the pressure within the droplet being higher than the pressure outside. This pressure jump at every location on the interface is proportional to the surface tension γ and the local mean curvature κ (see [16] for a detailed discussion). Forcing the droplet through an aperture therefore increases the pressure within the drop in the location where it is squeezed since the curvature increases, which requires the external fluid to apply this extra pressure. The drop will therefore remain trapped as long as the external pressure difference is below the pressure necessary to deform the interface.

This reasoning gives a local criterion for the ability to hold a drop stationary, based on the local flow rates and viscosities of the different fluids, as well as on the value of surface tension and constriction size. However, a different way to think about these problems is to consider the surface energy of the drop as it deforms. Indeed, the pressure field arguments above can be reformulated in terms of surface energies and the two approaches are equivalent [16]. The surface energy arguments however provide new insights on droplet manipulation, in particular when considering energy gradients, as discussed below.

Indeed, once a droplet is formed, its volume is essentially fixed (if the effects of dissolution or Ostwald ripening are negligible). However, the surface area of the interface that separates the droplet fluid from the surrounding medium can vary, as the geometry of the drop changes. The minimum surface area for a given volume is a sphere, which is the typical shape of a small, unconfined droplet, and any drop shape that departs from a perfect sphere corresponds to an increase in surface area. Moreover, creating this surface area corresponds to an energetic cost associated with the free energy of the interface, which can be written as

$$E = \gamma A, \tag{2.1}$$

where E is the surface energy, γ is the surface tension, and A is the surface area of the immiscible interface.[1] An increase of surface area therefore leads to an increase of free energy, such that squeezing a droplet between two plates increases the surface energy of the droplet, as shown in Fig. 2.2. This therefore requires a force to be applied by the plates on the droplet.

A more subtle consequence of the dependence of surface energy on confinement occurs when the droplet feels a gradient of confinement. The simplest such gradient is sketched in Fig. 2.3, where a drop can release some of its surface energy by migrating from the left to the right of the confining chamber, as expected intuitively. Indeed, the force that acts on the droplet and generates this motion can be written as the

[1] We will always consider to be constant in this discussion

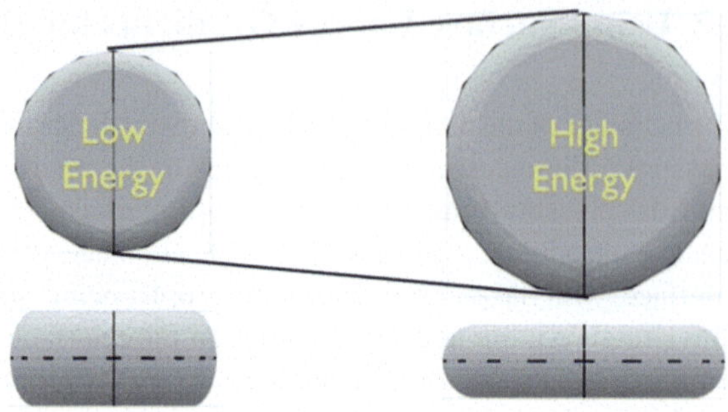

Fig. 2.2 Squeezing a droplet does not change its volume but it does change the surface area of the immiscible interface. Therefore the drop sketched here has a lower surface energy when it is less squeezed (*left*) than when it is more squeezed (*right*)

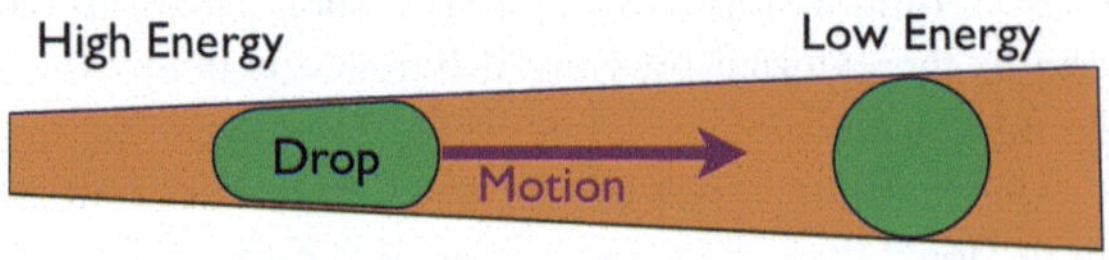

Fig. 2.3 A gradient of confinement corresponds to a gradient of surface energy and therefore a force that pushes the drop from the more to the less confined regions

gradient of the surface energy:

$$\vec{F}_y = -\vec{\nabla} E = -\gamma \vec{\nabla} A \tag{2.2}$$

if the surface tension is considered constant.

This migration of liquid drops as a result of gradients of energy has been known since the 18th century. Indeed, Hauksbee [31] observed that a drop of wetting oil migrates towards the more confined end between two non-parallel plates, thus reducing the total surface energy of the liquid-gas-solid system. This phenomenon has recently been studied in different geometries for both wetting and non-wetting drops and bubbles. For instance, a drop of wetting liquid on a conical needle [32] or inside a pipette cone [33] migrate to minimize the total surface energy, as does non-wetting bubble in a tapered channel [34, 35]. The travel direction for the wetting drop is towards the more confined end, while the non-wetting drop or bubble will migrate towards the less confined end. In the latter case, the migration stops either when the drop reaches a region in which it is no longer confined, i.e. when it becomes spherical, or when it reaches an obstacle that blocks its advance. The first case corresponds to a global energy minimum and the second case to a local energy minimum.

2.4 Rails and Anchors

2.4.1 Principle of Droplet Anchors

In the context of microfluidics, lithography methods allow confinement gradients to be produced locally, for example by etching a small groove in the surface of a microchannel. Therefore if a spherical drop enters a microchannel whose height is smaller than the sphere diameter, the drop must squeeze and thus depart from its spherical shape. The resulting deformation leads to an increase in its interfacial area and a corresponding increase in free energy. By this mechanism, the drop can store and transport this extra energy as it travels down the channel. Given the chance, it will tend to decrease its surface area in order to reduce its free energy. These confinement variations can be in the form of a local indentation in one of the channel boundaries. The simplest implementation is therefore to make a single hole in the surface of the microchannel and to lead the drops to this position with the outer flow. For a wide range of hole and droplet sizes, this leads to the "anchoring" of the drop at the location of the hole, even when the outer phase is flowing past the drop location, as shown in Fig. 2.4.

Moreover, lithographic microfabrication methods allow the production of complex two-dimensional shapes. Therefore etching a linear groove into the surface of an otherwise flat channel can lead the drops to follow the direction of these so-called "rails". When they are directed at an angle compared with the mean flow, the rail can be used to guide the droplets sideways within the wide section, thus allowing droplet guidance in the absence of rigid lateral walls. As an example, a sinusoidal rail is shown in Fig. 2.5, where drops are shown to follow the wavy path imposed by this etched structure. Naturally, any shape can be drawn in order to create rails of simpler or more complex structures, as we shall see in later sections.

The surface energy landscape for a traveling droplet can therefore be fashioned with a complex pattern of energy barriers and wells which then lead the drop to follow the path of least resistance. In the case of dilute flux of drops, predicting the motion of each of them is relatively straightforward, as we shall see in the sections below. This can be understood by first calculating the magnitude of the forces that come into play on the droplets.

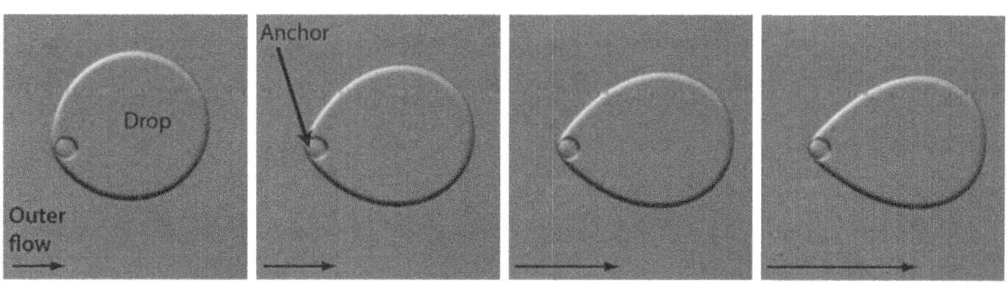

Fig. 2.4 A drop that is initially squeezed between the top and bottom surfaces of a Hele-Shaw cell will be attracted to the location of a local hole in the microchannel wall. As the strength of the outer fluid flow is increased, the drop can deform without detaching from the anchor

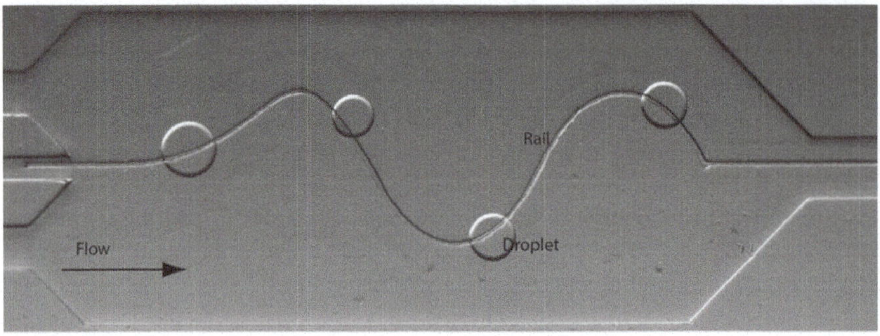

Fig. 2.5 Drops enter into a wide region which is patterned with a sinusoidal rail. The flow is from *left* to *right*. The drops follow the trajectories imposed by the rail

2.4.2 The Anchor Strength

Before tuning to practical applications of the rails and anchors, we will first consider the physical parameters that determine the anchor behavior and the strength of a particular anchor. The detailed discussion of the anchor strength is described in Refs. [30, 36]. Here we will limit ourselves to intuitive physical reasoning. Since the force that attracts a drop to an anchor is due to the gradient of surface energy, estimating it must begin by calculating the energy difference between the situation when the drop is above the anchor or far away, as shown in Fig. 2.6. In the case of an axisymmetric droplet confined between the top and the bottom planes, the shape of the two droplets (a) and (b) of Fig. 2.6 can be obtained from geometric considerations [37].

For the case sketched here, i.e. that $\varepsilon = h/R_b << 1$ and $\zeta = d/h \leq 2$, the droplet penetrates only slightly into the anchor. This can be understood by recalling that the interface curvature must be equilibrated at every point on the drop surface, except where the droplet is confined by the top and bottom walls. This implies that the interface curvature above the hole must be equal to the curvature on the edge of the droplet. The latter can be estimated as a function of the curvature in the plane $(1/R_b)$ and the curvature in the perpendicular direction $(2/h)$, as $\kappa \approx 2/h + \pi/(4R_b) \approx 2/h$ when $\varepsilon << 1$. Therefore the curvature at any point of the interface must have this value, including above the hole, so that the Laplace pressure is equilibrated inside

Fig. 2.6 Sketch of the droplet above the anchor and far away. The channel height is h, the hole diameter is d and its depth is e

the drop. If we model the interface shape above the hole as a section of a sphere of total curvature κ, we obtain that the radius of the corresponding sphere is $R = 2h$. Therefore the drop will enter into the hole until it reaches this radius and, if the hole diameter is small ($\zeta \leq 2$), the sphere will never touch the bottom of the hole.

In this case, the difference in surface area between the case when the drop is above the hole and the case where it is far away, $\Delta A = A_b - A_a$, can be written as [36]

$$\Delta A = -\left[\frac{\pi}{2}\zeta G(\zeta) + O(\varepsilon)\right] h^2 \tag{2.3}$$

where

$$G(\zeta) \equiv \frac{\zeta}{2} - \frac{4}{3\zeta}\left[1 - \left(1 - \left(\frac{\zeta}{2}\right)^2\right)^{\frac{3}{2}}\right] \tag{2.4}$$

is a parameter that depends only on the microchannel geometry.

The force that attracts the drop to the hole can then be calculated by estimating Eq. 2.2. This requires us first to transform the difference in surface area into a difference in surface energy $\Delta E = \gamma \Delta A$. Then the gradient of surface energy can be estimated by diving ΔE by characteristic length scale over which the drop surface area changes. In the limit of a small hole discussed here, a reasonable length scale is the hole diameter d. This yields the maximum force due to the change in surface area as:

$$F_\gamma \cong \frac{\gamma \Delta A}{d} = \gamma \frac{\pi}{2} h G(\zeta) \tag{2.5}$$

Equation 2.3 shows that the drop surface energy above the anchor is always lower than its surface energy far away, regardless of the geometric details or physical parameters of the fluids. This implies that the presence of a hole will always serve to reduce the surface energy of the droplet, thus applying an attractive force. This can be understood intuitively by considering that, as the drop penetrates into the anchor, its in-plane radius is reduced ($R_b <$ R$_a$) such that the net effect is to reduce the surface area. Nevertheless, this reduction of surface area can be very small in the case of large droplets. However, since it takes place over a small distance as well (small d), it can still lead to a large energy gradient and an equivalently large attractive force towards the anchor position. The net effect is that F_γ does not depend on the drop size. It is completely determined by the channel and hole geometries and by the value of the interfacial tension.

In the presence of a flow of the carrier phase, this anchoring force must balance the drag force due to the external flow in order to keep the drop in place. This drag force can also be estimated from physical arguments. In the geometry considered here, it is dominated by the pressure drag, i.e. the pressure difference between the upstream and downstream directions, applied on the cross-sectional surface of the droplet. A detailed calculation yields the drag force

$$\vec{F}_d = 24\pi \mu u \frac{R_b^2}{h} \tag{2.6}$$

where μ is the viscosity of the outer fluid, u is the mean velocity of the outer fluid, and Rb and h are drop radius and channel height, respectively [36]. Contrary to the gradient force, F_d depends on the viscosity of the carrier phase as well as on the drop radius.

Equating the expressions for F_d and F_γ gives the maximum velocity of the outer fluid for which the anchor is still able to hold the droplet stationary. This velocity can be expressed in terms of the dimensionless Capillary number $Ca = \mu U/\gamma$, which represents the relative strength of the viscous and interfacial effects. The prediction for the citrical value $Ca = Ca^*$ can therefore be written as

$$Ca^* = \alpha G \frac{h^2}{R_b^2} \tag{2.7}$$

where α is a proportionality constant that can be extracted from experiments. This result was validated for different channel and hole geometries, fluid pairs, and drop sizes. The experiments show a very good agreement with the theoretical prediction over two orders of magnitude in Ca^*.

In the case when $\zeta > 2$ the droplet goes into the hole until reaches the bottom. This adds an additional parameter to the geometric analysis, since now the hole depth will play a role in the strength of the anchor. Finally, as the hole size becomes comparable with the droplet size, the "characteristic length scale" that is necessary for the estimate of the energy gradient becomes less obvious. Such analysis has not yet been done but should be possible with some simple physical arguments.

2.4.3 Parking Versus Buffering Modes

In practical applications, multiple drops may be following the inside of the microchannel such that mobile droplets will collide with anchored ones. This raises the issue of droplet interactions that modify the simple force balance. Indeed, when two drops are in contact, the total force acting on the droplet pair is larger than the force acting on a single drop, since the pressure drop across the pair is larger. On the other hand, the anchor strength is not increased, in particular when the anchor is smaller than the drop size, as discussed above. For low flow rates of the carrier phase, the anchoring force is sufficiently strong for the anchored drop to remain stationary even when it is in contact with a second drop. This mode is called the "parking" mode. On the other hand, when the flow rate is increased, a "buffering" of droplets is observed, where each successive drop replaces the previous one by bumping it out of the anchor position. These two modes are illustrated in Fig. 2.7.

The buffering mode exists in a wide range of velocities and is robustly reproducible. As we shall see below, this makes it very useful in order to manipulate

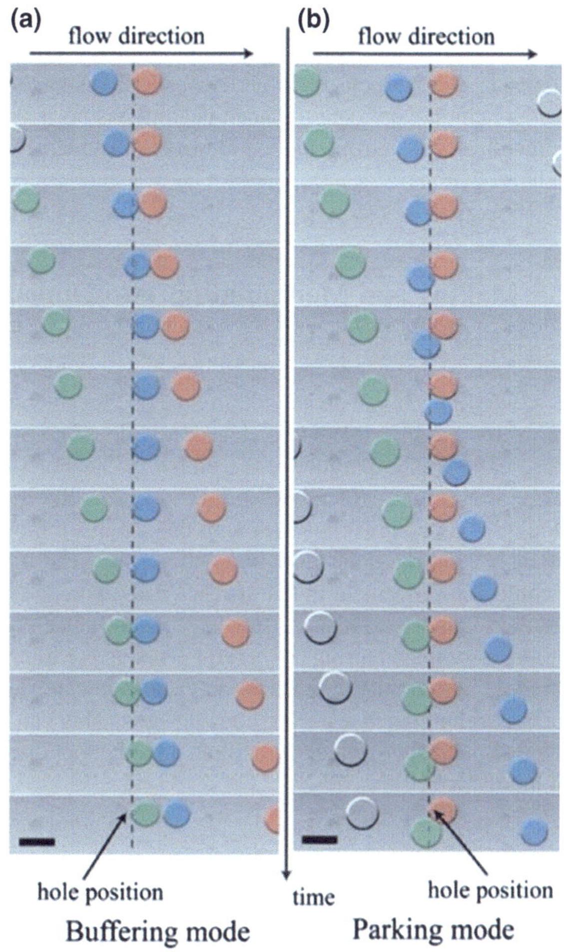

Fig. 2.7 *Two* modes of capture of individual droplets. When the flow rate is very low, a captured drop remains in place even when it is in contact with a different drop (*parking mode*). Alternatively, when the flow rate is increased slightly, each new droplet replaces the stationary drop (*Buffering mode*). Reprinted from Ref. [28]

drops in large arrays and to produce complex sequences of operations that involve filling then emptying anchors. In contrast, the parking mode is useful for long term observations of drop contents.

2.4.4 Forces Due to External Fields

This description of the anchoring mechanism, based on a simple force balance, allows us to imagine the possibility of adding active forcing mechanisms to the system which can be turned on and off as desired. For instance, an external force can be applied by using electric fields [38], focused heating from a laser [10, 39], acoustic fields

[40], etc. The force due to these external fields can then be added to the force balance written here to determine if a droplet remains anchored or if it can be removed on demand. Although electrical and acoustic fields have not yet been demonstrated in such situations, optically-induced heating has been shown to be sufficient to derail or un trap droplets from their equilibrium movement. This will be discussed in detail in later sections.

2.5 Making and Manipulating Two-Dimensional Arrays

The "rails and anchors" approach therefore can be used to guide and trap droplets in a two-dimensional area, without the need for microchannel side walls. Since any geometry can in principle be etched, a wide range of operations can be imagined. The simplest is to parallelize the individual anchor sites, in order to create a large array of droplets. This allows similar functionality to the traps shown by Huebner et al. [24]. In contrast however with those traps, removing the droplets can be done by increasing the outer flow rate without the need to reverse it, providing a slight technical improvement in that respect. The two approaches of pockets and anchors are not mutually exclusive. They have recently been combined together on order to produce complex pairings of several droplets per trapping site [41].

More importantly, the density of drops that can be anchored per unit area can be increased compared with previous designs. This can be achieved by reducing the scale of the device, which is facilitated by the fact that its performance independent of the scale.

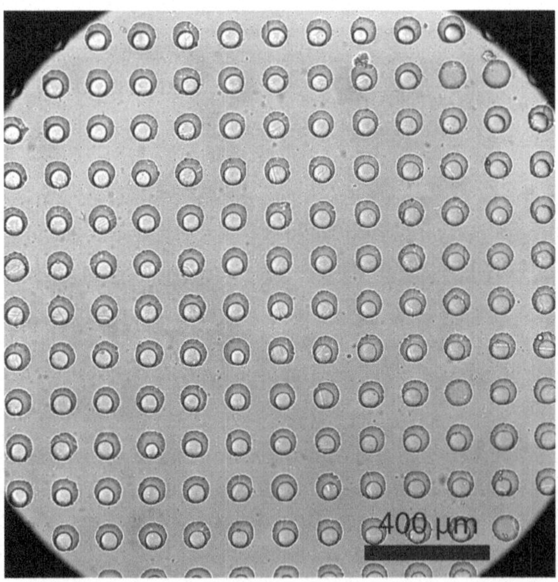

Fig. 2.8 A high-density array containing 5,000 drops/cm^2. The drops are visible in this picture as *small circles* and they are trapped inside the anchors (*the larger circles*). The anchor diameter is 75 μm and they are separated by a distance of 75 μm

Indeed, the critical capillary number Ca^* of Eq. 2.7 is independent of the scale of the problem, since multiplying all lengths by a constant would lead to an unchanged value of Ca^*. In practice, low cost photo-lithography masks can easily produce features around 75 μm. By allowing a distance between anchors equal to the anchor size, in order for the drops to flow between anchors even once an initial drop is stored, we obtain a density or nearly 5,000 drops/cm^2, as shown for instance in Fig. 2.8 which shows a microscope image of droplets trapped in a high density array. A matrix with nearly 10,000 anchors can be filled in under 10 min and the drop contents can be observed for several hours.

2.6 Active Manipulation in Two-Dimensional Geometries

2.6.1 Actuation by Laser Beams

In order to go beyond random filling and emptying of droplet arrays, an active force must be applied to perform operations on demand. A few operations that use the forcing from a focused laser have recently been demonstrated by Fradet et al. [30], who have shown how an active forcing can be combined with the rails and anchors. The general philosophy was to rely on the passive behavior of the microchannel design for most operations and to use the laser forcing as a localized perturbation that pushes the drop past an energetic barrier, from one state to another.

Indeed, such an optical setup had already been used to control droplet motion in linear or quasi-linear channels, in addition to forcing their fusion [10, 39, 41, 42]. Similar operations have also been demonstrated for drops outside microchannels by Faris and coworkers [4, 43]. The mechanism for the laser actuation is through a localized heating of the interface between the droplet and its surroundings, which induces a local depletion of the surfactant and a convection pattern due to spatial variations of surface tension [44]. The advantage offered by the laser heating is that it can be focused anywhere in the field of view of the microscope and that it provides an additional force that can be applied on the droplet. This force then complements the two forces described in Sect. 2.4.2 and can be applied at will.

Three basic operations were demonstrated by Fradet et al [30], by combining passive manipulations of the rails and anchors with an active laser forcing: 1. Selectively removing drops from an array, 2. Determining the anchoring positions of a train of drops, and 3. Inducing a chemical reaction by merging two droplets.

2.6.2 Removing a Drop From an Anchor

The operation of removing a drop from an anchor is the simplest to describe: If we consider that the laser adds a supplementary force to the force balance described in

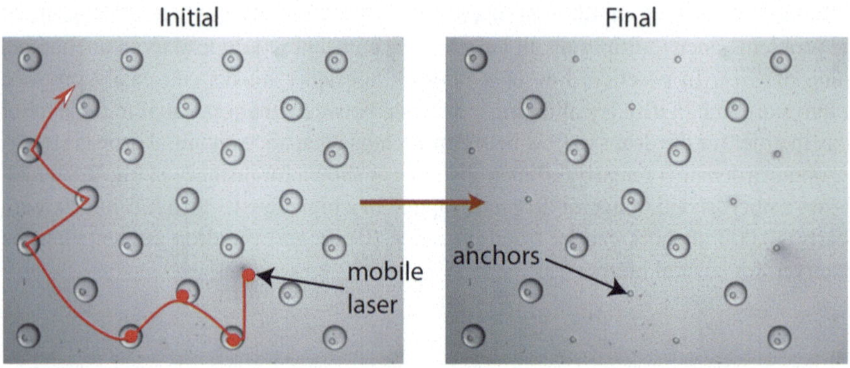

Fig. 2.9 The array is initially randomly filled with droplets. Then placing the laser focus on selected drops untrap them within a few seconds. This can lead to any droplet being removed from the array, for example to keep the X shape in the final state

Sect. 2.4.2, focusing the laser on a particular droplet can be used to remove it from its anchor. This is shown in Fig. 2.9, where a series of drops is initially trapped in a regular array. Then by positioning the laser at particular locations in the array, selected drops can be extracted. These operations can be applied in order to selectively recover certain drops, depending on their contents, or in order to free up certain anchor sites for new drops to be captured.

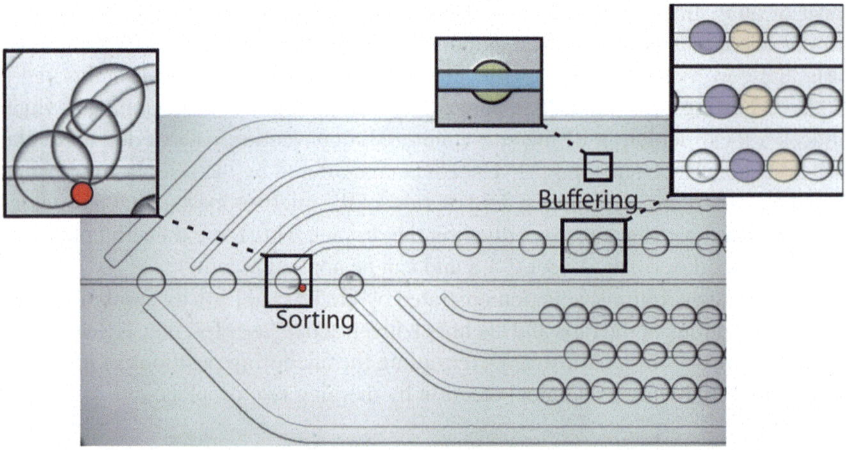

Fig. 2.10 Example of a device that allows the placement of individual drops in particular rails. The device contains a default rail in the center that the drops follow if they are not actively deviated. The storage rails that emanate from this central rail contain anchor sites that block drops in place. Finally, wide gutter rails drain any large drops or bubbles that may appear and protect the region or interest from being damaged

2.6.3 Selectively Filling an Array

Conversely, an array can be selectively filled by placing the desired drops at the desired locations. For this, a device was developed that integrates three basic operations, as shown in Fig. 2.10. First, a default rail goes through the center of the device from left to right. Droplets that enter into the test region follow this rail if they are not deviated by an external force. The second feature consists of a sequence of "storage rails" that begin near the default rail then curve back to the direction of the flow. These rails are patterned with traps along their lengths and serve to store the droplets. Droplets will follow these storage rails of they are deviated by an external force, such as the focused laser. Indeed, although the laser here does not completely derail the droplets, it deflects them sufficiently so that they feel the presence of the side rails, which they then follow as shown in the top-left inset of Fig. 2.10. Finally, large side-rails are placed upstream of the region of interest, in order to drain any large bubbles or drops that may appear into the system. These "gutter rails" were found to be very useful during the device operation, mainly in the periods of flow adjustment when the rates were modified or stopped. During these periods, large droplets can appear in the device and can destroy the matrix of drops that has been produced. For this reason the gutters offer a way to remove these unwanted drops upstream of the region of interest.

An important feature of this device is the relative width of the different structures. A drop that feels the presence of different rails will follow the path of largest energy gradient. Since a narrow rail provides a small energy gain while a wide rail provides a large energy gain, a drop that is near both rails follows the larger rail. Therefore, the storage and gutter rails are made wider than the central default rail, in order to ensure that a drop under the influence of a side-rail always derails from the default rail and follows the side path. Finally, the storage rails contain wide regions that correspond to anchor sites. However, these anchors are made weak enough that a second drop arriving at the location of an anchor will kick the initial droplet out. This leads to successive drops replacing the previous ones through "buffering".

2.6.4 Initiating a Chemical Reaction on Demand by Laser-Controlled Droplet Fusion

Finally, chemical reactions between two droplets can be induced on demand by implementing the time dependent protocol shown in Fig. 2.11. In this device, the anchors were made large enough to fit two drops within a single site, so that the key step in this protocol is to ensure that each anchor site contains a single drop of each of the two reagents (A and B). Once this is achieved, the laser heating is used to induce fusion of the two drops and thus the chemical reaction (shown in Fig. 2.12).

The protocol is as follows: drops containing the two reagents (A and B) are produced from two independent flow-focusing devices [7] upstream of the test region

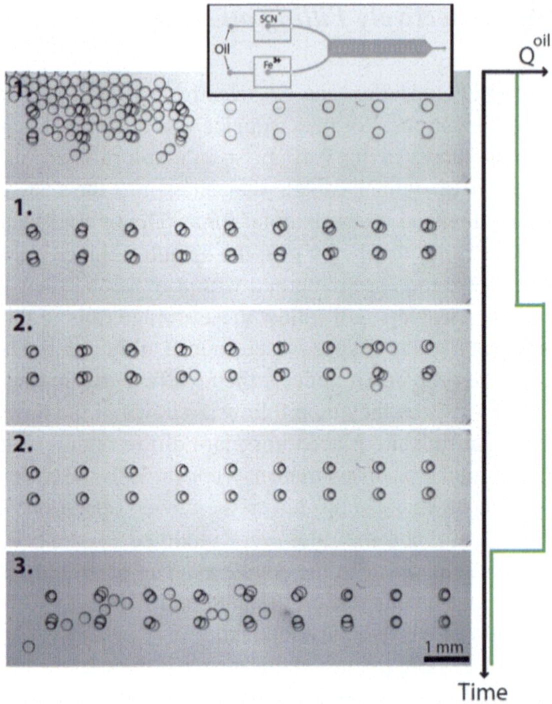

Fig. 2.11 Drops containing SCN^- and drops containing Fe^{3+} are produced at independent flow focusing junctions. They are sequentially flown into the test region that contains a series of large anchors. By modulating the oil flow rate, we ensure that each anchor contains a single drop of *A* and a single drop of *B*. See text for details

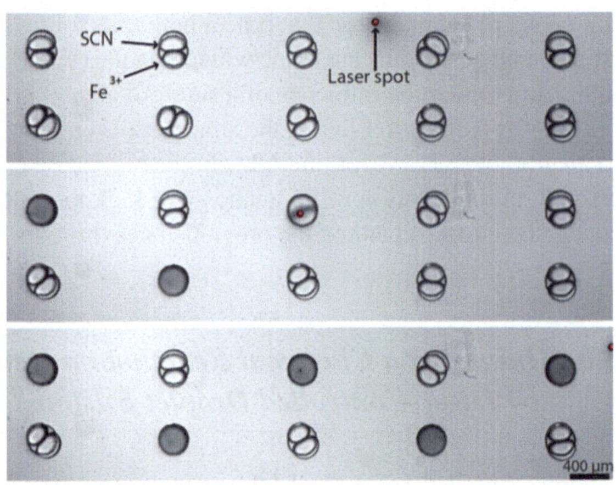

Fig. 2.12 Individual droplet pairs are merged on demand by local heating from a focused laser. The reaction produces a *deep red color*, which is seen here as *dark gray*. The laser path is chosen to produce a "W" pattern

where they meet at anchor sites. First drops containing species A are produced and flow into the test region, as shown in the two top panels of Fig. 2.11 (step 1). In this step, each anchor is filled with two drops of A. Once all of the sites are full, the oil flow rate is increased in order to place the anchors in the buffering mode (step 2), in which they are capable of holding a single drop but not two drops. This leads to one of the two droplets leaving each anchor, which leaves each site containing a single drop of A. At this stage, the oil flow rate is reduced and drops of B are produced in the second flow focusing junction (step 3). They reach the test region where they are held in the empty spots of the different anchors. Note that the simplicity of the physical mechanisms ensures that the operation is well reproducible and that all of the anchors behave in the same manner.

Finally, the chemical reaction can be triggered by merging the drops together. This can be done at once by using an electric field [29], or it can be done on the level of a single pair of drops by using the laser-induced merging. This second method is shown in Fig. 2.12, where the laser path is chosen in a way to merge individual pairs of drops on demand. Once merged, the drops of SCN^- and Fe^{3+} produce $Fe(SCN)_3$ which is deep red, and the color invades the merged droplet.

2.7 Using Surface Energy Gradients Without a Mean Flow

The examples shown above introduce the possibilities that are available when microfluidic devices are designed with height variations. They show that by removing the channel side walls, the coupling between drops in different regions of the channel is reduced, which makes operations on individual droplets easier to implement. These operations can be as simple as holding a drop stationary or increasingly complex by adding further guidance and trapping blocks, more advanced experimental protocols, and active forcing mechanisms to the experimental design. Nevertheless, the presence of a mean flow of the carrier phase, leading from the channel entrance to its exit, leads to some coupling in the droplet behavior since they are all pushed in the main direction. This motivates thinking about microfluidic devices that function without the need for a flow of the carrier phase. In addition to allowing drops to be directed in any direction in a two-dimensional device, removing the mean flow implies that the drop movement is determined locally by the drop's immediate neighborhood and not by the externally imposed mean flow.

Two steps must be added in order to implement flow-less droplet devices: the drop production and their transport . Indeed, the geometries that have become standard for the production of droplets in microchannels all rely on hydrodynamic forces to detach the drops at a channel junction. All three methods (T-junctions [6], flow focusing devices [7], and co-flow devices [45]) allow the production of a well calibrated train of droplets by continuously injecting the droplet and carrier fluids through a well-designed microfluidic geometry. At the junction where the fluids meet, drops detach due to the hydrodynamic forces that are determined by a coupling of the flow rates with the geometric parameters and fluid properties. This fixes the size, volume fraction, transport velocity, and production frequency of the droplets [16, 17, 48]. It is not possible to vary one of these parameters without affecting the others, expect by using active external forcing [10, 46–49].

This contrasts with non-microfluidic methods to produce drops, such as micropipetting or ink-jet printing, where only the dispersed phase is injected and the drop detachment is due to a local loss of equilibrium between the force due to surface tension and a body force: weight or inertia, respectively. The simplicity of the physical mechanisms of these methods yields a high degree of flexibility and stability, which explains their widespread use from the production of a single drop on demand to highly parallel automated platforms. However these techniques cannot be applied at microfluidic scales, since body forces become negligible as the drop size decreases. At microfluidic scales, surface energy effects become dominant. Then recalling the analogy between confinement gradients and gravitational potential energy, one can imagine ways of using these gradients both for the production and transport of droplets.

The production of drops at a step change in the microchannel height has been known for some time, in the form of "step emulsification", which was developed as a microfabricated equivalent of membrane emulsification techniques [50]. Since then, a series of articles (e.g. [51]) have shown that this technique is widely applicable and robust and that it can be used to produce monodisperse drops over a wide range of sizes, dictated mainly by the microchannel and step geometries. More recently, different groups have used steps in their microchannel design in order to produce drops of a well calibrated size [52, 53], usually combining them with the classical T-junction of flow-focusing junction. Although most previous studies have used a

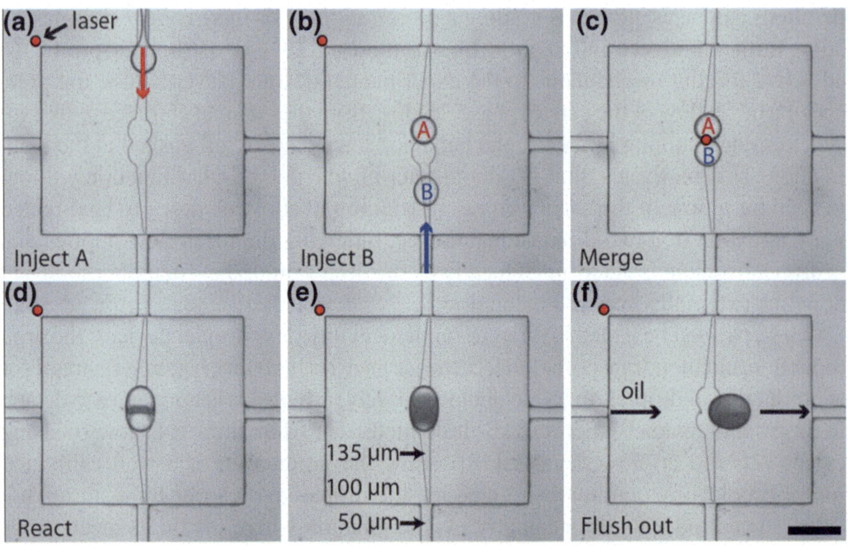

Fig. 2.13 Individual droplets are produced on demand at a step change in channel height. The different heights are labeled on *panel e*: The inlet channels are 50 μm high, the test section has 100 μm height and it is patterned with a V-shaped rail of height 135 μm. *A* corresponds to SCN^- and *B* corresponds to Fe^{3+}. The device is initially filled with the oil, which is then kept stationary. Then sample *A* is introduced from the *top channel* and allowed to detach into a droplet that automatically moves to the anchor position in the central region. This is followed by the injection of *B*, which does the same. Once both drops are in place, the laser is used to trigger their fusion and a reaction takes place (*panel c*). Finally, once the reagents are consumed, the drop is removed by flowing oil from *left* to *right*, The scale bar is 1 mm

mean flow of the external carrier phase to remove the drops from the nozzle, this flow is not necessary for the detachment of the droplet, as shown recently [54]. Instead, the drop production is dominated by the presence of the step; its transport can be ensured by a gradual reduction in confinement as shown below.

One implementation of such an approach is shown in Fig. 2.13, which combines droplet production at a step with V-shaped rails for their transport [55]. This device consists of a reservoir in which a chemical reaction will take place by merging two droplets that contain two reagents and which are produced and guided to the meeting point without using a flow of the external phase. The central reservoir, which contains regions of variable heights as shown on Fig. 2.13, is connected to four microchannels, two for oil and two for aqueous solutions.

When using these devices, the reservoir is initially filled with the liquid that will form the continuous phase and this liquid can be left at rest for the remainder of the experiment. Then the liquid that will form the droplet phase is injected through the channels for the aqueous which meet the reservoir at a step change in height and width. The change in geometry leads to drop detachment. Here, the step is designed such that the drops that are produced remain flattened by the top and bottom walls, such that they store a large amount of surface energy.

Their migration away from the nozzle is ensured by providing a gradient of confinement, as described and sketched in Fig. 2.3. Here, the confinement is released by a V-shaped rail which is etched in the top surface of the reservoir and which leads to a gradual release of the surface energy. This pushes the drop away from the step and to a central region where it is held by a local energy minimum.

This device allows on-demand droplet production and passive guidance to a meeting location, where two drops containing two reagents can be merged together to produce a chemical reaction [55]. An example reaction is shown in Fig. 2.13, which shows drops of SCN^- and Fe^{3+} that are merged together to produce $Fe(SCN)_3$ which appear dark on the grayscale images. In contrast with the devices of Figs. 2.11 and 2.12, the current device produces a single droplet of A and a single droplet of B, which are then passively transported to a meeting position and merged on demand. The total sample use for this device is therefore much lower, corresponding to the drop volume (around 10 nL) and the volume of the connecting channels (around 1 μL).

The device of Fig. 2.13 consists of three layers of different heights, which can be simply produced by multilayer soft lithography. First the four inlet channels have height of 50 μm. They lead to a square reservoir whose height is 100 μm, which is patterned with a deeper rail (height 135 μm). The left and right channels on the image are used as oil inlet and outlet, respectively, while the top and bottom channels are used to introduce the two reagents in aqueous solutions. The total operation shown in Fig. 2.13 takes place in nearly 40s, after which a flow of oil is induced to extract the reacted droplet and to reset the device for further reactions.

Similar operations can be performed if the confinement gradient is implemented differently. Dangla et al. [56] showed that a sloping reservoir ceiling can also lead to the production of drops on demand, through similar physical mechanism as steps. In contrast with the step emulsification however, the sloping ceiling automatically provides a confinement gradient that transports the drop away from the nozzle [35] and no further confinement gradients are required. A high frequency generation of emulsion drops can be achieved in this way, by parallelizing the inlet channels into the central reservoir and extracting the drops from a central hole.

2.8 Summary and Conclusions on Droplet Manipulation by Surface Forces

Many approaches have been developed for the production, transport, manipulation, and observation of micro-droplets and fluid segments. The results shown here demonstrate methods for manipulating drops in parallel, by removing the constraints due to the channel side walls. Removing this constraint reduces the interactions between different drops in the device, such that operations on a particular drop have a minimal influence on other drops elsewhere. This allows operations to be performed on individual drops without perturbing the behavior of the rest of the population. In this way, drops can be held stationary even when their neighbors are flowing.

In addition to providing added functionality with respect to other methods, this approach also simplifies the device operations. Indeed, reducing the interactions between far away droplets allows the devices to function "as desired" in a wider range of parameters, such as flow rates, drop frequencies, or fluid pairs. For this reason, many of the devices shown here work simply without the need for long optimization periods.

Shown here is a range of operations, ranging from droplet production, sorting, holding, and fusion. These operations now allow the implementation of a wide range of applications, such as the measurement of chemical kinetics, the manipulation of cells, or the exploration of fluid flow properties. More generally, this approach provides a platform that is simple to scale up to a large number of operations, with very low overhead cost. The physical scalings for the rails and anchors, as well as for the laser forcing, are favorable to miniaturization. This implies that further reduction in droplet size and increase in density are possible.

Acknowledgments The results described here summarize the work of several gifted PhD students and postdocs in my lab. I am particularly grateful to Rémi Dangla, Etienne Fradet, Sungyon Lee, and Paul Abbyad for the bulk of the results shown here. Caroline Frot provided important technical support. Figures 2.2, 2.3, 2.4, and 2.5 were provided by Rémi Dangla. Figure 2.8 was provided by Gabriel Amselem. Figures 2.9, 2.10, 2.11, 2.12, and 2.13 were provided by Etienne Fradet. Some of the research leading to these results has received funding from the European Research Council under the European Union's Seventh Framework Programme (FP/2007-2013) / ERC Grant Agreement n. 278248 "Multicell".

References

1. M.G. Pollack, R.B. Fair, A.D. Shenderov, Electrowetting-based actuation of liquid droplets for microfluidic applications. Appl. Phys. Lett. **77**(11), 1725–1726 (2000)
2. S.K. Cho, H. Moon, C.J. Kim, Creating, transporting, cutting, and merging liquid droplets by electrowetting-based actuation for digital microfluidic circuits. J. Microelectromech. Syst. **12**(1), 70–80 (2003)
3. A.A. Darhuber, J.P. Valentino, J.M. Davis, S.M. Troian, S. Wagner, Microfluidic actuation by modulation of surface stresses. Appl. Phys. Lett. **82**(4), 657–659 (2003)
4. K.T. Kotz, K.A. Noble, G.W. Faris, Optical microfluidics. Appl. Phys. Lett. **85**(13), 2658–2660 (2004)
5. A. Wixforth, Acoustically driven planar microfluidics. Superlattices Microstruct. **33**(5–6), 389–396 (2003)
6. T. Thorsen, R.W. Roberts, F.H. Arnold, S.R. Quake, Dynamic pattern formation in a vesicle-generating microfluidic device. Phys. Rev. Lett. **86**(18), 4163–4166 (2001)
7. S.L. Anna, N. Bontoux, H.A. Stone, Formation of dispersions using 'flow-focusing' in microchannels. Appl. Phys. Lett. **82**(3), 364–366 (2003)
8. R. Dreyfus, P. Tabeling, H. Willaime, Ordered and discordered patterns in two phase flows in microchannels. Phys. Rev. Lett. **90**, 144505 (2003)
9. K. Ahn, C. Kerbage, T. Hynt, R.M. Westervelt, D.R. Link, D.A. Weitz, Dielectrophoretic manipulation of drops for high-speed microfluidic sorting devices. Appl. Phys. Lett. **88**, 024104 (2006)
10. C.N. Baroud, M.R. de Saint Vincent, J-P. Delville, An optical toolbox for total control of droplet microfluidics. Lab Chip **7**, 1029–1033 (2007)
11. K. Ahn, J. Agresti, H. Chong, M. Marquez, D.A. Weitz, Electrocoalescence of drops synchronized by size-dependent flow in microfluidic channels. Appl. Phys. Lett. **88**, 264105 (2006)
12. H. Song, J.D. Tice, R.F. Ismagilov, A microfluidic system for controlling reaction networks in time. Angew. Chem. Int. Ed. **42**(7), 768–772 (2003)
13. H. Song, D.L. Chen, R.F. Ismagilov, Reactions in droplets in microfluidic channels. Angew. Chem. Int. Ed. **45**, 7336–7356 (2006)
14. S.-Y. Teh, R. Lin, L.-H. Hung, A.P. Lee, Droplet microfluidics. Lab Chip **8**, 198–220 (2008)
15. A.B. Theberge, F. Courtois, Y. Schaerli, M. Fischlechner, C. Abell, F. Hollfelder, W.T.S. Huck, Microdroplets in microfluidics: an evolving platform for discoveries in chemistry and biology. Angew. Chem. Int. Ed. **49**(34), 5846–5868 (2010)
16. C.N. Baroud, F. Gallaire, R. Dangla, Dynamics of microfluidic droplets. Lab Chip **10**(16), 2032–2045 (2010)
17. R. Seemann, M. Brinkmann, T. Pfohl, S. Herminghaus, Droplet based microfluidics. Reports on progress in physics **75**(1), 016601 (2012)
18. I. Barbulovic-Nad, S.H. Au, A.R. Wheeler, A microfluidic platform for complete mammalian cell culture. Lab Chip **10**(12), 1536–1542 (2010)
19. N.R. Beer, B.J. Hindson, E.K. Wheeler, B. Sara, K.A. Rose, I.M. Kennedy, and B.W. Colston. On-chip, real-time, single-copy polymerase chain reaction in picoliter droplets. Anal. Chem. **79**(22), 8471–8475 (2007) (Quantalife people)
20. J. Shim, G. Cristobal, D.R. Link, T. Thorsen, Y. Jia, K. Piattelli, S. Fraden, Control and measurement of the phase behavior of aqueous solutions using microfluidics. J. Am. Chem. Soc. **129**(28), 8825–8835 (2007)
21. W. Shi, J. Qin, N. Ye, B. Lin, Droplet-based microfluidic system for individual Caenorhabditis elegans assay. Lab Chip **8**(9), 1432–1435 (2008)
22. H. Boukellal, Š. Selimović, Y. Jia, G. Cristobal, and S. Fraden. Simple, robust storage of drops and fluids in a microfluidic device. Lab Chip **9**(2), 331–338 (2008)
23. C.H.J. Schmitz, A.C. Rowat, S. Koester, D.A. Weitz, Dropspots: a picoliter array in a microfluidic device. Lab Chip **9**(1), 44–49 (2009)

24. A. Huebner, D. Bratton, G. Whyte, M. Yang, A.J. deMello, C. Abell, F. Hollfelder, Static microdroplet arrays: a microfluidic device for droplet trapping, incubation and release for enzymatic and cell-based assays. Lab Chip **9**(5), 692–698 (2009)
25. M. Sun, S.S. Bithi, S.A. Vanapalli, Microfluidic static droplet arrays with tuneable gradients in material composition. Lab Chip **11**(23), 3949–3952 (2011)
26. W.H. Tan, S. Takeuchi, A trap-and-release integrated microfluidic system for dynamic microarray applications. Proc. Nat. Acad. Sci. **104**(4), 1146–1151 (2007)
27. D. Di Carlo, L.Y. Wu, L.P. Lee, Dynamic single cell culture array. Lab Chip **6**(11), 1445–1449 (2006)
28. P. Abbyad, R. Dangla, A. Alexandrou, C.N. Baroud, Rails and anchors: guiding and trapping droplet microreactors in two dimensions. Lab Chip **11**, 813–821 (2011)
29. A. Huebner, C. Abell, W.T.S. Huck, C.N. Baroud, F. Hollfelder, Monitoring a reaction at submillisecond resolution in picoliter volumes. Anal. Chem. **83**, 1462–1468 (2011)
30. E. Fradet, C. McDougall, P. Abbyad, R. Dangla, D. McGloin, C.N. Baroud, Combining rails and anchors with laser forcing for selective manipulation within 2D droplet arrays. Lab Chip **11**(24), 4228–4234 (2011)
31. Fr. Hauksbee. An account of an experiment touching the direction of a drop of oil of oranges, between two glass planes, towards any side of them that is neareste press'd together. Philos. Trans. R. Soc. **27**, 395–396 (1710)
32. E. Lorenceau, D. Quéré, Drops on a conical wire. J. Fluid Mech. **510**, 29–45 (2004)
33. M. Reyssat, L. Courbin, E. Reyssat, H.A. Stone et al., Imbibition in geometries with axial variations. J. Fluid Mech. **615**, 335 (2008)
34. T. Metz, N. Paust, C. Müller, R. Zengerle, and P. Koltay. Passive water removal in fuel cells by capillary droplet actuation. Sens. Actuators, A Phys. **143**(1), 49–57 (2008)
35. T. Metz, N. Paust, R. Zengerle, P. Koltay, Capillary driven movement of gas bubbles in tapered structures. Microfluid. Nanofluid. **9**, 341–355 (2010)
36. R. Dangla, S. Lee, C.N. Baroud, Trapping microfluidic drops in wells of surface energy. Phys. Rev. Lett. **107**(12), 124501 (2011)
37. S. Lee, C.N. Baroud, *Drop shape in a hele-shaw cell with a localized indentation* (In preparation, 2013)
38. F. Mugele, J.C. Baret, Electrowetting: from basics to applications. J. Phys.: Condens. Matter **17**, R705 (2005)
39. M.L. Cordero, D.R. Burnham, C.N. Baroud, D. McGloin, Thermocapillary manipula- tion of droplets using holographic beam shaping: microfluidic pin ball. Appl. Phys. Lett. **93**(3), 034107 (2008)
40. H. Bruus, Acoustofluidics 10: scaling laws in acoustophoresis. Lab Chip **12**(9), 1578–1586 (2012)
41. J. Xu, B. Ahn, H. Lee, L. Xu, K. Lee, R. Panchapakesan, K.W. Oh, Droplet-based microfluidic device for multiple-droplet clustering. Lab Chip **12**(4), 725–730 (2012)
42. M. Robert de Saint Vincent, R. Wunenburger, J.P. Delville, Laser switching and sorting for high speed digital microfluidics. Appl. Phys. Lett. **92**(15), 154105 (2008)
43. K.T. Kotz, Y. Gu, G.W. Faris, Optically addressed droplet-based protein assay. J. Am. Chem. Soc. **127**, 5736–5737 (2005)
44. E. Verneuil, M.L. Cordero, F. Gallaire, C.N. Baroud, Laser-induced force on a microfluidic drop: Origin and magnitude. Langmuir **25**(9), 5127–5134 (2009)
45. A.S. Utada, E. Lorenceau, D.R. Link, P.D. Kaplan, H.A. Stone, D.A. Weitz, Monidisperse double emulsions generated from a microcapillary device. Science **308**, 537–541 (2005)
46. G.F. Christopher, S.L. Anna, Microfluidic methods for generating continuous droplet streams. J. Phys. D: Appl. Phys. **40**, R319–R336 (2007)
47. J.C. Galas, D. Bartolo, V. Studer, Active connectors for microfluidic drops on demand. New J. Phys. **11**, 075027 (2009)
48. J. Guzowski, P.M. Korczyk, S. Jakiela, P. Garstecki, Automated high-throughput gener- ation of droplets. Lab Chip **11**(21), 3603–3608 (2011)

49. H. Gu, C.U. Murade, M.H.G. Duits, F. Mugele, A microfluidic platform for on-demand formation and merging of microdroplets using electric control. Biomicrofluidics **5**(1), 011101 (2011)
50. S. Sugiura, M. Nakajima, S. Iwamoto, M. Seki, Interfacial tension driven monodispersed droplet formation from microfabricated channel array. Langmuir **17**(18), 5562–5566 (2001)
51. I. Kobayashi, S. Mukataka, M. Nakajima, Novel asymmetric through-hole array microfabricated on a silicon plate for formulating monodisperse emulsions. Langmuir **21**(17), 7629–7632 (2005)
52. C. Priest, S. Herminghaus, R. Seemann, Generation of monodisperse gel emulsions in a microfluidic device. Appl. Phys. Lett. **88**, 024106 (2006)
53. F. Malloggi, N. Pannacci, R. Attia, F. Monti, P. Mary, H. Willaime, P. Tabeling, B. Cabane, P. Poncet, Monodisperse colloids synthesized with nanofluidic technology. Langmuir, **26**, 10–1441 (2010)
54. R. Dangla, E. Fradet, Y. Lopez, C.N. Baroud, The physical mechanisms of step emulsification. J. Phys. D: Appl. Phys. **46**, 114003 (2013)
55. E. Fradet, P. Abbyad, M. Vos, C.N. Baroud, *On demand pairing of microfluidic droplets for chemical kinetics* (Submitted 2013)
56. R. Dangla, S.C. Kayi, C.N. Baroud, Droplet microfluidics driven by gradients of confinement. Proc. Nat. Acad. Sci. **110**(3), 853–858 (2013)

Chapter 3
Electrical Switching of Droplets and Fluid Segments

Matthias Budden, Steffen Schneider, J. Michael Köhler and Brian P. Cahill

Abstract Switching operations in transport of microfluidic compartments are of high interest in miniaturized biotechnology, cell cultivation and screening programs as well as for future applications in miniaturized and automated diagnostics and in particular for automated experiments in ultraminiaturized combinatorial chemistry and combinatorial screenings in multidimensional parameter spaces. In addition to switching by laser actuation, surface forces, by centrifugal forces and electrowetting, electrical switching using electrostatic or dielectric manipulation represents an important class of microfluidic actuation principles. Electrical operations are of particular interest for fast switching and for addressing single selected fluid segments. Thus, they can be used for defining distances and orders of fluid segments and for sorting of droplets and segments in dependence on individual properties. Principles of electrostatic manipulation and the specific conditions for multi phase systems with strong differences in the electrical conductivity and electrochemical behaviour of the involved liquids are described in this chapter. The manipulation by DC fields is compared with the manipulation using AC fields by positive and negative dielectrophoresis. The manipulation by potential switching in Y-shaped micro channels is an example for efficient electrical manipulation of segments without galvanic contact. It can be shown that simple segment manipulation without any electrochemical changing of liquid composition is possible by switching under non-galvanic conditions. It is demonstrated that electronic data sets can be converted into a fluid pattern proving the high reliability of segment operations by potential switching. The robustness related to chemical composition and the applicability to cell suspensions will

M. Budden (✉) · S. Schneider · J. M. Köhler
Ilmenau University of Technology, Institute of Chemistry and Biotechnology,
PF 10 05 65, 98684 Ilmenau, Germany
e-mail: matthias.budden@tu ilmenau.de

B. Cahill
Institute for Bioprocessing and Analytical Measurement Techniques,
Rosenhof, 37308 Heilbad Heiligenstadt, Germany
e-mail: brian.cahill@iba-heiligenstadt.de

J. M. Köhler and B. P. Cahill (eds.), *Micro-Segmented Flow*,
Biological and Medical Physics, Biomedical Engineering,
DOI: 10.1007/978-3-642-38780-7_3,

be shown and the potential for cell cultivation and miniaturized screenings will be discussed. For further development, the possibilities and requirements related to the successive reduction of volumes of fluid segments, downscaling of functional elements and enhancement of switching frequencies are treated. Finally, future tasks for switching coming both from applications with different segment composition, involving cell cultures, multicellular systems, single cell operations, biomolecular diagnostics and from organic and inorganic synthesis, generation and application of nanomaterials, catalysis, single particle processes and supermolecular chemistry are discussed.

3.1 Introduction on Electrical Switching of Droplets

The idea of lab-on-a-chip systems (LOC) is based on the analogy between miniaturization and integration of electronic functions and microfluidic operations. Downscaling of transistors, electrical resistances, capacitors and other elementary functional units and their logical interconnection into integrated circuits became the model for the vision of highly integrated fluidic devices for operating chemical substances and biological entities—smaller molecules as well as biomacromolecules, supermolecules, cells and organisms. Some approaches in the last two decades tried to look for fields in which microelectronic networks—information processing chips—could be substituted by microfluidic networks. But, the insight in the behaviour of fluids and the experiences of restrictions in the development of integrated fluidic systems has taught us the important differences between the electronic and the fluidic worlds and the trials of microfluidic computing failed in many cases, in general due to lower degrees of integration, lower reliability and lower operation rates.

Nevertheless, the analogy still has a fascination. After some developments in microfluidics of the 1990s using homogeneous phase systems, the rediscovery of droplet based fluidic systems in the first decade of the 21st century led to the automation of operations on small samples of liquids in microchannels. In the field of droplet-based micro fluidics, small liquid samples are not regarded as alternatives to electron clouds or electronic states in electronic devices, but are mainly understood as small vessels or reactors for manipulation of substances or cells to which an electrical actuation could be applied. Microelectronics and microfluidics are not considered to be different competing concepts for miniaturized devices but are regarded as two different principles. The cooperative application of their specific and complementary power could be used for information processing in miniaturized chemical and biotechnological systems as well as for the development of interfaces between biological, chemical and digital-electronic operation units.

The recent state of the art of electronic chip development offers a huge field of possibilities of electronic measurement and actuation. So, it seemed to be very attractive to use this power for direct read-out of information and manipulating of fluidic systems by "fluido-electronic" interfaces. Two main tasks are related to the connection between fluidic systems and electrical transducers: (i) the exchange of informa-

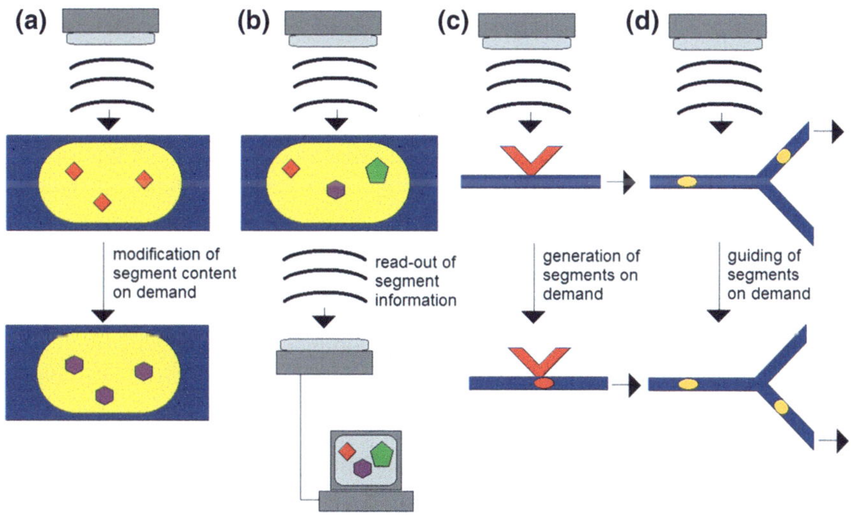

Fig. 3.1 Actuation of micro fluid segments and sensing (general): **(a)** initiation of conversion of segment content (biological, chemical processes), **(b)** non-contact read-out of content information (for example: chemical species, concentrations, cell content, physiological activity), **(c)** generation of segments on demand using switching by liquid/liquid interface actuation, **(d)** guiding of segments on demand by actuation in a branching region in a Y-channel

tion between components inside the liquid and an external information processing system, (ii) the generation, identification, addressing and manipulation of small liquid portions. Sensing and actuation are of interest for both types of interactions (Fig. 3.1).

In the following chapter, the application of electrical actuators for manipulation of liquid compartments, micro fluidic devices and systems will be described. Therefore, the different principles of individual addressing of small liquid samples will be presented and examples of the experimental realization and fields of applications will be discussed.

3.2 Droplets and Segments

3.2.1 Droplets

Droplets as well as microfluid segments are fluid compartments. They are formed by releasing a certain volume of a liquid into a non-miscible carrier medium. The formation of droplets of aqueous solutions containing chemicals, biomolecules or cells, by injecting these liquids into a carrier stream of a liquid alkane, e.g. tetradecane, or a perfluorocarbon liquid, e.g. perfluorodecalin, is a typical example of this strategy for formation of liquid micro compartments. Droplets and segments can be separated from each other by gas bubbles or by a carrier liquid. Both types of compartments can be realized over a large range of volumes reaching typically from microliters down to picoliters or even femtoliters. This corresponds to liquid compartments between millimeter and micrometer size. The deciding difference between both types is not given by the individual volume but by the relation to a wall, a guiding channel or interface.

Droplets are released into and transported inside a vessel which is significantly larger than the droplet size. In this way, individual droplets do not contact the wall and cannot be distinguished from other droplets by its position (Fig. 3.2a). The normal case is an "anonymous" situation of a single droplet inside a larger ensemble of droplets. There is no well-defined order of segments which is preserved over longer time spans. The situation of each droplet inside the microfluidic systems corresponds with the situation of droplets in a classical emulsion. The whole ensemble of micro fluid compartments is characterized by "anonymity". The advantage of this "freedom" is the high mobility of droplets and the high degree of freedom for droplet manipulation.

Switching and sorting of droplets is only reasonable in the case of a close temporal relation between a characterization of a droplet and the decision about a following actuation. Short-time loops are required for switching and sorting procedures for droplets with fluctuating relative position in a given volume containing many droplets.

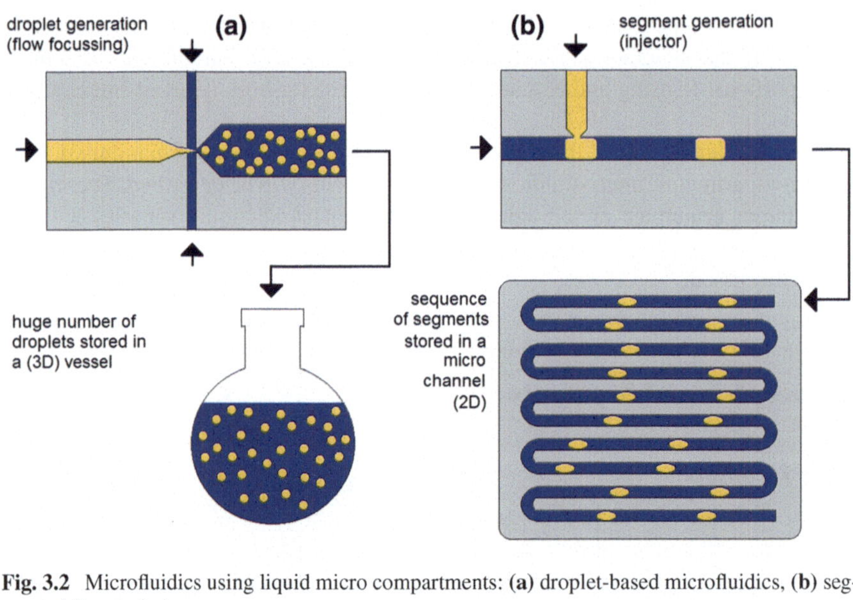

Fig. 3.2 Microfluidics using liquid micro compartments: (**a**) droplet-based microfluidics, (**b**) segmented flow technique

The only strategy for overcoming the anonymity in such droplet ensembles is individual labelling. Such labels could be realized by colour coding or by introduction of barcoded micro particles into the droplets.

3.2.2 Micro Fluid Segments and Their Manipulation Without Electrical Actuation

Micro fluid segments are fluid compartments with a strong relation to the wall or surface of a guiding material. All compartments are generated and manipulated in a strict order. It is not possible that one segment spontaneously overtakes one of the other segments. The position of a segment inside each large sequence of segments allows individual segments to be identified (Fig. 3.2b). Each segment is addressable by its position. The whole ensemble of micro fluidic compartments is characterized by "individuality". Single segments can be recognized and addressed by their position inside the segment sequence. The price of this order is a reduction of mobility and freedom of manipulation.

This principle order in the sequence of segments is conserved during segment operations, in general as far as they are not released in a larger volume and lose the guidance by a suited capillary or micro channel. In general stacking of segments from two or more input sequences into one output sequence as well as segment transformation, splitting and segment fusion preserve the sequence information, unless segment patterns become changed.

A conversion of segment shape can be of interest in order to influence the behaviour of the segment in mixing and splitting operations. A reduction of channel radius r from r_1 to r_2 must be accompanied by an increase in segment length l from l_1 to l_2 (stretching) because the volume must remain constant. In a simple cylinder model, this transformation can easily be described:

$$V_1 = \pi \cdot r_1^2 \cdot l_1 = V_2 = \pi \cdot r_2^2 \cdot l_2 \tag{3.1}$$

or

$$l_2 = \left(\frac{r_1}{r_2}\right)^2 \cdot l_1 \tag{3.2}$$

Stretching is always related to an increase of interface energy (Fig. 3.3: scheme of segment shape and enhancement of interface energy). This interface energy is only determined by the liquid/liquid interface tension σ and the liquid/liquid interface area A_{IF} if the wall is not wetted by the segment liquid. From a thermodynamic point of view, the segment becomes more sensitive to fluidic operations if the interface area is enhanced. The increase of interface energy can be described by the difference between the surface energy in the final γ_∞ and the initial state γ_0:

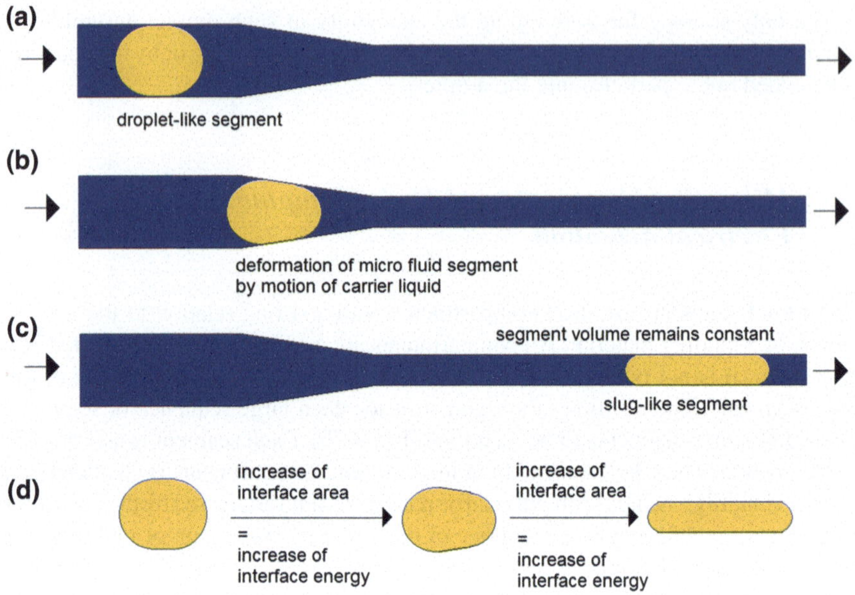

Fig. 3.3 Transformation of fluid segments (from droplet to slug): induction of additional interface energy by segment transformation inside a conical channel

$$\gamma_{\infty} - \gamma_0 = A_{\infty} \cdot \sigma - A_0 \cdot \sigma = (A_{\infty} - A_0) \cdot \sigma \tag{3.3}$$

It is difficult to split segments of nearly spherical size regularly. Normally these segments will leave the splitting position—for example a Y-junction—completely in one or in the other of two possible directions (Fig. 3.4: Scheme of droplet behaviour at Y-junction). In contrast, regular splitting can be performed robustly, if spherical segments are transformed into slug-like segments in order to have a preformed splitting shape. This transformation introduces the required additional interface for splitting. From a thermodynamic point of view, this enhanced interface energy is the driving force for the segment splitting. The principle of control of segment motion by controlling the interface energy corresponds completely with the interface-controlled operation of droplets using microfluidic anchors and rails (compare Chap. 2).

3.3 Electrostatic Manipulation of Droplets in a Liquid Carrier

3.3.1 Droplet Charging

Droplets can be actuated by a static electrical field, if they are charged. This actuation principle is based on the normal electrophoretic effect. The principle is well

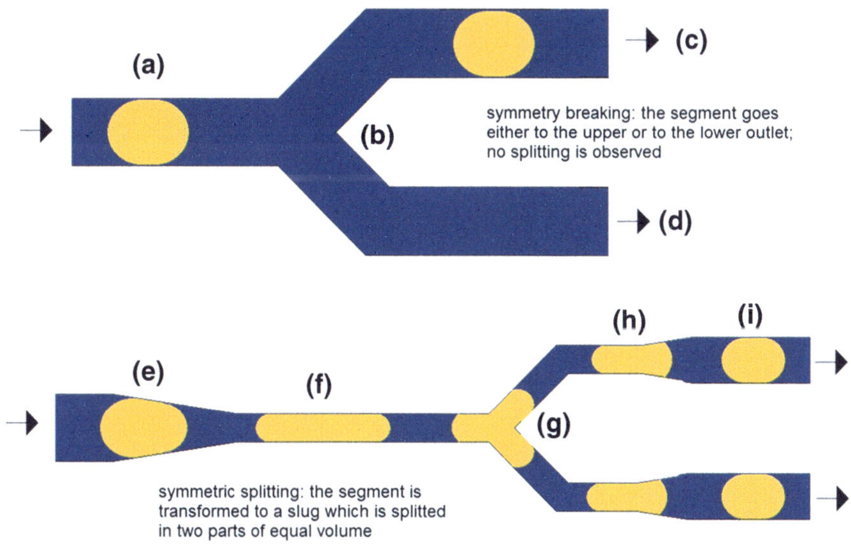

Fig. 3.4 Behaviour of segments at Y-junctions at symmetric flow rates: droplet-like segments (**a**) will have 50 % probability to choose the one or the other direction at the branching point (**b**) and leave a Y-shaped channel either by the upper (**c**) or the lower outlet (**d**); in contrast a segment splitting is normally observed if the droplet-like segment (**e**) is transformed into a slug (**f**) before reaching the branching position (**g**); by invers conical channel structures (**h**) the parts of the splitted segment can be reshaped in to droplet-like segments (**i**)

known from sorting of particles and small droplets in the so-called FACS devices for Fluorescence-Activated Cell Sorting. Charged particles are deflected in an electrical field after measurement of fluorescence signals.

The electrical charging of droplets can be achieved by contact with electrodes. So, the liquid can pass an electrode before forming droplets. In this case, charged droplets are generated from the charged liquid. Alternatively, droplets formed by a neutral liquid can be charged after the formation of droplets. Such a charging can also be performed by a contact with an electrode. But, in this case, it is important to avoid a destruction of the droplets by the electrode contact. Alternatively, the application of a focussed beam of small charged particles can be applied for droplet charging. Charging by particle beams is a non-contact technique. Negative charging can be realized by an electrode beam, positive charging by an ion beam. A low absorption of particles by the surrounding medium—for example by using vacuum—is a precondition for the application of particle beams for charging. Charging of droplets can also be applied in liquid/liquid two-phase systems, if the droplet is formed by an electrolyte and the carrier liquid has a dielectric character.

3.3.2 Actuation of Droplets by Static Electrical Fields

The electrophoretic force F during actuation is given by the product of particle or droplet charge Q and the electrical field strength E. The field strength is defined by the electrode potential U and the electrode distance b, if the dielectric medium is homogeneous:

$$F = Q \cdot E = Q \cdot \frac{U}{b} \tag{3.4}$$

For particles, the friction between the surrounding liquid and the particle has to be taken into account for estimating the motion in case of actuation. It can be calculated by using the Stokes Law:

$$|F| = 6 \cdot \pi \cdot \eta \cdot v \cdot r \tag{3.5}$$

The actuation velocity v of particles with a radius r can be calculated by combination of Eqs. (3.1) and (3.2):

$$v = Q \cdot \frac{U}{6 \cdot \pi \cdot \eta \cdot r \cdot b} \tag{3.6}$$

or for a given volume $V = \frac{4}{3} \cdot \pi \cdot r^3$:

$$v = Q \cdot \frac{U}{b \cdot \eta \cdot \sqrt[3]{162 \cdot \pi^2 \cdot V}} \tag{3.7}$$

These equations are a good approximation for the transport of droplets in an immiscible liquid medium, too. Instead of energy dissipation by viscous shear forces in the carrier liquid alone, a combination of kinetic energy loss by viscous flow in the carrier liquid and in the droplet liquid has to be considered. The ratio of both contributions is depending on the ratio of the both viscosities.

Charging of particles or droplets can be regarded as the formation of an electrochemical potential U_{ec}. This potential can be estimated by a balance between electrochemical energy and approximated internal electrostatic energy of the droplet or particle:

$$\frac{e^2}{2 \cdot \pi \cdot \varepsilon \cdot r} \cdot \sqrt{\frac{z^5}{\pi}} = U_{ec} \cdot z \cdot e \tag{3.8}$$

whereby the particle charge was substituted by the product of the number of elementary charges z and the elementary charge e:

$$Q = z \cdot e \tag{3.9}$$

resulting in:

$$U_{ec} = \frac{1}{2 \cdot \varepsilon \cdot r} \cdot \sqrt{\frac{1}{e} \cdot \left(\frac{Q}{\pi}\right)^3} \tag{3.10}$$

or in the velocity v for a given electrochemical potential for particle charging U_{ec} in dependence on particle radius r, electrode distance b and switching voltage U:

$$v = \frac{U}{6 \cdot b \cdot \eta} \cdot \sqrt[3]{\frac{(U_{ec} \cdot \varepsilon)^2 \cdot e}{r}} \tag{3.11}$$

For practical applications in chemical operations or in miniaturized biotechnology, it is important to avoid intensive electrochemical processes during the particle charging. Such processes can seriously affect the screenings, tests or experiments. Therefore, the electrochemical potentials for droplet charging should be kept below the order of magnitude of one volt.

An estimation of switching frequencies f in dependence on switching voltages U and droplet size is possible under the assumption that the elongation of droplet should be at least 1/3 of the droplet radius:

$$f = \frac{2 \cdot U}{b \cdot \eta} \cdot \sqrt[3]{\frac{e \cdot (U_{ec} \cdot \varepsilon)^2}{r^4}} \tag{3.12}$$

This clearly indicates that this method can be advantageously used for small droplets (diameters below 50 μm) only. In case of larger droplet volumes (several ten to several hundred nanoliters), very high charging voltages have to be applied or the flow rate and the switching frequency must be low.

3.3.3 Droplet Sorting by Electrostatic Electrical Manipulation

Sorting of droplets always requires four process steps:

(1) measurement or recognition of a characteristic feature,
(2) data processing and decision about switching,
(3) preparation of droplet for switching and finally
(4) actuation of droplets.

The typical process chains include selective fluorescence labelling for generation of distinguishable droplets. Binding of fluorescence-labelled antibodies is a typical way for a selective labelling in cellular or biomolecular systems. Differences in fluorescence intensity can easily be used for fast differentiation between two types of micro objects by miniaturized flow-through fluorimeters. The switching decision is based, in general, on the definition of a threshold in fluorescence intensity. Normally, the fluorescence labelling has very low or no effect on the sensitivity of droplets to the electrostatic field. Therefore it is necessary to charge the droplets in order to sensitize them for the electrostatic actuation. The charge transfer can be realized by bringing the droplets into contact with an electrode. This charging principle is frequently applied for charging of droplets for separation in the gas phase (like FACS devices), but can also be applied for aqueous droplets or droplets of other conductive liquids in a dielectric environment, for example if the droplets are embedded in an alkane or perfluorocarbon phase. The separation of droplets is simply performed by choosing the polarity of the electrical field in the switching region.

Jung and Kang [1] described the transport of droplets between uninsulated electrodes on a plane. The droplets acquired a charge by direct contact with an electrode and were transported between the electrodes by the Coulombic force acting on charge objects in an electric field. Ahn et al. [2, 3] developed a chip-based droplet-sorting device that relied on the active charging of the droplet surface by the injection of charge from an electrode in contact with the droplet. As droplets must be in contact with an uninsulated and an insulated electrode the droplet must be squeezed through a constriction. A constant sorting voltage is applied that switches negatively charged droplets into the upper channel, positively charged droplets into the lower channels and allows non charged droplets to pass straight through to the middle channel. Im et al. [4–6] have investigated similar charging and subsequent transport of droplets in electric fields and have noted that field strengths of 1-3 kV/cm do not affect biological cells suspended in such droplets adversely [4].

3.4 Dielectric Manipulation of Droplets by Alternating Fields in a Liquid Carrier

3.4.1 Trapping of Droplets in Field Cages

Droplets and segments are distinguished from the carrier solution by the differences in the dielectric constant. Therefore, any inhomogeneous electrical field induces a force on the droplets. This force is directed to a minimization of the energy of the whole system. So, droplets of relatively lower polarizability in a carrier liquid of higher polarizability tend to move in the direction of decreasing field strength. This effect is called "negative dielectrophoresis" . Droplets of higher polarizability tend to move in the direction of increasing field strength. This effect is called "positive dielectrophoresis". Both effects have to be distinguished from the electrophoresis effect, where a motion of charged particles or droplets is caused by Coulomb forces in a directed electrical field. Dielectrophoresis acts on objects that do not have any net charge because the applied field can polarize neutral objects. If an object is polarized, it obtains a dipole moment. Due to inhomogeneities in the electric field, the electric field gradient across the polarized object may interact with its dipole moment so that a net force acts on the object. Alternating potentials (AC fields) can be applied for the actuation because the induced dipole can reverse direction as the electric field reverses its polarity. Thus the direction of the force depends on the geometry of the system and, in particular, the electrode geometry. In this way, the transport of particles and droplets can be controlled without any dependence on their electrical charge.

The maximum dielectric force (perpendicular to the applied capacitor arrangement) acting on a particle can be calculated by a dipole approximation [7]:

$$F_{EDP} = \frac{27}{32} \cdot \pi \cdot {U_{rms}}^2 \cdot \left(\frac{R}{b}\right)^3 \cdot \varepsilon \cdot Re\,(f_{CM}) \tag{3.13}$$

with the Clausius-Mosotti factor being given by:

$$f_{CM} = \frac{\varepsilon_1^* - \varepsilon_2^*}{\varepsilon_1^* + 2 \cdot \varepsilon_2^*} \tag{3.14}$$

where b is the distance between the electrodes, R is the particle radius, U_{rms} is the applied voltage and ε^* is the complex dielectric constant of and the subscript denotes medium 1 or 2 respectively.

Particles and droplets can be localized in a well-defined position if they are trapped in a volume of carrier liquid with a field extremum in the three-dimensional space. Such field traps can be realized by multipole electrode arrangements [7].

Droplets of aqueous solution which are dispersed in an immiscible organic carrier solution have much higher electrical permittivity than the surrounding medium. They have to be trapped by a field cage using positive dielectrophoresis. Vice versa, polymer particles or droplets of organic solvents in an aqueous environment can be trapped in a field cage by negative electrophoresis. Octopole thin film electrode arrangements have been introduced successfully for trapping and for sorting of particles and cells by negative dielectrophoresis in microfluidic devices [7]. An overview on the recent status of dielectrophoresis on transport and manipulation of liquids and particles was given by Pethig [8].

3.4.2 Dielectric Actuation of Droplets by Dielectrophoresis

In droplet-based microfluidics, dielectrophoresis can be used for the generation of droplets controlled by an electrical field. Switching-on the field causes the formation of a liquid finger inside a microchannel filled with immiscible carrier liquid. This technique can be used for the generation of droplets on demand in lab on a chip systems [9, 10].

Velev et al. [11] used dielectrophoresis to transport aqueous droplets floating at the interface between a perfluorinated oil and water. As the perfluorinated oil is denser than water, water droplets float at the oil-water interface. It was noted that although the transport mechanism relied on dielectrophoresis, response of droplets to applied DC fields was observed and the authors related these to charging effects in the perfluorinated oil. This method was further developed for application of droplets as bioreactors: for particle synthesis [12] and bioassays [13].

Schwartz et al. [14] realized a multielectrode chip for dielectric manipulation of water droplets embedded in a liquid hydrocarbon carrier. They were able to demonstrate the manipulation of droplets in the device without any mechanically moving parts. The controlled motion of droplets can be used for pair-wise droplet fusion, for mixing and, therefore, for initiation of chemical reactions. In addition, it is an interesting principle for the realization of biochemical assays requiring the mixing of reactants.

The principle of positive dielectrophoresis can be applied for a large spectrum of solvents—including aqueous solutions as well as different organic liquids—if the electrical permittivity of the environment of a droplet can be kept low. This is in particular the case if the droplets are manipulated in air or another surrounding gas. Chatterjee et al. [15] showed that electrical fields can be used for actuation of droplets of different liquids guided through air-filled micro channels.

Baret et al. [16] showed how dielectrophoresis is very well suited for manipulation of small free-moving droplets through an application in a micro channel for continuous sorting of cells. Thes device was used for the realization of a fast fluorescence activity dependent cell sorting ("FAD") in a liquid environment. There was achieved a sorting efficiency above 99.99 % for droplets with a volume of about 12 pL. Ahn et al. [17] describe how they could separate 8 pL droplets generated at a rate of 1.6 kHz using dielectrophoresis.

3.5 Manipulation of Fluid Segments by Potential Switching

Valve-less switching is very important for higher switching rates and parallelized operations in microfluidics. In particular, an urgent need for valve-less switching exists in case of further reduction of volumes of liquid compartments. So, this type of switching operations will dominate fluidic manipulation in future. In contrast to free flowing droplets in an emulsion-like environment, the degrees of freedom for lateral motion are strictly reduced in micro segmented flow. The fixed position and the close relation to the wall of the channels or capillary are essentially for the reliable order of segments in a segment sequence. The well-defined order and the fixed positions of individual segments seem to be in contradiction to the desire for segment manipulation with an individual addressing of single fluid segments. Therefore, a device is required for switching which allows a sufficient lateral droplet motion without loss of the segment order.

Several droplet switching techniques have been described in the literature that rely on the action of a DC field on an aqueous droplet suspended in oil. These droplets have not been actively charged by contact with an electrode such that the droplet moves depending on the polarity of the applied field. Niu et al. [18] presented the switching of droplets at a Y junction based on the switching of the polarity of the electric field at the junction without elaborating on the switching mechanism very clearly. Guo et al. [19] developed a droplet sorter that could switch 65 pL droplets at a rate of 100 Hz. The droplets were not precharged and the switching direction

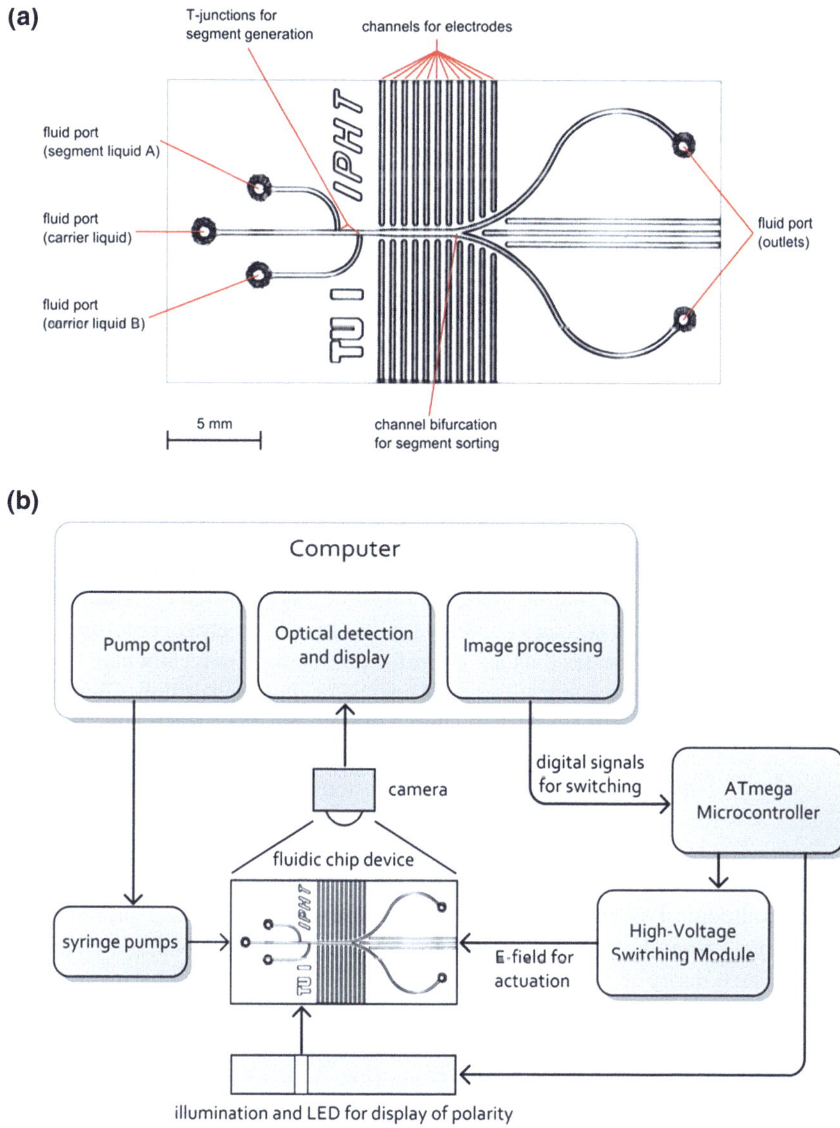

Fig. 3.5 Chip for electrical segment switching (a) and flow chart of the experimental arrangement (b)

depended on the polarity of the DC field. Although the switching mechanism was not explained, it was suggested that the mechanism may depend on charges induced on the droplet surface by the electric field [20] or on preferential ion adsorption at the oil water interface [21].

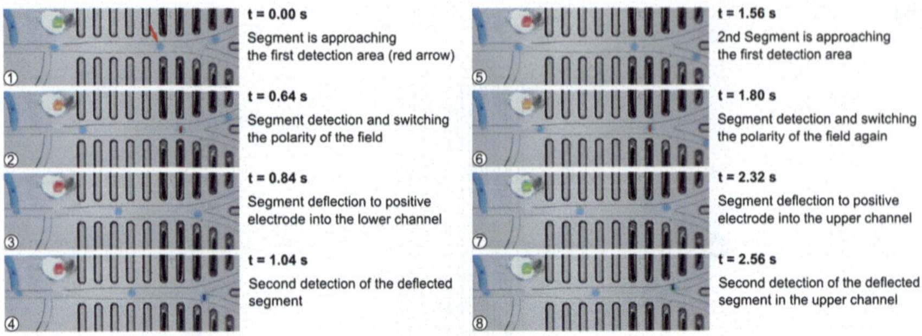

Fig. 3.6 Switching of aqueous segments in a micro fluidic chip that has a moderate lateral extension of the channel in the switching region

This section describes an example for the realization of such a device that can switch droplets depending on electric-field polarity without any active precharging of the droplet. The microfluidic device consists of microchannels, fluid ports and channels for the insertion of electrode wires. A chip was fabricated by photolithography and subsequent wet etching of a glass substrate that delivered rounded channel profiles and smooth wall surfaces combined with a high geometrical precision. This glass-wafer structure was used as a pattern for producing a polycarbonate disk by injection moulding. Fluid connections were opened by precision drilling. Figure 3.5a shows the droplet generation/switching chip. Oil is pumped into the system through the fluidic port on the far left. Aqueous solutions are pumped into the system through the next two input ports; this allows droplets of various compositions to be generated. The channels for electrodes can accommodate metal wires that are used to generate the electric field to switch droplets into the upper or lower channel on the right-hand-side of the chip and exit the system through either output ports. Figure 3.5b shows the external control system used to control switching. The computer processes an optical image from the camera to control the pumping and high-voltage switching module. The polarity of the electric field is recorded by LEDs connected to an optical fiber below the chip.

The key issue of this separation device is the moderate lateral extension of the micro fluidic channel in the switching region (Fig. 3.6). The slight enlargement of channel width over a limited channel length preserves the segment order and at the same time makes lateral motion possible. A switching of a laterally directed electrical field breaks the symmetry of segment motion. The field switching causes a lateral motion and brings the segment in direction of one or the other outlet branches of the Y-junction as required.

Investigations with aqueous micro fluid segments embedded in tetradecane in a polycarbonate micro device prove that a very reliable switching is possible [22]. Fluid segments of about 12 nL volume were generated at a micro injector and guided through a microchannel with rounded cross section and dimensions of about 0.3 mm width and 0.26 mm height. The device was equipped with lateral sets of five electrodes in a distance of about 0.5 mm. An electrical droplet switching was possible down to

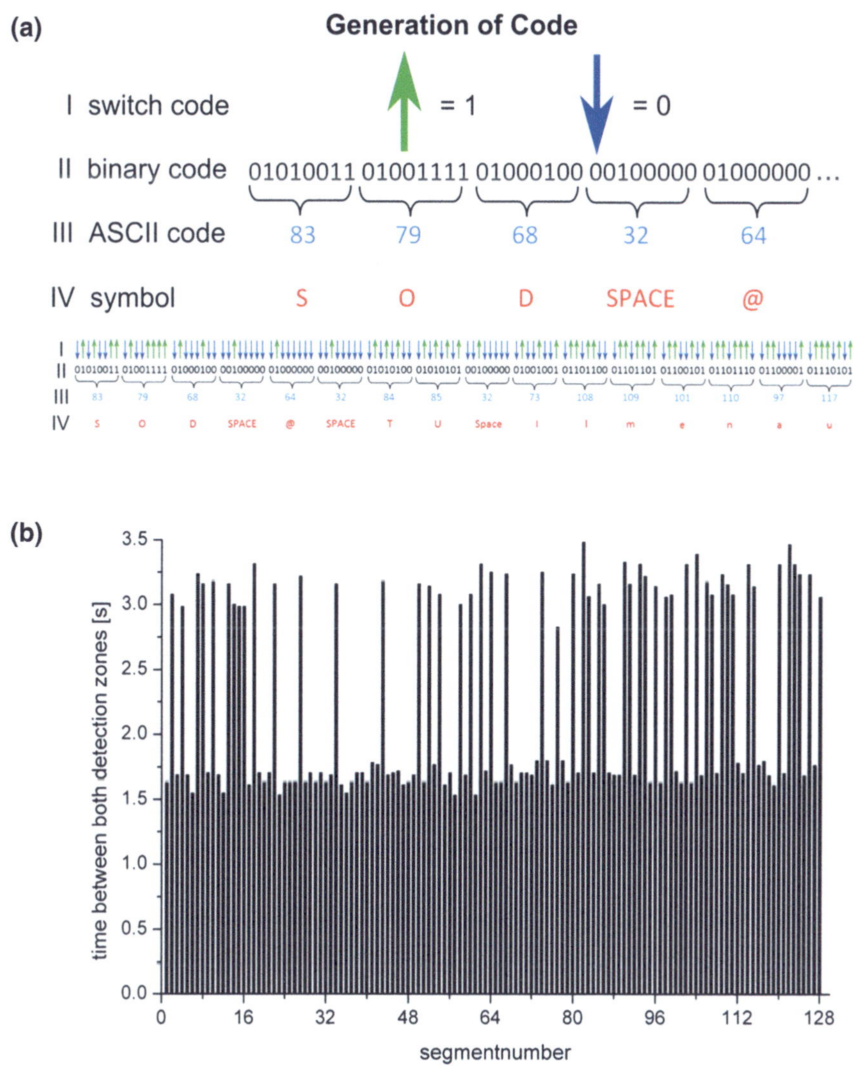

Fig. 3.7 Principle of microfluidic coding by segment switching: (**a**) binary coding by fluidic switching direction, (**b**) measurement segments with visualization of code by different delay in the optical signals in the upper and the lower channel

voltages of 200 V corresponding to an electric field strength of about 4 kV/cm. The electrical signal for switching can be applied across a section of the micro fluidic channel of about 1.4–2.6 mm. In this region the success of actuation is not affected by the actual position of the fluid segment. So, the switching principle is robust against fluctuations in segment/segment distances in larger segment sequences. The

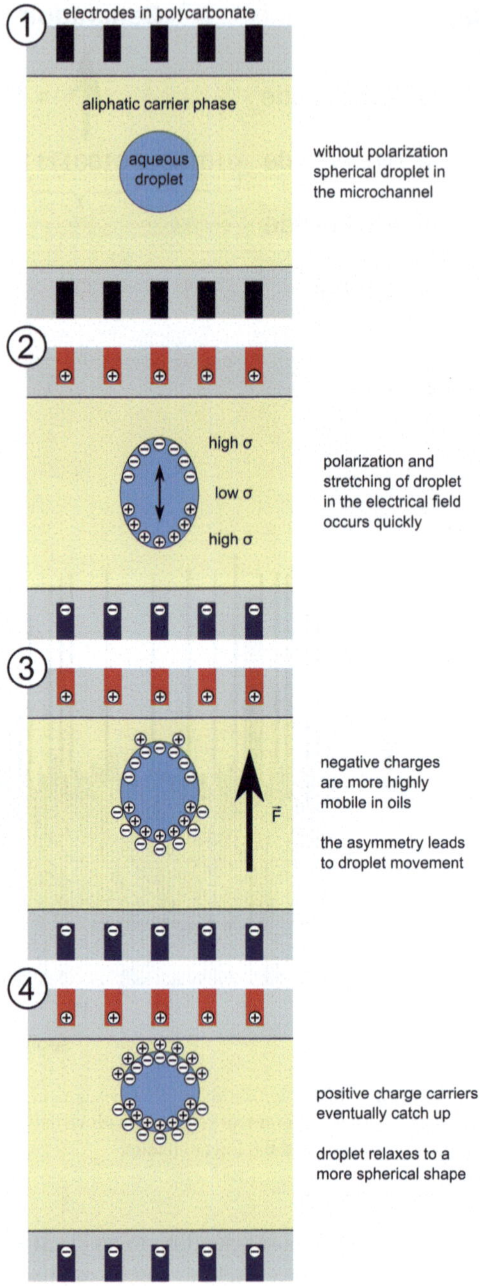

Fig. 3.8 Polarization and stretching of a droplet in a micro channel

switching principle can be used for sorting of segments [23] and can also be applied for generation of arbitrary segment sequences (Fig. 3.7) and liquid encoding [24].

The exact mechanism of the switching process is not clear up to now. But, model experiments [24] have shown that the switching is neither caused by electrowetting nor by a displacement in a static electrical field or pure dielectric actuation. Slow-motion microscope video recordings of the response of droplet on the switching of the polarity of the directed field show that the droplet motion in channels as well as in case of droplets with no wall contact is initiated by a lateral stretching of the droplet. This enlargement of liquid/liquid interface is probably induced by a polarization of the droplet in the electrical field. This stretching can be understood as a transition of the droplet shape into a new—temporary—state of equilibrium of interface forces. This effect can be explained by an increase of charge density and subsequently the interface tension in the both pole regions of the stretched droplet (Fig. 3.8). This deformation relaxes during a time interval of about 0.2–0.4 s. The actuation force is effective within this time frame and disappears afterwards.

The reasons for the relaxation and the asymmetry related to the field polarity are still a matter of research and an explanation based on induced charge is presented here. Although oils are generally regarded as non-conductive, they possess a low but finite conductivity [25]. The induction of mobile charges at interfaces has been used as the basis of pumping mechanisms in oil [26–29] and in water [30–32]. This deformation relaxes during a time interval of about 0.2–0.4 s; this compares relatively closely to the charge relaxation time in oil. The charge relaxation time is a measure of the time scale necessary for the formation of the electric double layer at a surface in a particular liquid. It is given by $\tau = \varepsilon/\sigma$, where σ and ε are the conductivity and dielectric constant of the liquid in question [25, 33]. In aqueous solutions, the charge relaxation time is significantly shorter due to the high concentration and mobility of salt ions in water. Hughes et al. [34] showed how the medium conductivity influences the polarization of mobile charge in the electric double layer in a time-dependent electric field and how this affects dielectrophoretic forces on charged particles. At low frequency (less than 1 kHz), ionic polarization tends to predominate over dielectric polarization. If the droplet is much more conductive than the medium, the aqueous droplet will experience rapid ionic polarization. As the electric field is not inhomogeneous, this dielectrophoretic force causes the droplet to be stretched in the direction of the electric field but does not result in net motion. The polarization of the droplet tends to generate an image charge in the oil. Thus counter charges are drawn from the oil. The anions and cations in hydrocarbons can have significantly different mobilities from each other [35, 36]. The conductivity at any given point in a liquid containing one species of positive ion and one species of negative ions is given by $\sigma = \rho_+ u_+ + \rho_- u_-$, where ρ_+ and ρ_- are the charge densities of the positive and negative ions at any point in space and u_+ and u_- are the mobilities of the positive and negative ions. In this way, the counter charge layer on one side of the droplet can form more quickly, so that the imbalance can cause the net motion described schematically in Fig. 3.8. As the counter charge of the slower ions catches up with the quicker ions, the imbalance vanishes and motion ceases.

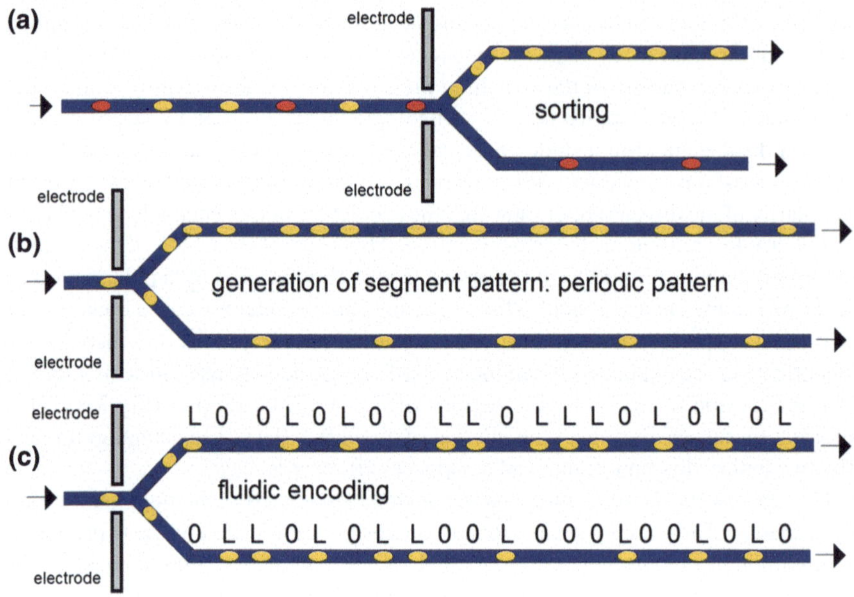

Fig. 3.9 Different applications of fluidic switching at a Y-junction: (**a**) sorting, (**b**) generation of a periodic pattern, (**c**) fluidic encoding

3.6 Applications and Challenges for Electrical Switching of Droplets and Segments

The sorting of droplets is probably the most important type of application of switching mechanisms for liquid segments. Such sorting is of interest for a large spectrum of biological applications, for example for the sorting of cells or cell ensembles. The formation of larger sets of different monoclonal cell cultures by cultivation of single cells is a typical application of droplet-based microfluidics. The automated recognition of certain cell types and switching of segments allows a selection of identified clones and their use in further biological procedures.

A sorting is also very interesting for larger screenings in biology as well as in chemistry (Fig. 3.9a). The micro segmented flow is very promising for the screening in multidimensional parameter spaces [37]. This strategy is applied for the identification of combination effects in microtoxicological screenings [38] as well as for the realization of nanomaterials with specific physical properties, for example in plasmonics [39]. The switching of segments enables the selection of segments of interest for individual manipulation, detailed characterization and other experiments.

In addition, the electrical control of switching inside networks with moving segments allows any rearrangement to preserve the complete information of the prehistory of each single compartment. The simplest form of patterning is the modulation of segment distances. So, regular periodic segment patterns can be generated, for

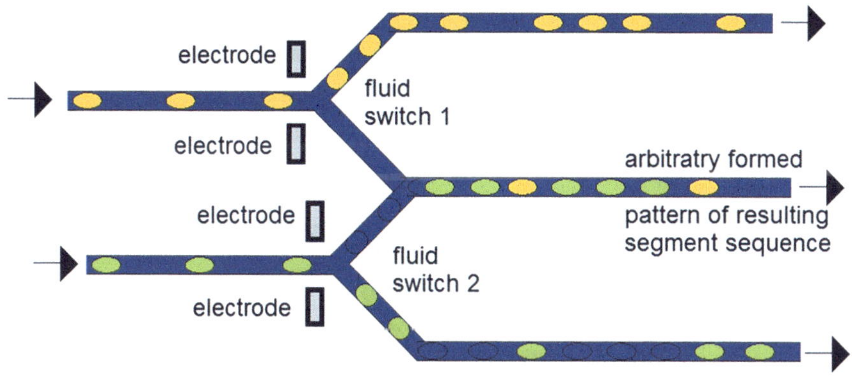

Fig. 3.10 Formation of an arbitrary binary segment pattern by electrical switching at a double-Y junction

example (Fig. 3.9b). The guiding of segments to one or the other outlet channel can be used for the generation of codes as described above and was realized in the same type of Y-branched switching unit (Fig. 3.9c)

Further, it becomes possible to generate new segment sequences by selective merging of parts of two or more original sequences. An example for the design of a micro fluidic device with two switching units for generation for binary segment sequences is given by a branched system with two input channels and three outlet channels (Fig. 3.10). This type of liquid operations is of interest, for example, for the simple inclusion of fluidic segments for reference and calibration tests as well as for inclusion of labelling segments in larger segment sequences for screenings. Further, it can be used for a selective sequence merging based on comparative measurements.

It is expected that electrical switching will become a key topic for logical operations in microfluidic applications. It will contribute to an enlargement of application fields and will become particularly important for automated procedures in experiments, screenings and diagnostic processes with large sample sets.

Challenges for further developments are presented by the diversity of compositions of fluid segments and carrier solutions. It is necessary to understand the mechanisms of switching in order to know the parameter ranges for reliable and robust switching in the case of varied liquid compositions. The switching conditions in screenings including cell cultivation have to respect very large differences in cell density and in the surface activity of involved cell cultures. Problems might arise in particular from the deposition of adherent cells on the walls of micro channels and by formation of biofilms. In addition, cellular growth causes an increase of viscosity. The release of proteins, carbohydrates and other biomacromolecules from cells into the segment liquid can lead to the formation of a gel like state and to non-Newtonian flow behaviour. A specific challenge arises from miniaturized screenings with multicellular systems like cell assemblies, spheroids, tissue fragments, embryos and small multicellular organisms. On the one hand, sidewall interaction,

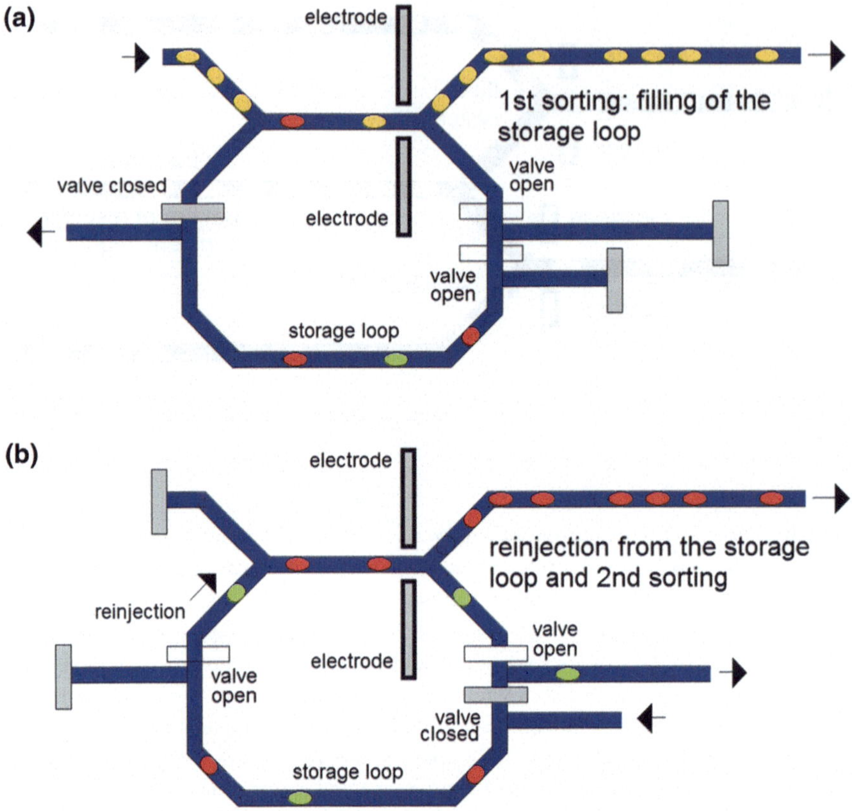

Fig. 3.11 Concept of a microfluidic realization of two sorting steps by use of a channel loop for segment storage: **(a)** first sorting step and incubation of selected segments in the storage loop, **(b)** reinjection of selected segments and second sorting step

enhanced viscosity and the danger of channel blocking demand for robust concepts for microfluidic switching. On the other hand galvanic processes, considerable heat release and strong electrical fields have to be avoided in order to exclude damages to the biological objects.

The switching principle can also be applied for controlling segment transport in microfluidic systems with channel loops. So, for example, a two-fold sorting process can be applied with one switching unit (Fig. 3.11). The integration of additional channels for fluid actuation and valves would also allow the design of microfluidic systems with loop structures for multi-fold repeated sorting operations (Fig. 3.12).

Further, the micro fluid segment technique is a very powerful strategy for single cell operations, and for biomolecular diagnostics and molecular biological operations. Small cell ensembles, single cells and molecular ensembles of low numbers of biomolecules can be operated at moderate concentrations due to the small volume

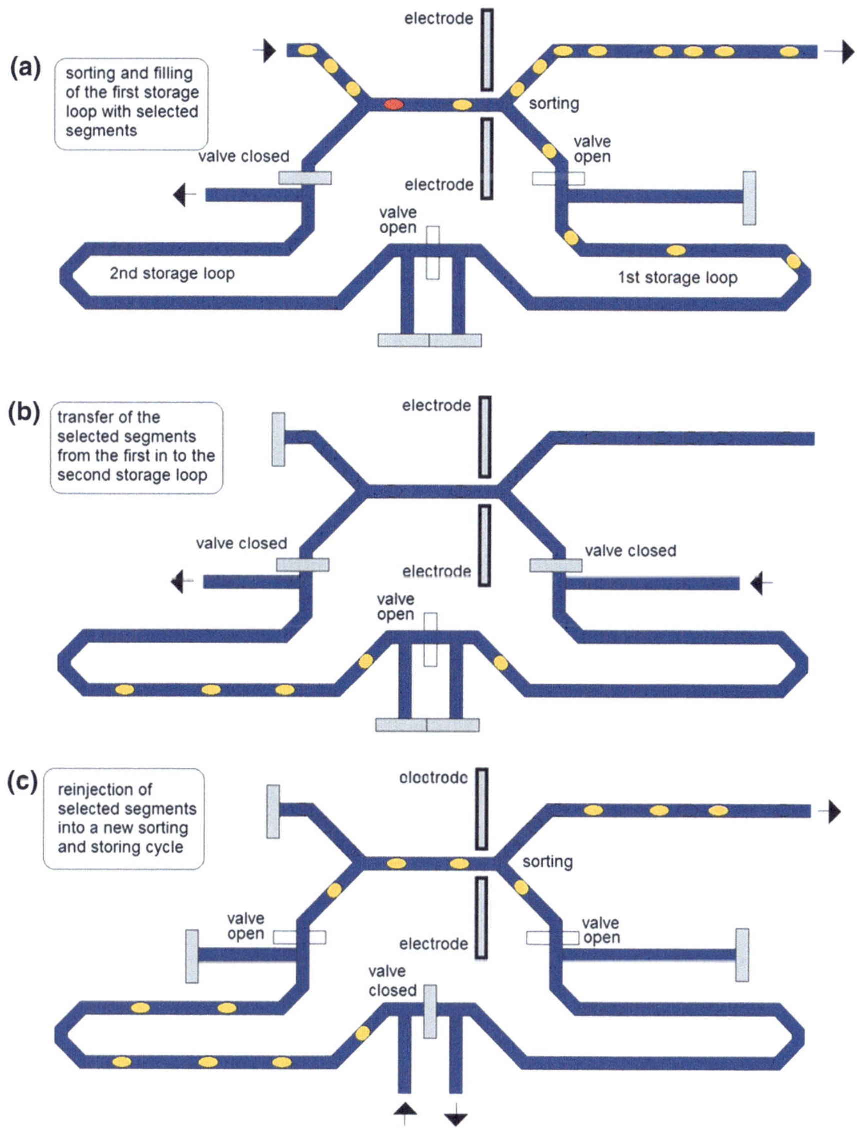

Fig. 3.12 Concept of a microfluidic double loop arrangement for realization of several repeated cycles of segment storage and electrical sorting: (**a**) first sorting and storage of selected segments in the right storage loop, (**b**) transfer of segments from the fist to the second storage loop, (**c**) reinjection and next sorting of segments

of liquid compartments. Single cell techniques and, in particular single molecule techniques demand an ultimate down-scaling of volume.

Apart from biological operations, the switching of fluid segments is also of interest for the synthesis and screening of new organic and inorganic substances and for the generation and application of new types of nanomaterials. Highly parallelized screenings for experimental combination of molecular modules and nanosized particles, the controlled nanoassembly and the realization of combinatorial coupling of different nanoparticles could become an exciting field for micro fluid segment technique and for the application of electrical switching in microfluidic systems. An accurate control of transport of particles and molecules in micro fluid segments could also support new developments in supramolecular chemistry.

The method of droplet switching presented in Sect. 5 has several advantages over the previously presented droplet switching techniques. Firstly, there is no need to inject charge by direct contact with an electrode: contact with an electrode that is not hydrophobic can lead to fouling of the surface that will lead to cross-contamination of droplets or imperfect droplet transport across the surface. In the presented switching method there is no need for surface contact or mechanical deformation of the droplet in a constricted microfluidic channel. Secondly, the switching operation can be compactly performed with one set of electrodes being responsible for both droplet charging (through induced charge) and switching. Thirdly, the charge is induced on the surface of the droplet which means the charge disperses after switching and subsequent switching operations can be performed unproblematically. These advantages are also shared by dielectrophoretic droplet switching.

In this context, the down-scaling of segmented-flow technique is a further promising field for future developments. Whereas the electrical sorting of free movable droplets with volumes in the picoliter range has already been established, there is a recent need for fast switching of fluid segments in the picoliter range in order to realize logical fluidic operations in chip-based micro fluidic networks.

Acknowledgments Financial support of the BMBF/VDI/VDE IT (project SOD-Kult, FKZ: 16SV5065) for the research on droplet switching in the frame of segment-on-demand technique for biological applications is gratefully acknowledged. Brian Cahill would like to thank the European Community for financially supporting the Marie Curie ERG project EWETDYNAM under reference number PERG05-GA-2009-247784.

References

1. Y.-M. Jung, I.S. Kang, A novel actuation method of transporting droplets by using electrical charging of droplet in a dielectric fluid. Biomicrofluidics **3**, 022402–11 (2009)
2. B. Ahn, K. Lee, R. Louge, K.W. Oh, Concurrent droplet charging and sorting by electrostatic actuation. Biomicrofluidics **3**, 044102 (2009)
3. B. Ahn, K. Lee, R. Panchapakesan, K.W. Oh, On-demand electrostatic droplet charging and sorting. Biomicrofluidics **5**, 024113 (2011)
4. D.J. Im, J. Noh, N.W. Yi, J. Park, I.S. Kang, Influences of electric field on living cells in a charged water-in-oil droplet under electrophoretic actuation. Biomicrofluidics **5**, 044112–10 (2011)

5. D.J. Im, M.M. Ahn, B.S. Yoo, D. Moon, D.W. Lee, I.S. Kang, Discrete electrostatic charge transfer by the electrophoresis of a charged droplet in a dielectric liquid. Langmuir **28**, 11656–11661 (2012)
6. D.W. Lee, D.J. Im, I.S. Kang, Electrophoretic motion of a charged water droplet near an oil-air interface. Appl. Phys. Lett. **100**, 221602-4 (2012)
7. S. Fiedler, S.G. Shirley, T. Schnelle, G. Fuhr, Dielectrophoretic sorting of particles and cells in a microsystem. Anal. Chem. **70**, 1909–1915 (1998)
8. R. Pethig, Review article–dielectrophoresis: status of the theory, technology, and applications. Biomicrofluidics **4**, 022811 (2010)
9. T.B. Jones, M. Gunji, M. Washizu, M.J. Feldman, Dielectrophoretic liquid actuation and nanodroplet formation. J. Appl.Phys. **89**, 1441–1448 (2001)
10. J. Zeng, T. Korsmeyer, Principles of droplet electrohydrodynamics for lab-on-a-chip. Lab on a Chip **4**, 265–277 (2004)
11. O.D. Velev, B.G. Prevo, K.H. Bhatt, On-chip manipulation of free droplets. Nature **426**, 515–516 (2003)
12. J.R. Millman, K.H. Bhatt, B.G. Prevo, O.D. Velev, Anisotropic particle synthesis in dielectrophoretically controlled microdroplet reactors. Nat. Mater. **4**, 98–102 (2005)
13. V. Rastogi, O.D. Velev, Development and evaluation of realistic microbioassays in freely suspended droplets on a chip. Biomicrofluidics **1**, 014107–17 (2007)
14. J. Schwartz, J. Vykoukal, P. Gascoyne, Droplet-based chemistry on a programmable microchip. Lab on a Chip **4**, 11–17 (2004)
15. D. Chatterjee, B. Hetayothin, A. Wheeler, D. King, R. Garrell, Droplet-based microfluidics with nonaqueous solvents and solutions. Lab on a Chip **6**, 199–206 (2006)
16. J.-C. Baret, O.J. Miller, V. Taly, M. Ryckelynck, A. El-Harrak, L. Frenz, C. Rick, M.L. Samuels, J.B. Hutchison, J.J. Agresti, D.R. Link, D.A. Weitz, A.D. Griffiths, Fluorescence-activated droplet sorting (FADS): efficient microfluidic cell sorting based on enzymatic activity. Lab on a Chip **9**, 1850–1858 (2009)
17. K. Ahn, C. Kerbage, T.P. Hunt, R.M. Westervelt, D.R. Link, D.A. Weitz, Dielectrophoretic manipulation of drops for high-speed microfluidic sorting devices. Appl. Phys. Lett. **88**, 024104-3 (2006)
18. X. Niu, M. Zhang, S. Peng, W. Wen, P. Sheng, Real-time detection, control, and sorting of microfluidic droplets. Biomicrofluidics **1**, 044101–12 (2007)
19. F. Guo, X.-H. Ji, K. Liu, R.-X. He, L.-B. Zhao, Z.-X. Guo, W. Liu, S.-S. Guo, X.-Z. Zhao, Droplet electric separator microfluidic device for cell sorting. Appl. Phys. Lett. **96**, 193701 (2010)
20. M.Z. Bazant, T.M. Squires, Induced-charge electrokinetic phenomena. Curr. Opin. Colloid Interface Sci. **15**, 203–213 (2010)
21. M.E. Leunissen, A. van Blaaderen, A.D. Hollingsworth, M.T. Sullivan, P.M. Chaikin, Electrostatics at the oil-water interface, stability, and order in emulsions and colloids. Proc. Nat. Academy Sci. U.S.A **104**, 2585–2590 (2007)
22. M. Budden, S. Schneider, P.M. Günther, T. Henkel, M. Kielpinski, J.M. Köhler (2012) Splitting and switching of micro fluid segments in closed channels for serial experiments, T3-P-14, *P*roceedings of IMRET XII (Lyon, 20–22 Febr. 2012), pp.291–292, Poster
23. M. Budden, S. Schneider, G. A. Groß, M. Kielpinski, T. Henkel and J. M. Köhler, Splitting and switching of micro fluid segments in closed channels for chemical operations in the segment-on demand technology". Chem. Eng. J.,(In Press). doi:10.1016/j.cej.2012.07.104
24. M. Budden, S. Schneider, G. A. Groß, M. Kielpinski, T. Henkel, B. Cahill and J. M. Köhler Microfluidic encoding: generation of arbitrary droplet sequences by electrical switching in microchannels. Sens. Actuators A: Phys.,(In Press). doi:10.1016/j.sna.2012.10.013
25. T. Harvey, R. Wood, G. Denuault, H. Powrie, Effect of oil quality on electrostatic charge generation and transport. J. Electrostat. **55**, 1–23 (2002)
26. R.M. Ehrlich, J.R. Melcher, Bipolar model for traveling-wave induced nonequilibrium double-layer streaming in insulating liquids. Phys. Fluids **25**, 1785–1793 (1982)

27. M. Zahn Washabaugh, J. Melcher, Electrohydrodynamic traveling-wave pumping of homogeneous semi-insulating liquids. IEEE Transact. Electr. Insulation **24**, 807–834 (1989)
28. S. Bart, L. Tavrow, M. Mehregany, J. Lang, Microfabricated electrohydrodynamic pumps. Sens. Actuators A: Phys. **21**, 193–197 (1990)
29. F. Moesner, P. Buhler, D. Politano, P. Prati, Electrohydrodynamic motor for tiny vessels. IEEE/ASME Int. Conf. Adv. Intell. Mechatron. **97**, 66–69 (1997)
30. Ajdari, Pumping liquids using asymmetric electrode arrays. Physical Review E **61**:R45 LP – R48 (2000)
31. B.P. Cahill, L.J. Heyderman, J. Gobrecht, A. Stemmer, Electro-osmotic streaming on application of traveling-wave electric fields. Phys. Rev. E **70**, 036305-14 (2004)
32. P. Ramos, A. Garcia, A. Gonzalez, AC electrokinetic pumping of liquids using arrays of microelectrodes. Bioeng. Bioinspired Syst. II **5839**, 305–313 (2005)
33. J.R. Melcher, *Continuum Electromechanics* (MIT Press, Cambridge, 1981)
34. M. Hughes, H. Morgan, M. Flynn, The Dielectrophoretic behavior of submicron latex spheres: influence of surface conductance. J. Colloid Interface Sci. **220**, 454–457 (1999)
35. I. Adamczewski, J.H. Calderwood, Viscosity and charge carrier mobility in the saturated hydrocarbons. J. Phys. D: Appl. Phys. **8**, 1211 (1975)
36. M. Yazdani and J. Seyed-Yagoobi (2009) Effect of charge mobility on electric conduction driven dielectric liquid flow. Electrostatics Joint Conference 2009
37. J.M. Köhler, A. Funfak, J. Cao, D. Kürsten, S. Schneider, P.M. Günther, Addressing of concentration spaces for bioscreenings by micro segmented flow with microphotometric and microfluorimetric detection. Opt. Nano- and Microsyst. for Bioanaly. **10**, 47–81 (2012)
38. Cao, D. Kursten, S. Schneider, A. Knauer, P.M. Gunther and J.M. Kohler, Uncovering toxicological complexity by multi-dimensional screenings in microsegmented flow: modulation of antibiotic interference by nanoparticles. Lab on a Chip,**12**, 474–484 (2012)
39. Knauer, S. Schneider, F. Möller, A. Csáki, W. Fritzsche and J.M. Köhler Screening of Plasmonic Properties of Composed Metal Nanoparticles by Combinatorial Synthesis in Micro Fluid Segment Sequences. Chem. Eng. J.(In press). doi:10.1016/j.cej.2012.10.008

Chapter 4
Chip-Integrated Solutions for Manipulation and Sorting of Micro Droplets and Fluid Segments by Electrical Actuation

Lars Dittrich and Martin Hoffmann

Abstract The following sections present the novel approach of combining continuous flow analysis (CFA) systems with the advantages of digital microfluidic (DMF) concepts. More precisely, the integration of electrostatic actuation into segmented flows (SF) is demonstrated. It is not focused on whether the actuation method should contingently be termed electrowetting rather than electrostatics, but in fact the intention is to show how well-known microfluidic concepts can be combined into an innovative and promising technique. The investigations show that a reasonable number of standard MEMS manufacturing processes are sufficient to fabricate the appropriate fluidic chips utilizing standard materials such as silicon and glass. The potentials emerging from the new systems are explicated, and a look ahead is given to possible applications as well as to research activities at issue.

4.1 Basics for Chip Integration of Droplet Actuators

4.1.1 Continuous Flow Analysis (CFA)

Current microfluidic analysis systems (often termed as µTAS–micro total analysis systems) can deliberately be classified into two fundamental approaches:

- Continuous flow analysis (CFA) and
- Digital microfluidics (DMF).

CFA systems process continuous streams of the analyte within closed channel structures whereas analyte flow is achieved by integrated micro pumps, external pumps or pressure gradients, or the utilization of capillarity enhanced by electrokinetic effects.

L. Dittrich (✉) · M. Hoffmann
Department Micromechanical Systems, Technische Universität Ilmenau, PF 10 05 65, 98684 Ilmenau, Germany
e-mail: lars.dittrich@tu-ilmenau.de

J. M. Köhler and B. P. Cahill (eds.), *Micro-Segmented Flow*, Biological and Medical Physics, Biomedical Engineering,
DOI: 10.1007/978-3-642-38780-7_4,

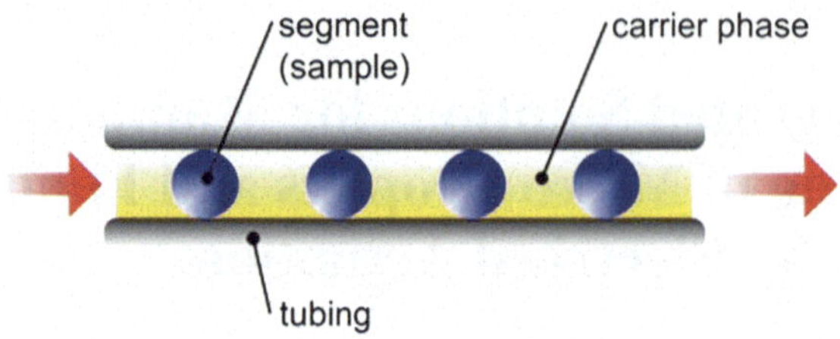

Fig. 4.1 Illustration of segmented flow (SF)

So-called segmented flows (SF; Fig. 4.1) are counted among CFA systems as well. In this special case, the continuous stream of the (liquid) analyte is compartmentalized into segments of a certain volume by systematically disrupting the analyte stream by well-defined "spacers" composed of a second fluid (i.e. a liquid or gas) being immiscible with the segments. CFA systems are designed for very specific tasks that impede their flexible application. The realization of more complex and challenging tasks in CFA systems such as mixing or separating certain media requires unreasonable efforts. On the other hand and inherent to the functional principle, the great advantages of CFA systems lie in the largely suppressed evaporation of liquid media as well as the protection of the analyte against environmental influences such as dust, harsh atmospheres etc. Even though CFA techniques like segmented flows indubitably provide the treatment of numerous uniform samples under similar conditions, the samples cannot be controlled and processed independently. Stopping the entire sample chain is inevitable in order to analyze or process selected samples making this method time-consuming and thus inefficient to some extent.

4.1.2 Digital Microfluidics (DMF)

Digital microfluidic (DMF) systems process discrete quantities of the analyte, e.g. droplets in case of a liquid. They usually cope without channel-like structures. The analyte compartments are controlled individually, directly, and independently, for example by moving them on an electrode array. In DMF systems, microfluidic functions are reduced to a number of basic fluidic operations (Fig. 4.2), e.g. the mixing of two media can be reduced to uniting two droplets of different media and subsequent agitation ("shaking"). This digitalization of fluidic functions facilitates hierarchical system architectures and thus the creation of flexible and scalable systems. By analogy, the DMF approach can be compared to programmable logic controls (PLC) whereas the CFA approach corresponds to hard-wired programmed logic controls. It is evident that for DMF systems all operations can be realized by software changes without hardware modifications. The resulting fluidic systems are often termed as labs on a chip (LoC) [2]. Anyway, DMF comprises planar (2D) assemblies whereas CFA rather refers to linear (1D) systems.

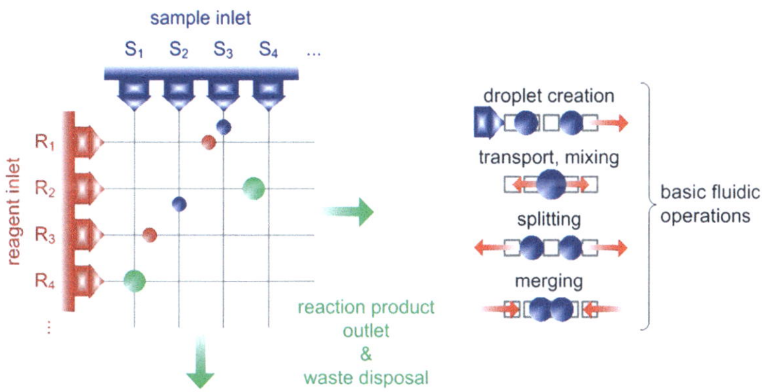

Fig. 4.2 Concept (*left*) and basic operations (*right*) of digital microfluidics (according to [1])

4.1.3 Labs on a Chip (LoC) and Micro Total Analysis Systems (μTAS)

The terms lab on a chip (LoC) and micro total analysis system (μTAS) are synonymously used for miniaturized fluidic systems providing the technical prerequisites for chemical or biological analyses [3, 4]. Miniaturization goes along with numerous advantages, amongst others [3, 5, 6]:

- Low reagent consumption and thus cost reduction,
- Low volume of waste materials,
- Integratability including coupling to sample preparation as well as to further analysis systems,
- High level of automation,
- Short sample response times enabling an improved process control,
- Short sample processing times and thus high throughput and the
- Option of task-specific interconnection of numerous systems, each one serially connected or in parallel portable systems.

μTAS usually comprise miniaturization of all functions of a conventional analysis method, e.g. pumps or dispensing systems, valves, branches, mixers, reaction chambers, separating devices, sieves and filters, detectors as well as electronics for communications and control. Holistic miniaturization is achieved by applying the manufacturing technologies known from microelectronics to the single components of conventional analysis systems [7–9].

Modular architectures are suggested in [10] for the design of exceptionally flexible μTAS that can be employed for many applications, e.g. following the partitioning:

1. biomedical functions (media identification, precision dispensing, analysis, and monitoring),
2. elementary functions (fluid transport, mixing, purging, filtration, analysis, detection, monitoring), and

3. basic components (buffers, channels, valves, mixers, branches, etc.).

Many microfluidic systems are based on continuous analyte streams within micro channels which are made of glass or transparent polymers for the purpose of surveillance [5]. Especially when dealing with SF, operations mentioned in the second hierarchical level are only realizable with restrictions. However, contrary to conventional microfluidics, digital microfluidics treats compartmentalized liquid agents ("droplets") whose typical diameters equal the dimensions of microchannels [5, 10]. In many cases, segments are embedded in (organic) solvents which are immiscible with the droplet medium. Within digital microfluidic LoC's, droplet manipulation takes place on planar surfaces. Important operations are droplet creation, transport, splitting, merging, and mixing (cf. Fig. 4.2). Several mechanisms can be utilized for droplet manipulation [10], e.g. electrowetting (on dielectrics), dielectrophoresis, thermocapillarity, or surface acoustic waves. Electrowetting is the most frequently used and most powerful actuation principle. So far, numerous contributions to the development of DMF LoC's utilizing electrowetting actuation have been published, e.g. [11].

4.1.4 Combining CFA Systems with DMF Concepts

It is obvious that the state of the art can be significantly extended by combining CFA systems with DMF concepts and thus to wed the two fundamental approaches to microfluidic problems. A considerable novelty is the manipulation of liquid segments within streaming two-phase segmented flows by applying electric fields. Figure 4.3 visualizes the idea.

Recent research works comprised the proof of concept. Heading from the idea towards novel LOCs, a number of challenges are to be mastered: Selecting appropriate material systems is essential for future applications, especially in view of disposable mass products (e.g. in home care diagnostics) and energy efficiency. Both demands open up the possibility of miniaturized fluidic systems whereby especially direct segment actuation utilizing electric fields renders the need for mechanical components such as pumps or valves redundant. The final goal of recent research works

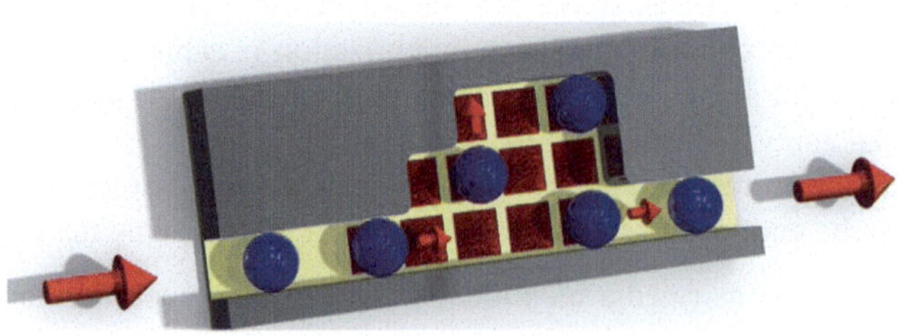

Fig. 4.3 Visualization of the novel approach: manipulation of liquid segments within streaming two-phase fluid flows

was to develop a shunting and sorting technique based on the electrowetting effect (i.e. electrostatic actuation) which allows changing the order of compartments within a CFA system. Two-phase systems in combination with the electrowetting effect are well-known, for example in liquid optical lenses [12]. The challenge was to develop the presented dynamical concept emanating from established stationary systems: single segments out of the serial order within a standard segmented flow shall be resorted (i.e. their order shall be changed) by shunting them on an electrode array. Automatically, the required arrangement enriches the system by the already mentioned functions and operations of DMF systems. Keeping the segments embedded within a carrier medium ensures a constant droplet volume by preventing evaporation. Both avoiding evaporation and nevertheless enabling all established features of CFA and DMF systems is the great advantage of the presented approach. In this manner, an LOC is created that combines the features and advantages of CFA with those of DMF systems.

A distinction is drawn between two basic application scenarios:

1. Processing station: Single segments are preselected (e.g. by a capable sensor by means of certain criteria) and decoupled from the continuously streaming segmented flow within the station in order to be processed (i.e. monitored, analyzed, modified, etc.) while being electrostatically detained at a stand-by position beyond

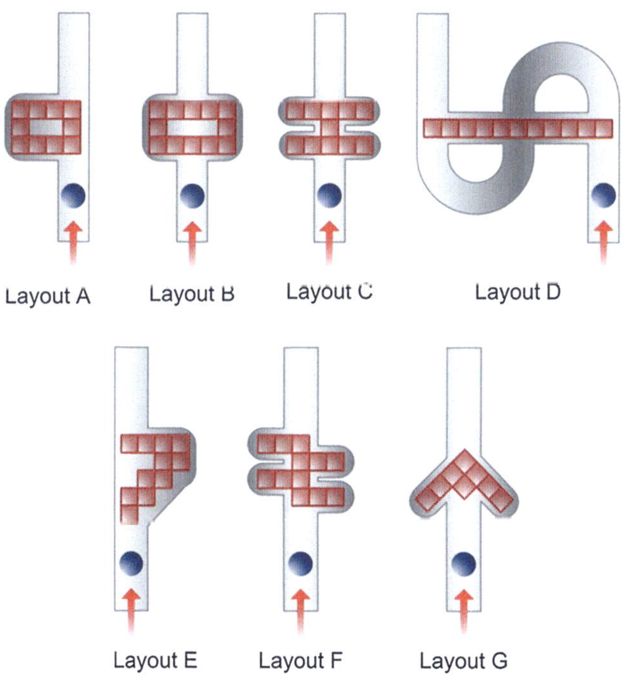

Fig. 4.4 Schematic chip layouts representing the two application scenarios processing station (layouts A to G) and sorting station (layouts A, B, D, and E)

the continuous flow. Once the processing is completed, the segment is sorted into the streaming chain again in order to be driven to the next station or to be disposed.
2. Sorting station: The order of segments can be changed within the station. On the one hand, this can be realized by decoupling selected segments from the continuously streaming segmented flow and sorting them into the SF at a later position. On the other hand, bypasses can be utilized in order to let single segments "overtake" some of their predecessors by electrostatic actuation. However, segment sorting might be quite helpful in order to preselect a number of segments which are relevant for a certain investigation.

Figure 4.4 shows sample layouts representing these two application scenarios.

4.2 Modeling and Simulation for Electrostatic Actuation in Integrated Devices

4.2.1 General Aspects of Modeling of Electrostatic Actuation

The magnitude of the electrostatic actuation forces is crucial for the determination of the flow velocity up to which a single segment can be reliably manipulated within the streaming carrier medium. In order to estimate the actuation forces that can be exerted on a single droplet by applying an electric field, a simple analytical model is introduced, and the configuration depicted in Fig. 4.6 was simulated in Ansoft Maxwell 14 and Comsol 4.2a, respectively. Afterwards, the derived electrostatic forces are compared to the flow forced exerted on a single segment that is electrostatically fixed within the streaming carrier medium.

Segments as used for the experiments and regarded here typically feature a volume of 60 nl. Since spherical segments exhibit the smallest surface-to-volume ratio of all geometrical bodies and thus represent an energetic minimum it is desirable to design the microfluidic chip in a way that the spherical shape can be maintained at any position within the microfluidic channels. Assuming spherical segments, a volume of 60 nl results in a segment diameter of 486 μm. The segmented flows have been provided within coils of Teflon® tubing featuring an inner diameter of 500 μm and an outer diameter of $\frac{1}{16}$ in. The channel should therefore not be smaller than 500 μm in diameter or its lateral and vertical dimension, respectively, throughout the whole microfluidic chip as well.

From here, the indices denote either the segment medium (s) or the carrier phase (c).

4.2.2 Modeling of Electrostatic Actuators

The simple analytical model of an electrostatic actuator containing a movable dielectric serves for a first estimation of the forces that can be exerted on a single segment by electrostatic actuation as well as for the verification of the simulation results which are introduced later.

Emanating from the virtual displacement principle, the total system energy depends on the amount of overlap between dielectric and electrode area. Mechanically displacing the dielectric within the capacitor by an infinitesimal distance **dx** is synonymous with the work done by the electric field. The electrodes of the introduced configuration are connected to a voltage source. Thus, not the charge but the voltage **U** is kept constant during dielectric displacement, and the work **U dQ** is done by the voltage source as well. With $\mathbf{dQ} = \mathbf{U\,dC}$, the energy balance reads

$$\mathbf{dW_{field}} = \frac{1}{2}\mathbf{U^2 dC} = -\mathbf{F_x dx} + \mathbf{U^2 dC}. \tag{4.1}$$

The force exerted on one segment in the x-direction can be derived from

$$\mathbf{F_x} = \frac{\mathbf{dW_{field}}}{\mathbf{dx}} = \frac{1}{2}\mathbf{U}^2\frac{\mathbf{dC}}{\mathbf{dx}} \tag{4.2}$$

As simplification for the calculation of the capacitance, a single segment is modeled as a cuboid exhibiting a sphere-equivalent volume $\mathbf{V_s} = 60\,\mathrm{nl} \stackrel{\mathrm{def}}{=} \mathbf{V_{cuboid}} = \mathbf{h_s\,a^2}$ (cf. Fig. 4.5). From electrostatic actuators it is well-known that the force generated is constant and independent on the amount of overlap between electrodes and dielectric [13]. Hence, the dimensions of the cuboid are approximated in z-direction in the first instance and $\mathbf{h_s}$ is set to 486 μm . Thus, the edge length **a** of the cuboid is 352 μm.

The total capacitance of the system (Fig. 4.5, left) can be derived from the equivalent network (Fig. 4.5, right) and depends on the amount of overlap only. The capacitance of a parallel-plate capacitor is derived from

$$\mathbf{C} = \varepsilon_0\varepsilon_\mathbf{r}\frac{\mathbf{A}}{\mathbf{d}}. \tag{4.3}$$

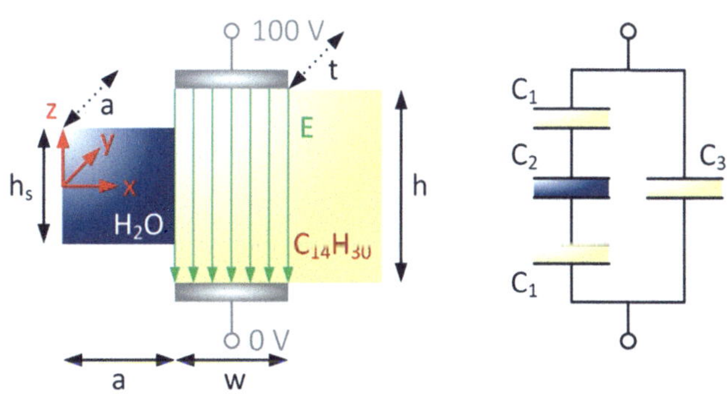

Fig. 4.5 Simple analytical model for the calculation of the capacitance and the actuation force–illustration (*left*) and equivalent circuit (*right*)

The total capacitance of the configuration is

$$\mathbf{C_{tot}} = \left(\frac{2}{\mathbf{C}_1} + \frac{1}{\mathbf{C}_2}\right)^{-1} + \mathbf{C}_3. \tag{4.4}$$

Since $\mathbf{a} > \mathbf{t}$, the single capacitances are

$$\mathbf{C}_1 = \varepsilon_0\, \varepsilon_\mathbf{c} \frac{\mathbf{2xt}}{\mathbf{h} - \mathbf{h_s}},\ \mathbf{C}_2 = \varepsilon_0\, \varepsilon_\mathbf{s} \frac{\mathbf{xt}}{\mathbf{h_s}},\ \text{and}\ \mathbf{C}_3 = \varepsilon_0\, \varepsilon_\mathbf{c} \frac{(\mathbf{w} - \mathbf{x})\mathbf{t}}{\mathbf{h}}, \tag{4.5}$$

respectively. Thus, the total capacitance sums up to

$$\mathbf{C_{tot}} = \frac{\varepsilon_0\, \varepsilon_\mathbf{c}\, \varepsilon_\mathbf{s}\, \mathbf{xt}}{\varepsilon_\mathbf{s}\,(\mathbf{h} - \mathbf{h_s}) + \varepsilon_\mathbf{c}\, \mathbf{h_s}} + \varepsilon_0\, \varepsilon_\mathbf{c} \frac{(\mathbf{w} - \mathbf{x})\, \mathbf{t}}{\mathbf{h}} \tag{4.6}$$

and the force

$$\mathbf{F_x} = \frac{1}{2}\mathbf{U}^2 \frac{\mathbf{dC}}{\mathbf{dx}} = \frac{1}{2}\left(\frac{\varepsilon_0\, \varepsilon_\mathbf{c}\, \varepsilon_\mathbf{s}\, \mathbf{t}}{\varepsilon_\mathbf{s}\,(\mathbf{h} - \mathbf{h_s}) + \varepsilon_\mathbf{c}\, \mathbf{h_s}} - \varepsilon_0\, \varepsilon_\mathbf{c} \frac{\mathbf{t}}{\mathbf{h}}\right)\mathbf{U}^2 \tag{4.7}$$

With the material properties from Table 4.1, the electrode depth and width $\mathbf{t} = \mathbf{w} = 300\,\mu\text{m}$, the channel height $\mathbf{h} = 500\,\mu\text{m}$, the dielectric edge length $\mathbf{a} = 352\,\mu\text{m}$ and height $\mathbf{h_s} = 486\,\mu\text{m}$, and the applied voltage $\mathbf{U} = 100\,\text{V}$, the force is calculated to be $\mathbf{F_x} = 979\,\text{nN}$.

For the simulations, the liquid droplet is now modeled as a solid and ideally dielectric (i.e. non-conductive) sphere between the two electrodes of a parallel-plate capacitor with the dielectric constant of pure water. The droplet is surrounded by tetradecane. Relevant material properties are compiled in Table 4.1.

The configuration of Fig. 4.6 represents a section of a microfluidic channel containing the segment chain. Based on a segment volume of $\mathbf{V_s} = 60\,\text{nl}$ and thus a droplet radius of $\mathbf{R_s} = 243\,\mu\text{m}$, a square channel cross section with an edge length of $500\,\mu\text{m}$ has been chosen. The channel is completely filled with tetradecane. A control electrode representing an element of a control electrode array is located at the

Table 4.1 Compilation of relevant material properties at room temperature and atmospheric pressure

Material	Property and source	Value
Water (H_2O)	Density [14]	$\rho_s = 998\,\frac{kg}{m^3}$
	Dynamic viscosity [14]	$\eta_s = 1.003 \cdot 10^{-3}\,\text{Pa} \cdot \text{s}$
	Static dielectric constant [15]	$\varepsilon_s = 80.29$
Tetradecane ($C_{14}H_{30}$)	Density [16]	$\rho_c = 762.9\,\frac{kg}{m^3}$
	Dynamic viscosity [17]	$\eta_c = 2.08 \cdot 10^{-3}\,\text{Pa} \cdot \text{s}$
	Static dielectric constant [18]	$\varepsilon_c = 2.06$

channel top at an electric potential of 100 V. In order to ensure an overlap for reliable droplet actuation, the horizontal section of the control electrode is 300 × 300 μm. The channel bottom forms a common ground electrode at zero potential with a horizontal section of 1300 × 1000 μm. Depending on the droplet position **x**, the force $\mathbf{F_x}$ exerted on the droplet in x-direction is to be determined. Figure 4.7 shows the simulation results.

In Fig. 4.7, the peaks of the electrostatic force in x-direction lie slightly below the result of the analytical model (979 nN). This can be explained by the fact that the simulations have been carried out using a spherical dielectric whereas a cuboid was employed for the analytical calculation whose z-dimension equaled the segment diameter. Further limitations of the modeling are discussed in Sect. 4.5. However, the result of the analytical model is of the same order of magnitude as the simulation results. The analytical model proves and confirms the simulation.

Further comparisons lend credence to the simulation results: the weight force of a single segment is $\mathbf{F_W} = m\mathbf{g} = \rho_s \mathbf{V_s g} = 587$ nN. Its buoyant force in tetradecane is $\mathbf{F_B} = \rho_c \mathbf{V_s g} = 449$ nN.

4.2.3 Electrostatic Forces in Relation to Flow Forces

Considering a single segment that is electrostatically retained within a sorting or processing station, respectively, the flow forces acting on this segment by the streaming carrier phase are to be estimated. Therefore, the REYNOLDS number is calculated [14] with the density ρ_c, the free-stream velocity $\mathbf{v}_\infty$ (i.e. far away from objects or boundaries), the characteristic dimension **d**, and the dynamic viscosity η_c:

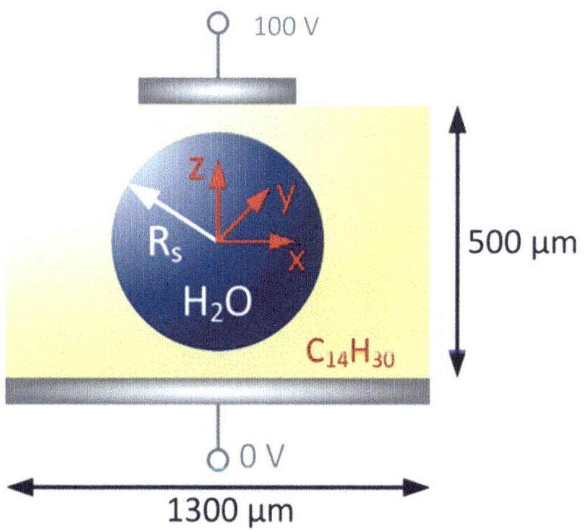

Fig. 4.6 Simulated configuration for the estimation of electrostatic actuation forces

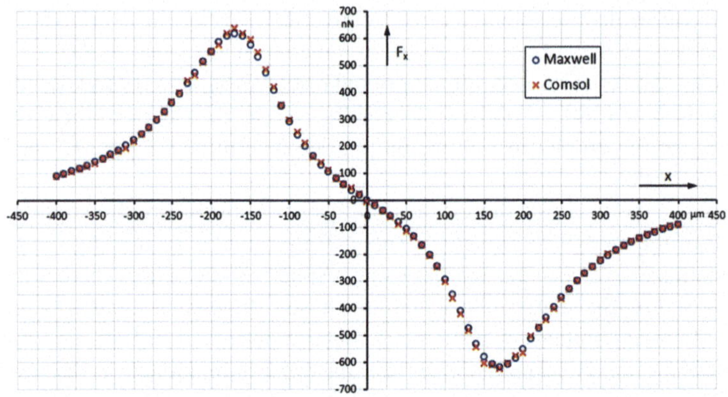

Fig. 4.7 Comparison of the simulation results from Maxwell 14 and Comsol 4.2a: electrostatic force in x-direction $\mathbf{F_x}$ versus droplet position $\mathbf{x}$

$$\mathbf{Re} = \frac{\rho_\mathbf{c}\,\mathbf{v}_\infty \mathbf{d}}{\eta_\mathbf{c}} \tag{4.8}$$

The flow rate has to be related to a square channel with an edge length of 500 µm so that the flow velocity can be calculated. For segmented flows, typical flow rates lie in the range between 30 and 50 $\frac{\mu l}{min}$. Entering the appropriate values and a maximum flow rate of 100 $\frac{\mu l}{min}$, one obtains a free-stream velocity of $\mathbf{v}_\infty = 6.7\,\frac{mm}{s}$ and a maximum Reynolds number of $\mathbf{Re} = 1.22$. As typical for microfluidic channel flows, no turbulence is to be expected since stationary ring-shaped turbulences occur for $\mathbf{Re} \geq 20$ only, [19].

The aqueous segment is now assumed to be immersed in the streaming carrier medium (tetradecane) whereas the segment itself is modeled as a rigid body which is assumed to be fixed to its position–in this case by electrostatic forces. The force $\mathbf{F_{flow}}$ exerted on one segment can be calculated with $\mathbf{A_p}$ being the projected area of the segment and $\xi(\mathbf{Re})$ being the drag coefficient of the flown body [20, 21]:

$$\mathbf{F_{flow}} = \xi\,(\mathbf{Re})\,\mathbf{A_p}\rho_\mathbf{c}\frac{\mathbf{v}_\infty^2}{2} \tag{4.9}$$

For $0 \leq \mathbf{Re} \leq 3 \cdot 10^5$ the drag coefficient can be calculated by [19]

$$\xi\,(\mathbf{Re}) = \frac{24}{\mathbf{Re}} + \frac{3.73}{\sqrt{\mathbf{Re}}} - \frac{4.83 \cdot 10^{-3}\sqrt{\mathbf{Re}}}{1 + 3 \cdot 10^{-6}\mathbf{Re}^{\frac{3}{2}}} + 0.49 \tag{4.10}$$

Inserting Re for the maximum flow rate (and thus the maximum flow velocity) results in a drag coefficient of $\xi\,(\mathbf{Re_{max}}) = 23.5$. Entering the material properties from Table 4.1 one obtains a flow force of $\underline{\mathbf{F_{flow}} = 74\,\mathrm{nN}}$. It can be concluded, that the force generated by the electric field can easily compensate the flow forces exerted on one segment when retained at a certain position within a sorting or processing station while the carrier medium stream is maintained.

4.3 Technology Considerations and Fabrication of Chip Devices for Electrostatic Actuation

4.3.1 Materials and Basic Concept

Silicon and glass as basic materials have been employed because of the following benefits:

- Materials used have to be resistant to the segmented flow liquids.
- The wide range of available standard MEMS technologies can be used for manufacturing the components.
- Thus, economically priced fluidic chips can be fabricated in view of single-use applications, e.g. for home care diagnostics etc.
- Glass as transparent material facilitates the optical monitoring of the aspired sorting movements of single segments (cf. [7]).

Figure 4.8 shows the functional layers of the chip.

For experiments, the segments consisted of water (H_2O). Tetradecane ($C_{14}H_{30}$) was used as continuous phase carrying the segments. Additionally, 2% (v/v) bromophenol blue ($C_{19}H_{10}Br_4O_5S$) was used as dye for the segments. (cf. Table 4.1).

4.3.2 Technology Concept and Manufacturing

Figure 4.9 shows a schematic drawing of the finished microfluidic chip. The intended microfluidic chip can be partitioned into an upper (cover plate, Fig. 4.10) and a lower part (substrate, Fig. 4.10). Both parts are processed independently and assembled in a final step. Bonding is realized by coating both parts with a 1 μm thick layer

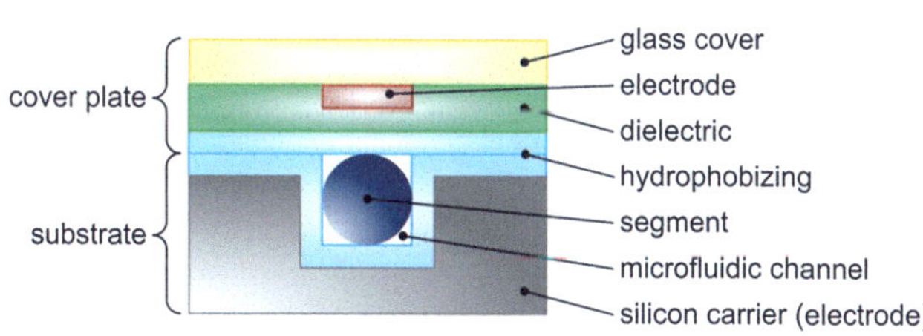

Fig. 4.8 Cross-sectional view through the microfluidic chip from a technological view

of DuPont™ Teflon® AF. The coated parts are aligned and exposed to elevated temperature in an oven to form the bond.

The process sequences for both parts are depicted in Fig. 4.10. Table 4.2 contains the relevant process parameters for the main fabrication steps.

The chrome masks for UV lithography featured a resolution of 5 μm. Several layouts representing the two basic application scenarios had been placed on the masks. Figure 4.4 illustrates the manufactured layouts. Each layout represents one of the mentioned application scenarios (sorting or/and processing station) in its specific manner and can be adapted to most different applications. Thus, the shown layouts should be considered as representative examples. The chips have been manufactured as described in Fig. 4.10 as well as in Table 4.2.

4.4 Experimental Realization of Chip-Integrated Electrostatic Actuators

A chip holding fixture clamps the fabricated microfluidic chips (Fig. 4.11). Within this fixture, the chip is forced down by two clamps so that resilient contact pins contact the ITO contact pads. The chip can be mechanically adjusted within the

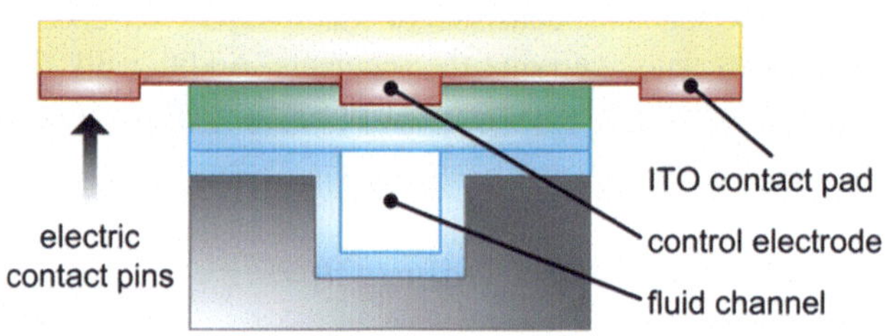

Fig. 4.9 Schematic of the fluidic chip, consisting of substrate (Fig. 4.10, *left*) and cover (Fig. 4.10, *right*)

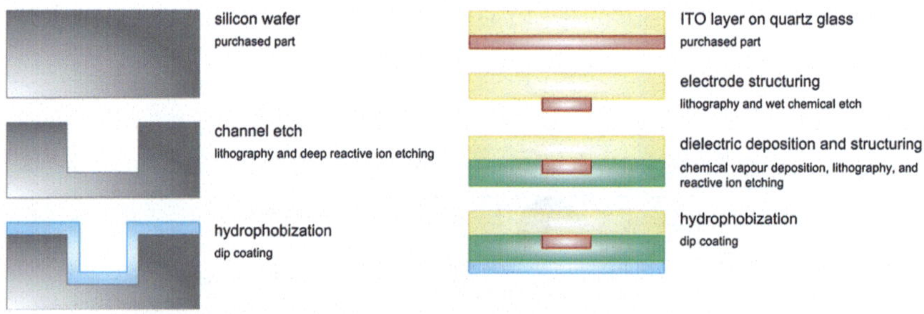

Fig. 4.10 Process sequence for the substrate (*left*) and the cover plate (*right*) of the microfluidic chip

Table 4.2 Process parameters for the main fabrication steps

No.	Step	Parameters
A–Substrate		
	Silicon Wafer	Ø100 mm thickness 1 mm double-side polished (DSP), resistivity $< 0.005\ \Omega$cm, p-doped (Boron)
A1	Silicon channel etch	Lithography AZ® 9260, 10 μm thickness Etch cycle at 600 W (ICP), 18 W(platen), 35 mTorr, 20 °C, and SF_6: 130 sccm, O_2: 13 sccm for 13 s; Deposition cycle at 600 W (ICP), 0 W(platen), 18 mTorr, 20 °C, and C_4F_8: 85 sccm for 7 s Altogether for 140 min; afterwards isotropic overetch for 2 min *Surface Technology Systems ICP Multiplex ASE*
A2	DuPont™ AF hydrophobization (Grade 601S2-100-6; 500 nm)	Dip coating extraction velocity 50 $\frac{mm}{min}$, 5min, drying at room temperature, bake at 40 °C for 15 min, subsequently at 160 °Cfor 180 min *Dip Coater KSV-D*, LOT-Oriel GmbH
B–Cover plate		
	Glass plate ITO layer (indium tin oxide)	Dimensions: 14 in × 16 in × 0.7 mm; diced to 100 mm × 100 mm × 0.7 mm Resistivity: 10 $\frac{\Omega}{\blacksquare}$
B1	ITO electrode structuring	Lithography AZ® 1518 Wet chemical etch @ 40 °C for 3 min in H_2O: HCl (37 %) : HNO_3 (65 %) = 20: 20: 1 $\left(\frac{V}{V}\right)$
B2	Silicon nitride dielectric deposition (400 nm)	Plasma enhanced chemical vapour deposition at 30 W, 500 mTorr, 350 °C and SiH_4: 20 sccm, NH_3: 14 sccm, N_2: 1500sccmfor 25 min *Surface Technology Systems 310/320 (PECVD)*
B3	Silicon nitride dielectric structuring	Lithography AZ® 1518 Reactive Ion Etching at 200 W, 75 mTorr, 20 °C and CF_4: 35 sccm, O_2: 3 sccm for 300 s *Oxford Instruments Plasmalab System100 (RIE)*
B4	DuPont™ Teflon® AF hydrophobization (Grade 601S2-100-6;500 nm)	See step 2 for substrate
C–Bonding		
		Chip dicing, manual alignment utilizing an optical microsope Heating to 305 °C within 30 min, dwell at 305 °C for 30 min, annealing to 40 °C within 30 min

fixture in order to precisely encounter the gently widened endings of the tube which comprises the segmented flows. Thus, the failure-free transfer of the segments from the tubing into the chips could be assured. Figure 4.12 shows a fabricated chip using the example of layout B. The device is controlled by a LabVIEW program (National Instruments). The following routines have been implemented:

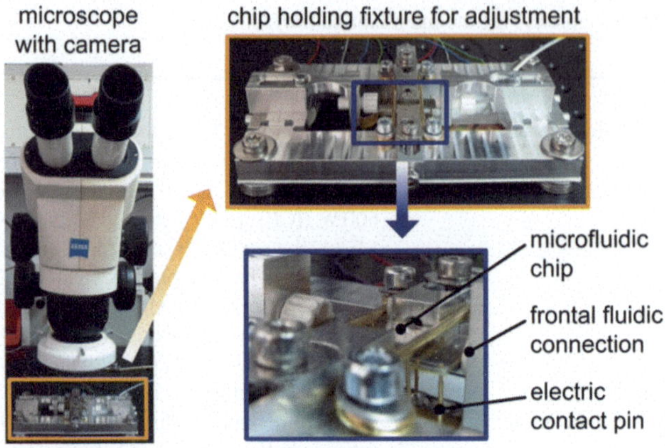

Fig. 4.11 Experimental setup for testing the fabricated chips

- Control of the syringe pump neMESYS (cetoni GmbH, Korbußen, Germany) in order to create the required flow rate of the segmented flow through the microfluidic chip, and
- Application of a preselected electric potential to selected electrodes in a programmable sequence.

The segments in the relevant region of the respective chip layout are observed with the trinocular microscope Stemi2000 (Carl Zeiss AG, Germany) on which the microscope camera Moticam 2500 (Motic, Wetzlar, Germany) is mounted. The pitch between two consecutive segments was several millimeters up to centimeters.

Utilizing the experimental setup described above, the feasibility of the described approach could be proven. Experimenting with several chips featuring the introduced layouts, it turned out that it is sufficient to actuate the segments by the electric (scatter) field between two adjacent electrodes of the ITO layer. The usage of the (conductive) silicon channel walls as common counter electrode became optional. 40 V (DC) were

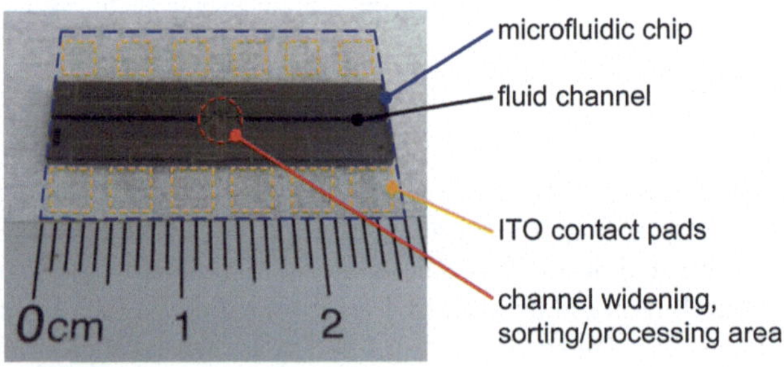

Fig. 4.12 Fabricated chip, layout B

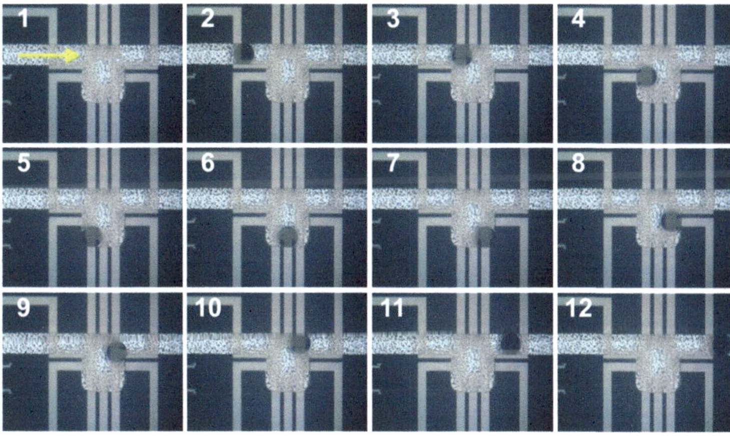

Fig. 4.13 Sequence of single images showing the electrostatic actuation of a water droplet within a processing station–Layout A; flow direction is from left to right

sufficient to move the water droplets within the flowing carrier medium ($C_{14}H_{30}$) whereas a flow velocity of e.g. 1 $\frac{\mu l}{min}$ was adjusted at the syringe pump.

4.5 Summarizing Conclusions on Modeling, Realization and Application Potential of Chip-Integrated Electrostatic Actuation of Micro Fluid Segments

In the previous sections, a modeling algorithm has been suggested in order to estimate the relevant forces acting on single segments when passing sorting or processing stations based on applied electrical fields. At this point, the approach is critically reviewed, its limits are discussed, and an outlook to possible refinements is given.

First of all, it has to be taken into account that the liquid segments have been treated as rigid spherical solid. Whereas external forces cause deformations and fluid flow for liquids, this behavior is excluded with the introduced approach. The parallel-plate capacitor (cf. Figs. 4.5 and 4.6) that has been used to simulate and calculate the electrostatic forces represents a simplification of the real situation: it is assumed that there is a uniform electric field between the two electrodes. The fringing field around the electrode edges is neglected. But de facto this fringing field component is responsible for the effect of attractive electrostatic forces [13]. Even in the region of the normally uniform field between the electrodes, any dielectric boundary surface causes field distortions which have been neglected. The channel side walls of the fabricated system can (partially) serve as ground electrode–they consist of doped silicon–that additionally reduces the region of a uniform field. The electrodes of the fabricated system additionally feature dielectric coatings, amongst others in order to prevent electrolysis.

The solution of the analytical model for the force in x-direction is independent on the segment position x. This reflects the fact that for a cube inserted into the region of a uniform electric field there are only dielectric boundary surfaces which are either perpendicular or parallel to the field lines. The combination of spherical segments and the consideration of fringing fields lead to the characteristic curve as shown in Fig. 4.7. The simulation additionally incorporates the fact that the energy of the electric field is converted into a force in z-direction as well.

Segments as used for applications in the fields of life sciences, pharmacy and cell biology usually consist of aqueous solutions or even of complex composed liquids. Hence, one can assume that the dielectric constant will decrease whereas the electric conductivity will increase for that case in comparison with the assumptions of the model. Moreover, it is the matter of a scientific discussion to which extent water satisfies the criteria of solid materials being classified into electric conductors and dielectrics on the other hand. There are investigations on the paraelectric properties of water which have not been incorporated so far, cf. [22, 23].

Regarding the fluid mechanics considerations it has to be noted that an electrostatically "fixed" segment partially blocks the microfluidic channel. Thus, the critical dimension is dramatically reduced and not as large as the 500 μm that have been employed for the calculation. At the same time, each sorting or processing station comes along with channel widening that again legitimates the approach.

From a more fundamental point of view it has to be mentioned that there are other approaches to model microfluidic capacitor-like assemblies as they have been described in the previous sections. For instance, electrohydrodynamic (EHD) forces can be derived reverting to dielectrophoresis, electrocapillarity, or electrowetting, respectively (cf. [24, 25]). However, the intention at this point was to show how one can benefit from simple engineering principles for the purpose of a straightforward chip layout routine.

The novel approach of combining continuous flow analysis (CFA) systems with the advantages of digital microfluidic (DMF) concepts has been presented. A simple process sequence has been utilized in order to fabricate adequate microfluidic chips. Leading single segments out of the continuous stream of the segmented flow by electrostatic actuation (scenario processing station) could be shown as well as changing the segment order within the segment chain (scenario sorting station). Thus, it could be proven that strict separation into CFA and DMF functionalities is no longer necessary. Depending on the respective application, the presented microfluidic chip and accordingly the chip layout can be adapted to many user-specific requirements. For example, silicon could be used as lateral channel wall and the cover plate is maintained as for the introduced design. But for the bottom channel wall, silicon could be substituted by Teflon®. The gas permeability of this material is much better than silicon. By this means, biological cells being carried (and kept alive) within an aqueous segment could be fed with several gases in case of a long-term retention within an analysis station of the microfluidic chip. Furthermore, the chip design can be simplified or adapted to certain application scenarios. For example, the channel walls can be made of ultraviolet-sensitive polymers such as SU-8 (MicroChem Corp.) by lithography processes.

Besides standard LoC tasks, the following conceivable applications of the presented approach are to be pointed out:

- Home care diagnostics,
- Cell cultivation, and
- Sample processing in health care and pharmacy

The following steps are scheduled for future works:

- Firstly, the physical limits of the chips are to be determined, e.g. maximum flow rate and pressure drop, maximum sorting velocity, minimum actuation voltage etc.
- In a second step and together with project partners it is planned to implement the proven functionalities into existing μTAS or LoC's, respectively.
- For future versions of the chip, bypasses will facilitate the bubble-free flooding of the microfluidic chip.
- For the case of employing the presented chip as stand-alone system (i.e. not as part of a μTAS or LoC, respectively), a suitable fluidic interface has to be found in order to reliably guide the SF's into the chip whereas the interface itself hermetically seals the segmented flows against the environment.

Acknowledgments The presented work has been funded by the German Federal Ministry of Education and research under contract 16SV5058. The authors are much obliged to the following partners and colleagues for their respective contributions to success:
• Boris Goj for the simulation in Maxwell and Yahia Cheriguen for his help with Comsol.
• The Bioprocess Engineering Department of the Institute for Bioprocessing and Analytical Measurement Techniques in Heilbad Heiligenstadt, Germany, for support and numerous segmented flows;
• Andrea Knauer and the Department Physical Chemistry and Microreaction Technology at the Ilmenau University of Technology, Germany;
• Michael Bertko, Alexander Groß, Stefan Hampl, Lothar Dressler, Manuela Breiter, Tobias Remdt, Birgitt Hartmann, Gabriele Harnisch, David Vopel, Claudia Krauße, and Judith Wolf.

References

1. S.K. Cho, H. Moon, C.-J. Kim, IEEE J. Microelectromech. Syst. **12**, 70–80 (2003)
2. Internet source: http://www.liquid-logic.com/technology; Accessed 06 Nov 2012
3. P.C.H. Li, *Microfluidic Lab-on-a-Chip for Chemical and Biological Analysis and Discovery*, 1st edn. (CRC Press, Taylor and Francis Group, Florida, 2006)
4. D.R. Reyes, D. Iossifidis, P.-A. Auroux, A. Manz, Anal. Chem. **74**, 2623–2636 (2002)
5. R.B. Fair, A. Khlystov, T.D. Tailor, V. Ivanov, R.D. Evans, P.B. Griffin, V.K. Vijay Srinivasan Pamula, M.G. Pollack, J. Zhou, IEEE Des. Test Comput. **24**, 10–24 (2007)
6. Internet source: http://en.wikipedia.org/wiki/Lab-on-a-chip; Accessed 06 Nov 2012
7. S.J. Haswell, Analyst **122**, 1R–10R (1997)
8. P. Bergveld, Biomed. Microdevices **2**, 185–195 (2000)
9. A.J. de Mello, N. Beard, Lab Chip **3**, 11N–19N (2003)

10. R.B. Fair, Microfluid. Nanofluid. **3**, 245–281 (2007)
11. V.K. Pamula, M.G. Pollack, P.Y. Paik, H. Ren, R.B. Fair, Apparatus for manipulating droplets by electrowetting-based techniques, 2005, United States Patent No. US 6911132 B2
12. Internet source: http://www.varioptic.com; Accessed 06 Nov 2012
13. D.J. Griffiths, *Introduction to Electrodynamics*, 3rd edn. (Prentice Hall, Inc., New Jersey, 1999), pp. 193–196
14. F.M. White, *Fluid Mechanics*, 4th edn. (WCB/MCGra-Hill, Singapore, 1999), pp. 24 and 771
15. C. Wohlfahrt, Springer Materials–The Landolt-Börnstein-Database, Springer Verlag, Berlin Heidelberg, 2008, vol. 17, supplement to IV/6, pp. 29–42
16. T.S. Khasanshin, A.P. Shchemelev, High Temp. **40**, 207–211 (2002)
17. C. Franjo, L. Segade, C.P. Menaut, J.M. Pico, E. Jiménez, J. Solution Chem. **30**, 995–1006 (2001)
18. C. Wohlfahrt, *Springer Materials—The Landolt-Börnstein-Database*. (Springer Verlag, Berlin Heidelberg, 2008), vol. IV/17, ch. 309
19. H. Brauer, D. Sucker, Chem. Ing. Tech. **48**, 665–736 (1976)
20. W.S. Janna, *Introduction to Fluid Mechanics*, 4th edn. (CRC Press, Taylor & Francis Group, Florida, 2010), p. 283
21. R.W. Fox, A.T. McDonald, *Introduction to Fluid Mechanics*, 4th edn. (John Wiley & Sons, Inc., New York Toronto, 1992), p. 439
22. J. Malsch, Über die Messung der Dielektrizitätskonstanten von Flüssigkeiten bei hohen elektrischen Feldstärken nach einer neuen Methode, Physikalische Zeitschrift XXIX, 1928, pp. 770–777
23. F. Booth, The dielectric constant of water and the saturation effect. J. Chem. Phys. **19**(4), 391–394 (1951)
24. J. Zeng, T. Korsmeyer, Principles of droplet electrohydrodynamics for lob-on-a-chip. Lab Chip **4**, 265–277 (2004)
25. T.B. Jones, J.D. Fowler, Y.S. Chang, C.-J. Kim, Frequency-based relationship of electrowetting and dielectrophoretic liquid microactuation. Langmuir **19**, 7646–7651 (2003)

Chapter 5
Electrical Sensing in Segmented Flow Microfluidics

Brian P. Cahill, Joerg Schemberg, Thomas Nacke and Gunter Gastrock

Abstract Microfluidic systems require measurement techniques to assess the bioprocesses taking place within these systems. In droplet-based microfluidics, sensing systems have thus far been dominated by the use of optical sensing techniques. Direct electrical measurement of droplets has been quite rare. This chapter describes the use of electrical sensors in droplet-based microfluidics. Electrical sensors offer a possibility to measure the presence/position of droplets but also their electrical properties.

5.1 Introduction in to Electrical Sensing of Droplets and Micro Fluid Segments

This contribution focusses on techniques for electrical sensing of droplets. Droplet sensing can be divided into two general categories (i) droplet presence sensing and (ii) droplet content sensing. The presence of a droplet between two electrodes or between a light source and a photodetector can lead to a change in the measured signal. Droplet presence allows the estimation of droplet position, droplet speed and droplet volume. Droplet content sensing is necessary for the analysis of the processes that take place within droplets. This may be the sensing of discrete objects within the droplet or the sensing of the concentration of a particular quantity in the droplet. In single-cell cytometry, it is necessary to ascertain if a single cell is present in a droplet and to ignore droplets that contain multiple or zero cells. Optical sensing generally measures concentration in droplets indirectly through the use of indicators, for example, droplets containing a pH indicator will change color if acid is titrated into the droplets at various concentrations. Optical sensing can also use the

B. P. Cahill (✉) · J. Schemberg · T. Nacke · G. Gastrock
Institute of Bioprocessing and Analytical Measurement Techniques, Heilbad Heiligenstadt, Germany
e-mail: brian.cahill@iba-heiligenstadt.de

J. M. Köhler and B. P. Cahill (eds.), *Micro-Segmented Flow*, Biological and Medical Physics, Biomedical Engineering,
DOI: 10.1007/978-3-642-38780-7_5,

percentage of light transmitted through the droplets to measure the optical density and in turn, for example, the cell count. Electrical sensing can distinguish between the dielectric constant and the conductivity and to the frequency dependence of these quantities. The use of impedance/capacitive sensing has increased significantly in recent years through the emergence of touch screen devices. In digital microfluidics, we can use electrical sensing to sense the presence or absence of a droplet through the large dielectric constant and conductivity differences between oil and water. In addition, differences in conductivity and permittivity can be also used to measure the content of individual droplets. The conductivity depends to a large extent on the ionic strength and the permittivity depends on the concentration of water-soluble solutes, such as, alcohol. In droplet-based cell culture, the concentrations of glucose, oxygen and carbon dioxide are of most importance.

5.2 Capacitive Sensing of Droplets

5.2.1 Principle of Capacitive Sensing

If a potential difference is applied between two conducting bodies separated by an insulating medium, electrical charges of equal and opposite magnitude accumulate on the surface of the conducting bodies. Capacitance is defined in terms of charge storage:

$$\mathrm{C} = \mathrm{Q}/\mathrm{V}, \tag{5.1}$$

where C is the capacitance in Farads, Q is the electrical charge in Coulombs, and V is the potential difference between the conductors in Volts. The capacitance depends on the geometry of the electrode system and the permittivity of the insulating medium between them. Applications of capacitive sensing are based on the variation of a parameter of interest being coupled to a change in capacitance.

Baxter [1] presents various schemes for optimizing capacitive sensing and to improve linearity, implement shielding, to avoid sensitivity to stray capacitance and to achieve adequate signal bandwidth. Heerens [2] provides a detailed analysis of how to implement a guard electrode in capacitive sensing and how to avoid adverse effects due to parasitic capacitance.

Taking a broad view of sensing technology, capacitive sensing has enjoyed a huge increase in applications in the last 5 years through the takeoff of touch-screen control in consumer electronics. Other applications of capacitive sensing involve mobile electrodes, whereby the geometry of the electrode system changes; this sort of measurement has found extensive application in micrometers, proximity detection and motion encoding [1] but is less relevant to measurement in segmented flow. This section will concentrate on applications of capacitive sensing in multiphase microfluidics. Van der Linden [3] described the construction of an inline capacitive sensor to measure the disperse water content in oil by measuring the difference in capacitance between the dispersion and dry oil at the same temperature.

Optical measurement has been the most prominent technique for the analysis of droplet content [4–7]. In many cases optical measurements require the addition of fluorescent markers and other molecules that act as indicators. Electrical measurements can often be performed without any need for modification of the object under test. The first electrical measurements of droplets that have emerged have aimed to measure the presence/absence of a drop or its position quantitatively [8–12]. Chen et al. [8] and Srivastava and Burns [9] used capacitive sensing of droplet position to automate further operations in a microfluidic system and to measure droplet size. Niu et al. [10] performed inline measurements of the capacitance differences of droplets flowing through a microfabricated chip and subsequently could direct the flow of droplets through a switching mechanism. The capacitive measurement of droplets senses changes in the dielectric constant of the liquid between the measurement electrodes; in short, these measurements sense the presence/absence of water between the measurement electrodes. This type of capacitive droplet measurement is also capable of measuring droplet velocity and volume [11].

Various research groups have used the electrodes of their electrowetting chips to measure the capacitance of the liquid between the electrodes. Nichols et al. [12] sensed the position of droplets in an electrowetting chip in order to direct the flow of droplets. Schertzer et al. [13] showed how the composition of droplets could be measured through the capacitive sensing of droplets. The resistance and capacitance variation was measured as a function of the concentration of soda-lime glass particles in droplets and the value of both was observed to be linearly dependent on concentration. In addition the progress of chemical reactions could be measured in situ. Previously they had related droplet capacitance to methanol concentration [14]. Shih et al. [15] implemented a sensing and feedback control system as part of an electrowetting chip to monitor and control droplet movement and tested droplets containing water, methanol, and cell culture medium containing fetal bovine serum. Murran and Najjaran [16] also measured impedance in an electrowetting chip.

Ernst et al. [17–20] developed a sensor to measure the volume of microdroplets ejected from a droplet dispenser as are commonly used in automatic pipette robots. Droplets are generated and pass through an open plate capacitor. The presence of the droplets between the plates of the capacitor alters the measured capacitance and this change of capacitance can be related to the volume of the water droplet. They used an alternating voltage of $V_{pp} = 20\,\text{V}$ at a constant sinusoidal frequency of $f = 156\,\text{kHz}$. The signal amplification and conditioning was performed on a PCB developed by the group itself.

The time dependence of the measured signal is shown below in Fig. 5.1. It is a little counterintuitive to observe that as the drop approaches the electrode the capacitance first drops before it rises again. Ernst et al. attributed this drop in capacitance to capacitive coupling through the pipette that leads to the presence of charge on the droplet surface that affects the measurement of capacitance [20].

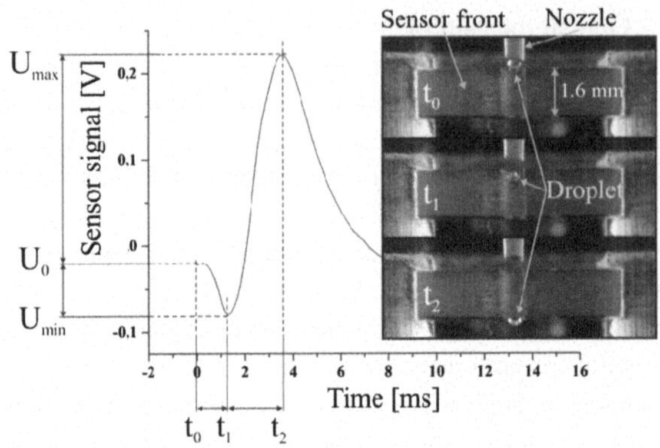

Fig. 5.1 Sensor signal as a function of time. The photo inset shows three frames of a video showing the passage of a droplet through the sensor. This image is from [20]

5.2.2 Experimental Example of Capacitive Measurements in Microfluid Segments Embedded in a Perfluorinated Carrier Liquid

Perfluorinated hydrocarbons are very attractive for use as the carrier liquid in micro segmented flow due to their specific chemical and physical properties. They are chemically inert, show high interface tensions against aqueous phases and also have high contact angles against other liquids and possess significant differences to many other liquids in several physical parameters.

Perfluorodecalin (PFD) is very often used in these systems. It is, at present, the only suitable perfluorinated hydrocarbon that has gained approval for use in medical experiments. PFD has the additional advantage of having a high capacity for adsorbing gases that facilitates gas exchange from/to droplet-based bioreactors. As the refractive index of PFD is 1.313, it is relatively difficult to distinguish optically between PFD and aqueous solutions (refractive index of 1.33) in comparison with segmented-flow systems using tetradecane or other oils. We present a capacitive measurement of the presence/absence of aqueous droplets separated by PFD that allows further operations to be performed. The authors required a droplet position sensor to automate medium exchange from droplet-based bioreactors and develop a capacitive sensor for this purpose. The droplets serve as bioreactors for the long-time cultivation (weeks) of multicellular aggregates. Medium exchange is necessary once per day.

Figure 5.2 shows the capacitive sensing module. The purpose of the system is to enable medium exchange to and from a droplet. Two main fluidic connections allow droplets to pass through the system. Two secondary fluidic connections facilitate medium exchange. Two spring-loaded electrodes serve to allow measurement of the capacitance signal.

The AD7746 Capacitance Measurement Chip was designed by Analog Devices to measure low capacitance due its very low parasitic capacitance. It has found several applications in microfluidics [21–24]. The chip communicates by means of the two-

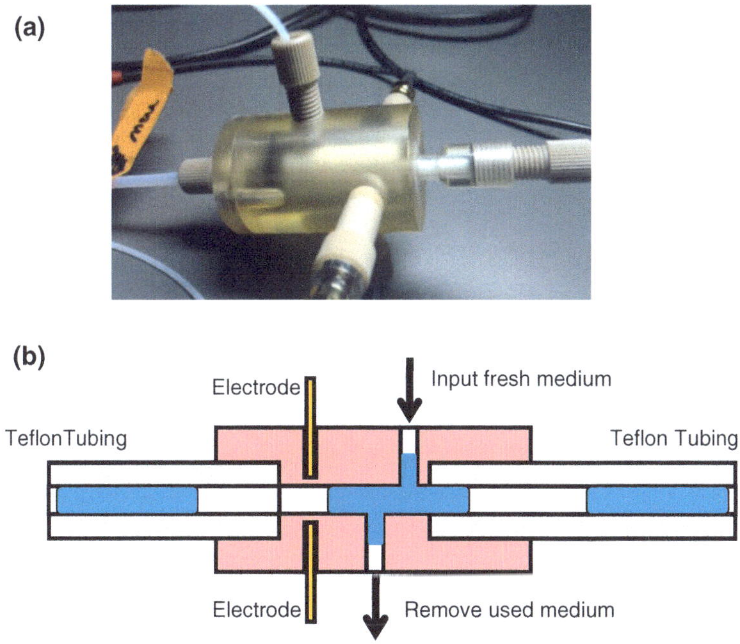

Fig. 5.2 **a** Photo of the capacitive sensing module and schematic diagram. **b** Schematic drawing of the capacitive sensing module showing medium exchange

wire serial interface that is compatible with the I^2C bus. An evaluation board is available from Analog Devices that connects the chip with a microcontroller system by means of a USB interface. Computer control is possible using a rather limited dedicated program supplied by Analog Devices or alternatively by means of LabView [11] or Visual Basic [24].

Figure 5.3a shows the capacitance traces for spherical droplets, that is, droplets that do not wet the wall of the tubing. The capacitance trace rises as the droplet passes between the electrodes. In contrast, Fig. 5.3b shows how the capacitance of larger droplets decreases at the phase boundary before rising in the centre of the droplet as expected. We attribute this behaviour to electrohydrodynamic effects related to interfacial charging at the oil-water interface.

Capacitative sensing is thus far the most widespread application of electrical sensing in droplet-based microfluidics and has been applied to segmented flow as well as droplets in electrowetting chips. It is generally used to sense the presence of a droplet between a pair of electrodes but can also be used to measure the change in dielectric constant.

Fig. 5.3 Capacitance trace for **a** spherical droplets and **b** elongated droplets with wall contact

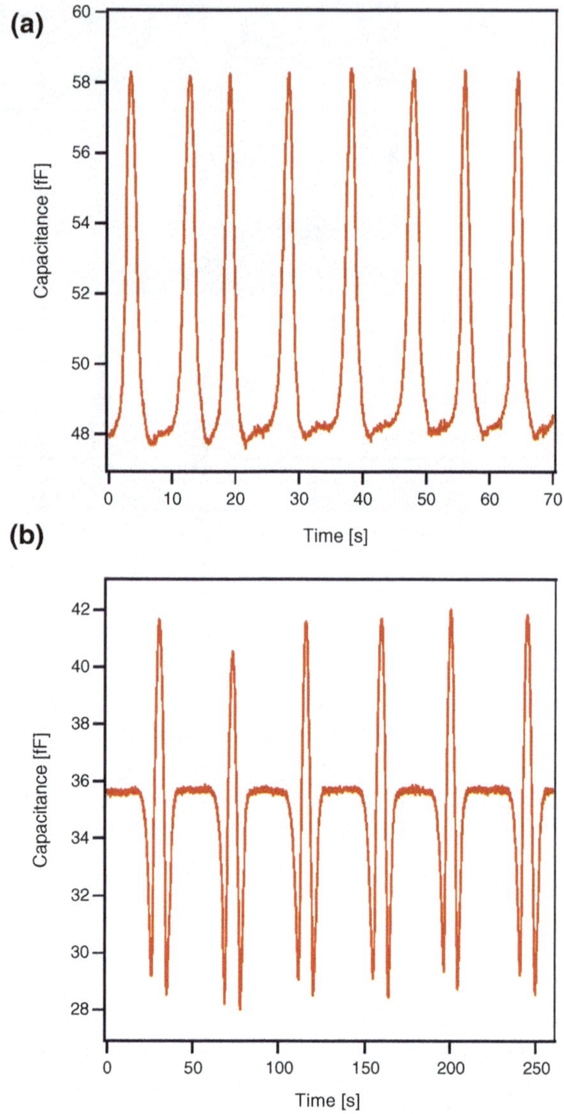

5.3 Impedimetric Measurement of Conductivity in Segmented Flow

5.3.1 Impedimetric Measurement Principle

For almost 15 years, contactless conductivity measurement of conductivity has been applied to measuring electrophoresis traces [25–27]. In electrophoresis, the variation of the ionic concentration is measured as a function of time. In recent years a couple of research groups have started using these techniques for measuring the content of droplets in segmented flow. Cahill et al. [28–30] presented the measurement of conductivity in segmented flow and Lederer et al. [31, 32] and Sadeghi et al. [33] described the measurement of droplet impedance in an electrowetting chip. These applications differ from the vast majority of the applications of impedance spectroscopy, where the electrodes make galvanic contact to the liquid that is to be measured [34–36] because the electrodes are insulated from the liquid to be measured. Direct contact between electrodes and the fluid medium minimizes the electrode impedance and greatly improves measurement sensitivity. The electrode impedance is adversely affected by the capacitance of insulation layers. In droplet-based microfluidic systems, the transport of droplets across the surface of the electrode without breaking up is aided by continuous and hydrophobic surfaces. To achieve this aim, Cahill et al. [28] chose to use electrodes attached to the outer surface of thin-walled glass capillaries; the surface of glass could be readily silanized with hydrophobic alkysilanes. The sensor measured the impedance spectrum over a range of frequencies and not just a single measurement at a fixed frequency as is common in electrophoresis. This facilitated the measurement of the conductivity over a wide range. As the thickness of the capillary walls is much lower than that in capillary electrophoresis, the measurement could be performed at lower excitation amplitudes. This complexity and cost of the electronics of any dedicated electronic circuits for performing such measurements is reduced.

Electrowetting on dielectric explicitly requires full isolation from the liquid in order to build up the capacitive energy required to move the droplet. The insulating layer prevents galvanic contact between the electrodes and the fluid being measured and although this complicates impedance measurement, it has some advantages. If the thickness of the insulation is much greater than the thickness of the electric double layer at the interface, the presence of the electric double-layer can be neglected. Lederer et al. [31] chose to use the electrodes of their electrowetting chip to measure droplet conductivity. This setup took advantage of thinner insulation based on thin-film insulation and as a consequence the analytical model needed to account for double layer capacitance [32]. Sadeghi et al. [33] developed an analytical technique to measure droplet size and concentration in an electrowetting chip. Shih et al. [37] recently measured the adsorption and growth of adherent cells at localized hydrophilic areas in an electrowetting chip by means of impedance measurement.

5.3.2 *Finite Element Model of Non-Contact Impedance Measurement*

Cahill et al. [29, 30] presented a numerical simulation of a contactless conductivity sensor with thin walls. The measurement setup shown below in Fig. 5.4 was modeled by means of finite element simulation using COMSOL 3.5a with Matlab used to externally control and log the results of the simulation. This automation allowed sweeps of parameters, such as, frequency and conductivity and presentation of the results in the form of results as acquired from a gain-phase analyzer, that is, impedance phase and impedance modulus as a function of frequency.

Geometry

The generalized geometry of the sensor is shown in Fig. 5.4a: a capillary contains two distinct phases with one phase positioned between the electrodes. The geometry of the capillary and the electrodes can be varied in the simulations below. A ground electrode is placed between the electrodes; this can be omitted if required. Axial symmetry of the capillary, droplet and electrodes greatly reduces the complexity and the time required to perform the simulation.

Finite Element Model

The electric potential U is described by a modified form of the Poisson equation:

$$-\nabla \cdot \left((\sigma + j\omega\varepsilon_0\varepsilon_r)\nabla U - \mathbf{J}^e\right) = Q_j \tag{5.2}$$

where σ is the conductivity, ω is the angular frequency, ε_0 is the permittivity of free space, ε_r is the relative dielectric constant, $\mathbf{J}^e$ is the external current density and Q_j is a current source.

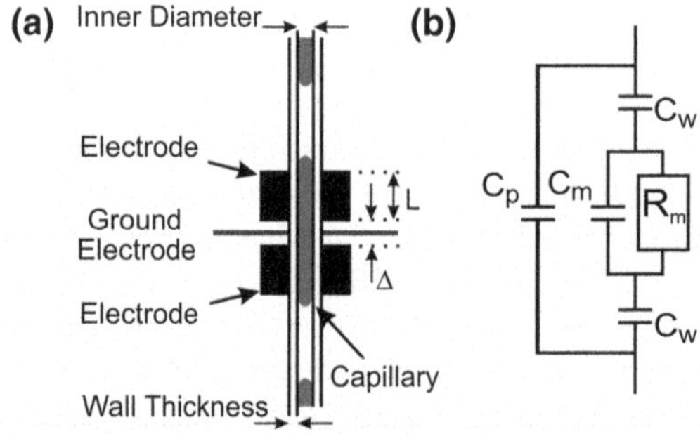

Fig. 5.4 **a** Schematic diagram of the impedance sensor showing the capillary with two ring electrodes and ground electrode. The geometric parameters that were adjusted as part of this study are shown: the electrode length L, the electrode spacing Δ, the wall thickness and the inner diameter of the capillary. **b** Equivalent circuit diagram of the sensor where C_w is the capacitance of the capillary wall, C_m is the capacitance of the medium in the capillary, C_p is the parasitic capacitance and R_m is the resistance of the medium in the capillary

The boundary conditions are:

$$n \cdot \mathbf{J} = 0 \tag{5.3}$$

$$n \cdot (\mathbf{J}_1 - \mathbf{J}_2) = 0 \tag{5.4}$$

where Eq. 5.3 uses the insulation boundary condition for external boundaries setting the current **J** to zero, Eq. 5.4 uses the continuity of current on either side of the boundary for all boundaries between dielectric media. The ground electrode and one of the measurement electrodes are set to ground potential while an AC potential of 1-Volt amplitude is applied to the other measurement electrode. The frequency of the applied potential was varied between 10 kHz and 30 MHz while the conductivity of the aqueous liquid was varied between 1 mS/m and 1 S/m. The conductivities and dielectric constants of all other media were set to constant values relating to the chosen materials. The cross-sectional area of the electrode was excluded from the simulation.

The voltage at the surface of the lower ring electrode in Fig. 5.4a was set to 1 V at various alternating frequencies and the voltage at the higher electrode to 0 V. The voltage in the plane of the ground electrode was either 0 V or was set to the continuity boundary condition (Eq. 5.4) to show how the sensor works without a ground electrode.

The current J passing between the electrodes was measured by integrating the current entering the higher electrode for one cycle of the AC signal and by integrating around the axis of symmetry. The real and imaginary parts of the current were computed for each frequency and conductivity value simulated. The impedance Z is given by $Z = V/J$, where the impedance modulus $|Z|$ and phase angle θ are given by $Z = |Z|e^{j\theta}$. A parasitic capacitance (see the equivalent circuit in Fig. 5.4b) of 75 fF was added to the impedance measured by the simulation. This value was obtained experimentally by measuring an impedance sweep with an HP 4194A Impedance/Gain-Phase Analyzer with only short unshielded cables of 2-cm length attached.

Table 5.1 details the default settings that were used in the simulation. The third column indicates the graphs where the simulation deviates from

Table 5.1 Default simulation settings

Quantity	Value	Exceptions
Medium conductivity	89.9 mS/m	Fig. 5.5b
Capillary wall thickness	10 μm	Figs. 5.6a, 5.9b
Capillary inner diameter	0,5 mm	Fig. 5.6b
Electrode length	2 mm	Fig. 5.7a
Electrode spacing	2 mm	Fig. 5.7b
Parasitic capacitance	75 fF	Fig. 5.8a
Ground electrode	Grounded	Fig. 5.8b
Medium dielectric constant	80	Fig. 5.9a, b
Wall dielectric constant (quartz)	4.2	
Wall conductivity (quartz)	1 pS/m	

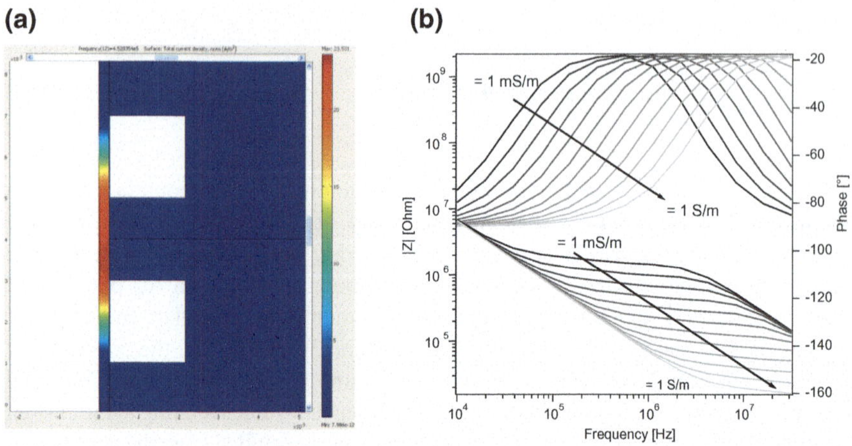

Fig. 5.5 **a** Example graphic of the simulation showing the current flow in the fluid. The area of the electrodes is *white* and was excluded from the simulation. **b** The *upper curves* show the dependence of the phase angle on frequency for a range of liquid conductivities from 1 mS/m to 1 S/m. The *lower curves* show the dependence of the impedance modulus on frequency for a range of liquid conductivities from 1 mS/m to 1 S/m

these values. Details of the exceptions are given below in the relevant figure captions.

Results

The primary parameter to be measured by this sensor is the medium conductivity. Secondary parameters that are of interest are the medium dielectric constant, system geometry and the electrical properties of the measurement system.

Figure 5.5a shows an example of the current flow between the two measurement electrodes. The centre of the two electrodes is shown in white and is not part of the area being simulated. Due to symmetry at $X = 0$, the area to the left of the y-axis is not part of the simulation and is shown in white. There is significant current flow in the area inside the capillary and between the electrodes. In the interior of the capillary adjacent to the electrode the current decays with increasing displacement from the counter-electrode.

Figure 5.5b shows how the impedance modulus and impedance phase angle depend on conductivity and frequency. At low frequencies, the capillary-wall capacitance C_w dominates the impedance measurement: the impedance phase θ is $-90°$ and the impedance modulus decreases with increasing frequency. At intermediate frequencies, $|Z|$ approaches a value approximately equal to R_m and θ is at its maximum value. At higher frequencies, the medium capacitance C_m and the parasitic capacitance C_p dominate the impedance measurement. The measurement of conductivity is optimized by maximizing the maximum phase in the intermediate frequency range and extending the bandwidth over which $|Z|$ takes a constant value.

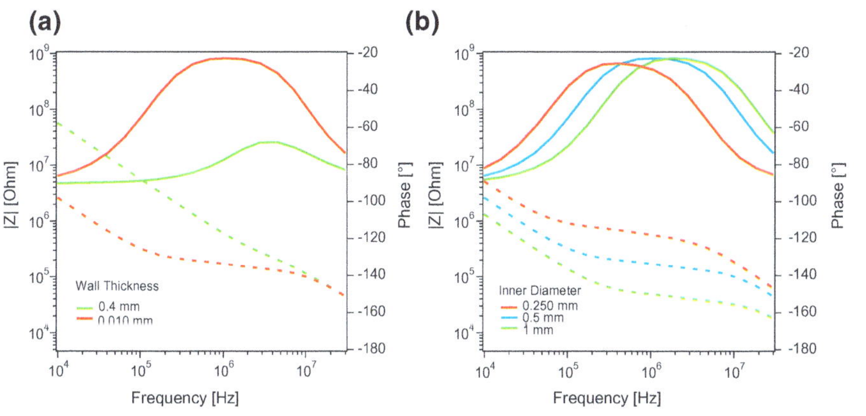

Fig. 5.6 **a** The *upper curves* show the dependence of the phase angle on frequency for wall thicknesses from 0.010 to 0.400 mm. The *lower curves* show the dependence of the impedance modulus on frequency for the same wall thicknesses. **b** The *upper curves* show the dependence of the phase angle on frequency for inner diameters of 0.25, 0.5 and 1.0 mm. The *lower curves* show the dependence of the impedance modulus on frequency for the same inner diameters

The effect of varying the geometry of the chosen capillary on the measurement of the impedance of the fluid within the capillary is shown in Fig. 5.6. Figure 5.6a compares a thin-walled quartz capillary with 10 μm wall thickness with typical PTFE tubing with 1.6 mm outer diameter and 0.5 mm inner diameter, that is, 0.4 mm wall thickness. The decrease of C_w shows how increasing wall thickness effectively damps the measurement of the impedance of the liquid in the capillary. The maximum impedance phase angle is much less than for the thin-walled capillary and the impedance modulus has a very minor inflection point at the frequency of maximum phase. In comparison, the thin-walled capillary has a bandwidth of around two orders of magnitude of frequency where the impedance modulus is relatively flat.

In Fig. 5.6b the impedance response for varying inner diameter of the capillary is displayed. The impedance of the capillary decreases with increasing diameter and the frequency of maximum phase shifts to higher frequency. The inner diameter determines to a great extent the measurement of medium resistance. Thicker capillaries have lower resistance. The decrease of this resistance shortens the RC relaxation time of the equivalent circuit shown in Fig. 5.4b and subsequently increases the frequency of maximum phase.

In Fig. 5.7a the impedance response for various electrode lengths is shown. The electrodes are 1–3 mm long and the electrode spacing is 2 mm. For longer electrodes, the frequency at which the impedance phase angle and modulus are first affected by the medium resistance decreases. This is due to the increase in C_w due to the increased contact area between the electrode and the capillary. In Fig. 5.7b the impedance response for various electrode spacings is shown. The performance of the sensor is affected by variation of both the medium resistance R_m and medium capacitance C_m. R_m increases at higher electrode spacings leading to a higher impedance modulus. C_m decreases at higher electrode spacings leading to a higher maximum phase angle

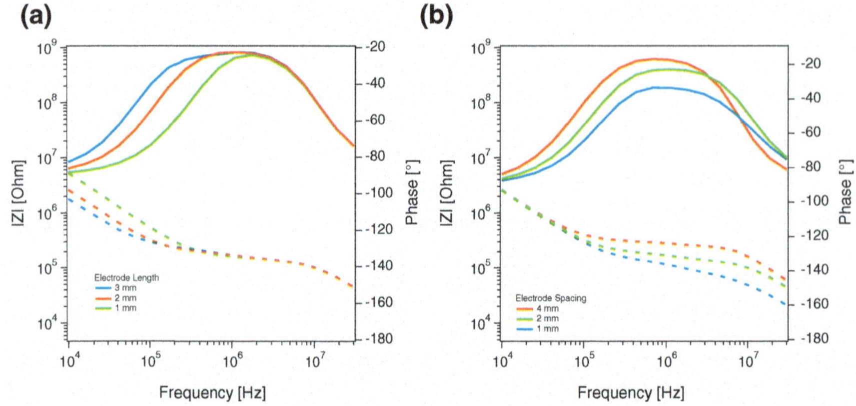

Fig. 5.7 **a** The *upper curves* show the dependence of the phase angle on frequency for electrode lengths of 1–3 mm. The *lower curves* show the dependence of the impedance modulus on frequency for the same electrode lengths. **b** The *upper curves* show the dependence of the phase angle on frequency for electrode spacings of 1, 2 and 4 mm. The *lower curves* show the dependence of the impedance modulus on frequency for the same electrode spacings

at higher electrode spacings. In general, Fig. 5.7 shows that larger electrode lengths and electrode spacings improve the sensor performance. This provides a challenge for the miniaturization of such impedance measurement in microsystem technology.

In Fig. 5.8 the effect of the electrical measurement instrumentation, cabling and shielding on the overall sensor performance is shown. For larger C_p the impedance measurement is damped as can be seen in Fig. 5.8a from the decrease in the maximum frequency at which can be measured and the decrease in the maximum phase angle. It is important that the design of the measurement instrumentation and cabling minimizes the value of C_p. Figure 5.8b shows how grounding of the electrode affects the measurement. Using a ground electrode reduces C_p and increases the measurement range and maximum phase angle slightly.

In Fig. 5.9a the effect of filling a thin-walled capillary with oil or water is shown; these are the two major liquids used in segmented-flow microfluidics. For oil the phase angle has a constant value of $-90°$ showing that capacitance dominates the measurement as oil has much lower conductivity than water. In Fig. 5.9b the effect of filling a typical PTFE tubing with 1.6 mm outer diameter and 0.5 mm inner diameter with oil or water is shown. For oil the phase angle also has a constant value of $-90°$ indicating that the measurement is dominated by capacitance. In contrast the response of the sensor to measuring water although much more highly damped than in the case of a thin-walled capillary, is still significantly different than for oil. It is feasible to consider the use of such a sensor in normal PTFE tubing to measure the presence of a droplet at the sensor. Although the optical measurement of the presence of droplets through PTFE tubing is possible, there are some difficulties, for example, PTFE is not perfectly optically transparent, sometimes the oil has a very similar refractive index to water and often biologists prefer to work without adding dyes or fluorescent molecules to cell media.

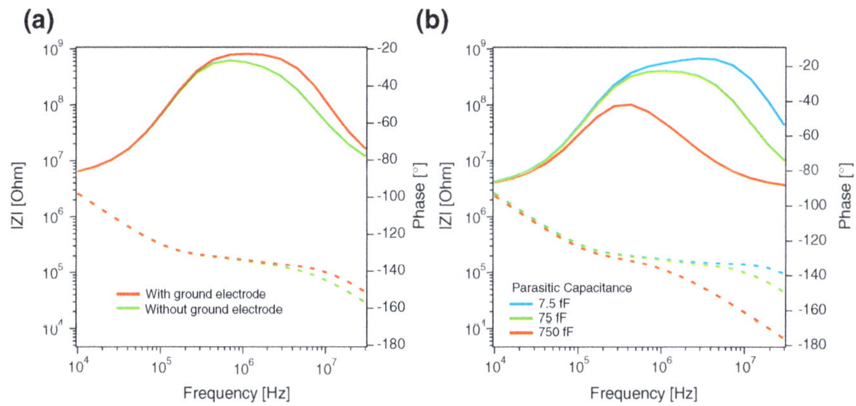

Fig. 5.8 **a** The *upper curves* show the dependence of the phase angle on frequency for parasitic capacitances of 7.5, 75 and 750 fF. The *lower curves* show the dependence of the impedance modulus on frequency for the same parasitic capacitances. **b** The *upper curves* show the dependence of the phase angle on frequency for electrode spacings with and without a ground electrode. The *lower curves* show the dependence of the impedance modulus on frequency with and without a ground electrode

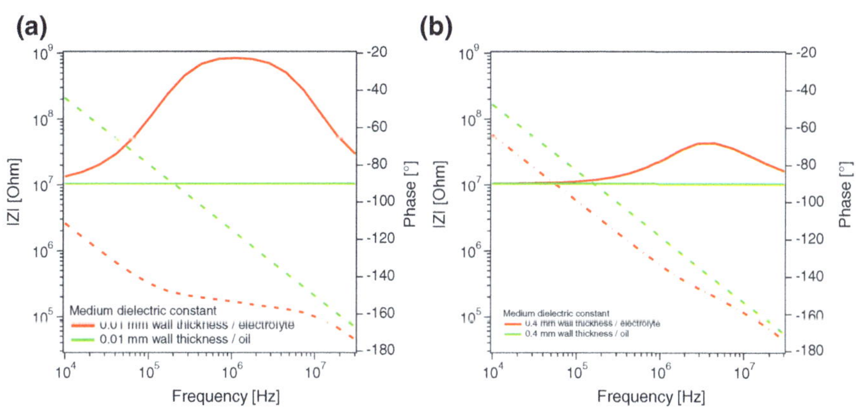

Fig. 5.9 **a** The *upper curves* show the dependence of the phase angle on frequency for dielectric constants of 80 and 2 for wall thickness of 0.01 mm. The *lower curves* show the dependence of the impedance modulus on frequency for the same criteria. **b** The *upper curves* show the dependence of the phase angle on frequency for dielectric constants of 80 and 2 for wall thickness of 0.01 mm. The *lower curves* show the dependence of the impedance modulus on frequency for the same criteria

5.3.3 Analytical Model of Non-Contact Impedance Measurement

Figure 5.4a shows an electrode system based on a thin-walled glass capillary. The general equivalent circuit describing this electrode system is shown in Fig. 5.4b. Analysis of the equivalent circuit shows how the sensor can be optimized. The system consists of a tube with two electrodes contacting the outside of the tube and with a grounded guard electrode shielding the two electrodes from each other. C_w the capacitance of the insulating walls is given by $C_w = \varepsilon_w A/d_w$, where ε_w is the dielectric constant of the wall, A is the area of the electrode and d_w is the thickness of the wall. C_p is the parallel capacitance resulting from parasitic capacitance between the electrodes and associated cabling and from the capacitance associated with the measurement system. The quantity that is to be measured is the impedance of the fluid in the capillary. This is determined by the geometry of the sensor and the conductivity and dielectric constant of the fluid and is described by the medium resistance R_m and medium capacitance C_m.

Although the insulated electrodes were chosen in order to facilitate improved droplet transit through the measurement system, a useful consequence of insulation is the fact that interference due to electrode polarization is minimized. The series addition of a capacitor corresponding to insulation with µm-scale thickness and the electric double layer with nm-scale thickness allows the formation of the electric double layer to be completely neglected.

In order to describe the behavior of the system so that we can construct a robust sensor, we will model the measurement response for the following sample values describing the equivalent circuit in Fig. 5.4b $R_s = 1\,\text{M}\Omega$, $C_w = 10$ pF, $C_p = 0.070$ pF and $C_m = 0.070$ pF. The total impedance of the equivalent circuit in Fig. 5.4b is given by:

$$Z = \frac{2 + j\omega R_m (C_w + 2C_m)}{j\omega (C_w + 2C_p) - \omega^2 R_m (C_m C_w + C_p C_w + 2C_p C_m)} \tag{5.5}$$

The general form of the curves in Fig. 5.10 consists of three distinct frequency regions for both impedance modulus and phase angle: the low-frequency range, the mid-frequency measurement range and the high-frequency range. For the impedance modulus the low-frequency and high-frequency ranges are decreasing curves indicative of the decrease in the impedance of a capacitor with increasing frequency. In the mid-frequency measurement range, the impedance modulus value is relatively flat indicative of the measurement of the frequency-independent impedance of a resistor. The impedance phase angle behavior shows the same general trend, in the first and third sections it is $-90°$ indicative of a pure capacitance and in the mid-frequency range it approaches a maximum value.

Figure 5.10a compares the impedance of the equivalent circuit for medium resistances R_m of 0.1, 1 and 100 MΩ respectively. In the mid-frequency range, the impedance modulus has a value close to that of R_m, this means in the mid-frequency range the resistance of the fluid in the capillary is the dominant element of the overall impedance. The mid-frequency range shifts to higher frequencies as the medium conductivity increases (lower medium impedance) but without a widening of the bandwidth relative to the centre frequency. Outside of the mid-frequency measurement range there is little or no difference between the three sample curves, thus at such frequencies the sensitivity to measurement of conductivity is very low.

Figure 5.10b compares the impedance of the equivalent circuit for various values of the parallel capacitance C_p. It is clear that the mid-frequency measurement region is much narrower for higher values of C_p and that as C_p decreases the mid-frequency range increases in width towards higher frequencies and the maximum phase angle increases. This shows the importance of minimizing C_p by implementing a grounded guard electrode between the measurement electrodes, by shielding cables or by otherwise optimizing the measurement electronics. Unshielded electrodes restrict the mid-frequency measurement region to a relatively small bandwidth. Decreasing the value of C_p is only effective if it is greater than C_m. For values of C_p less than C_m, the increase in measurement bandwidth is relatively small. The design of the sensor and the dielectric constant of fluid determine the value of C_m.

Figure 5.10c compares the impedance of the equivalent circuit with regard to the value of the insulation capacitance C_w. It can be seen that the mid-frequency measurement region is much smaller for lower values of C_w and that as C_w increases the mid-frequency range increases in width towards lower frequencies. It is clear from the graphs that lower values of C_w result in a larger change of phase angle and a wider bandwidth of measurement. For this reason, it is of primary importance to implement flow cells with thin electrode insulation d_w and of secondary importance to increase the area of the electrodes, A. The value of A is limited by the size of the droplet used in a particular experiment and the relation between the speed of measurement and the speed of transit of droplets across the electrodes.

5.4 Experimental Investigation of an Inline Noncontact Impedance Measurement Sensor

5.4.1 Impedance Measurement of Ionic Strength

The construction of a galvanically-decoupled through-flow sensor system for repeatable impedance measurements requires relatively high capacitance electrodes with thin electrode insulation. The measurement system should be relatively inexpensive and robust and enable inline measurements of aqueous droplets in segmented flow. The sensor used extremely thin-walled glass capillaries with nominal wall thickness of 10 μm with 500 μm inner diameter as are commercially available for use in X-ray crystallography (WJM-Glas, Berlin/Pankow, Germany). The measurement electrode rings were fabricated from Elastosil LR 3162 conductive silicone (Wacker AG, Burghausen, Germany) and could slide across the surface of the glass capillary making conformal contact to the capillary surface. The parasitic capacitance of the measurement system was reduced by using a thin copper sheet as shielding between

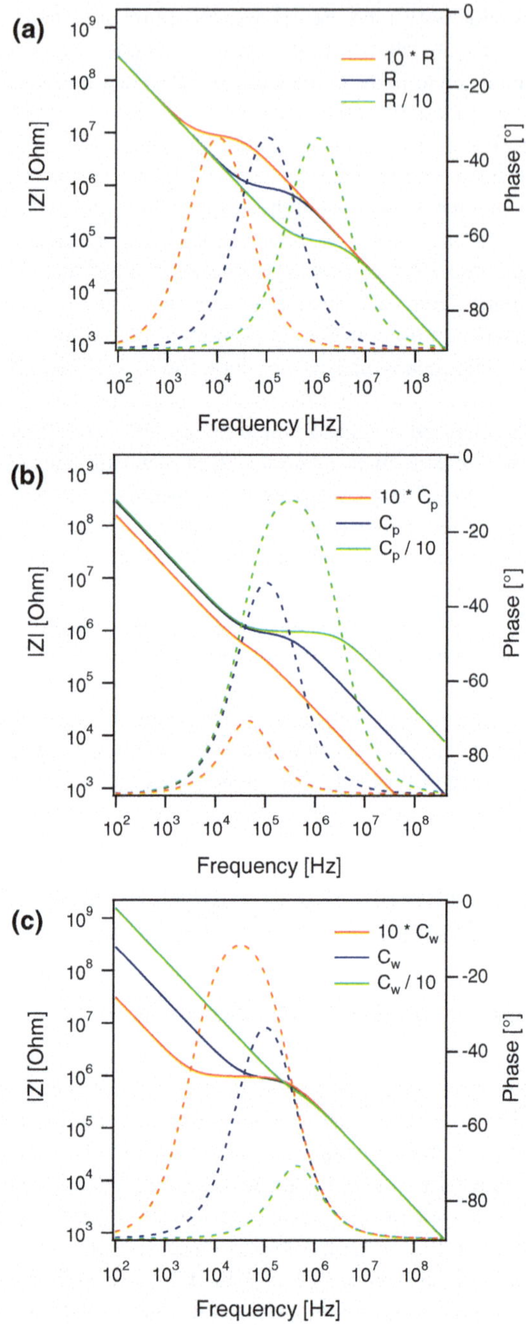

Fig. 5.10 **a** The dependence of the impedance modulus and impedance phase angle on medium conductivity. **b** The dependence of the impedance modulus and impedance phase angle on the parallel capacitance. **c** The dependence of the impedance modulus and impedance phase angle on the insulation capacitance. *Full lines* represent impedance modulus curves and *dashed lines* represent phase angle curves as modelled for the equivalent circuit displayed in Fig. 5.4b

the two electrodes [26]. The surface of the capillaries was rendered hydrophobic by coating with a monolayer of octadecyltrichlorosilane (ODTS) molecules (ABCR, Karlsruhe, Germany).

Impedance was measured using: (i) HP 4194A Impedance/Gain-Phase Analyzer and (ii) an IMPSPEC system (MEODAT, Ilmenau, Germany). The HP 4194A Impedance/Gain-Phase Analyzer scans the wavelength spectrally, that is, it sequentially applies a single excitation signal at a particular frequency until the preprogrammed frequency scan is completed. It is not particularly suited for integration into a process control setup where speed of measurement and cost become important factors. The IMPSPEC system is an impedance measurement device specifically designed for process control [38, 39] and consists of a 14-bit Digital to Analog Converter (DAC) for signal generation and 14-bit Analog to Digital Converters (ADC) for signal acquisition.

Aqueous solutions with ionic strengths between 0.39 and 200 mM KCl were prepared. Measurement of these samples frequencies between 100 Hz and 20 MHz allowed the effectiveness of the conductivity sensor to be clearly evaluated. Figure 5.11 shows the impedance characterization of an ionic strength series for the thin-walled capillary electrode structure with full and dashed lines representing data measured by the HP Impedance Analyzer and IMPSPEC system respectively. The form of the graphs shows a distinct resemblance to that simulated in Fig. 5.10b. The impedance modulus and phase angle both show a mid–frequency measurement region, the frequency of which varies with ionic strength. The impedance phase angle shows a peak value in the mid-frequency measurement region that increases in value slightly with increasing ionic strength. As the ionic strength increases, the frequency at which the peak value is observed also increases.

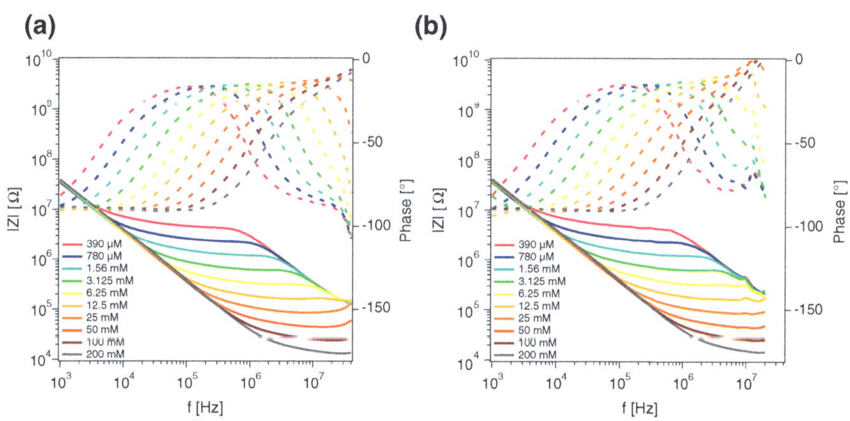

Fig. 5.11 **a** The dependence of the impedance modulus and impedance phase angle on signal frequency is displayed as measured by the HP Impedance Analyzer. **b** The dependence of the impedance modulus and impedance phase angle on signal frequency is displayed as measured by the IMPSPEC system. *Full lines* represent impedance modulus curves and *dashed lines* represent phase angle curves. The ionic strength of the aqueous solutions is indicated by the legend

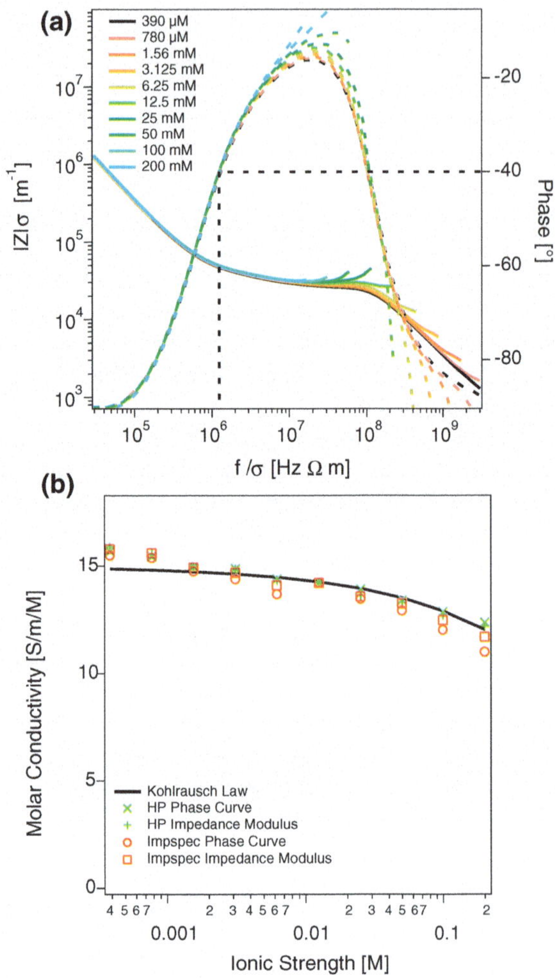

Fig. 5.12 **a** The results from Fig. 5.11a are normalized by multiplying the impedance modulus by the conductivity of each solution and dividing the frequency by the conductivity. *Full lines* represent impedance modulus curves and *dashed lines* represent phase angle curves. The ionic strength of the aqueous solutions is indicated by the legend. **b** The dependence of molar conductivity on molar concentration is displayed. Each *curve* displayed in Fig. 5.11 was used to calculate a molar conductivity value. The *solid line* is from Kohlrausch's Law

The data obtained from the HP 4194A Impedance Analyzer and the IMPSPEC system are quite similar in the measurement range corresponding to the impedance modulus plateau region but there is deviation from the analytical model described above at higher frequencies that is clear for the IMPSPEC system at frequencies above 3 MHz and for the HP Analyzer at frequencies above 10 MHz. These deviations are caused by irregularities in the transfer function of the amplifiers. The results acquired by the IMPSPEC system are acquired at a much higher rate because of the use of multisine excitation.

Figure 5.12a shows the normalized impedance characterization of the data shown in Fig. 5.11a. The frequency was normalized by dividing by the conductivity and the impedance modulus was normalized by multiplying by conductivity. For large frequency ranges, the data collapses to a single curve for both the impedance modulus and the impedance phase angle solely by means of a single variable, the conductivity. This coincidence of the normalization is most close for data at frequencies below the frequency of maximum phase angle. At the peak phase value, the slight increase of the measured phase with increasing ionic strength leads to a slight shift of the values from one another. At measurement frequencies above 10 MHz, the measurements deviate strongly from the model described above.

Both of the normalized values used for the axes of Fig. 5.12a can serve as cell constants. By using interpolation to measure the impedance modulus $|Z|_\theta$ and frequency f_θ at a particular phase value (in our case $-40°$ and increasing), it was possible to define two cell constants X and Y that each can be used to determine the conductivity:

$$\sigma = \frac{f_\theta}{X} \text{ and} \tag{5.6}$$

$$\sigma = \frac{Y}{|Z|_\theta} \tag{5.7}$$

Figure 5.12b displays the molar conductivity of all of the curves shown in Fig. 5.11. The data is displayed in terms of molar conductivity in order to clearly emphasize any deviation from the conductivity calculated for each molar concentration.

5.4.2 *Measurement of Droplets*

Figure 5.13 shows impedance measurements of droplets at transport rates of circa 4 droplets per minute as measured by the IMPSPEC system. Multi-sine wave excitation is used to apply five frequency values between 1 and 10 MHz simultaneously. The aqueous segments in Fig. 5.13b have clear plateau values for all phase angle measurements. The measurement is set up for the highest sensitivity as the aqueous droplet passes between the electrodes. For this reason significant noise is present when oil is located between the electrodes. Work is ongoing to develop signal-processing techniques that will enable high-throughput sensing of droplets in segmented flow [40].

5.5 Microwave Sensing in Micro Fluidic Segmented Flow

5.5.1 *Principle of Microwave Sensing in Microfluidics*

Through the use of microwave sensing it is possible to measure the variation of the dielectric properties (both conductivity and permittivity). Microwave sensing is well suited to the investigation of biological and chemical processes in microsystems, for example, cell growth, cell metabolism, and molecular and ionic concentrations. High frequency sensors can integrate RF/microwave detection in a microfluidic network

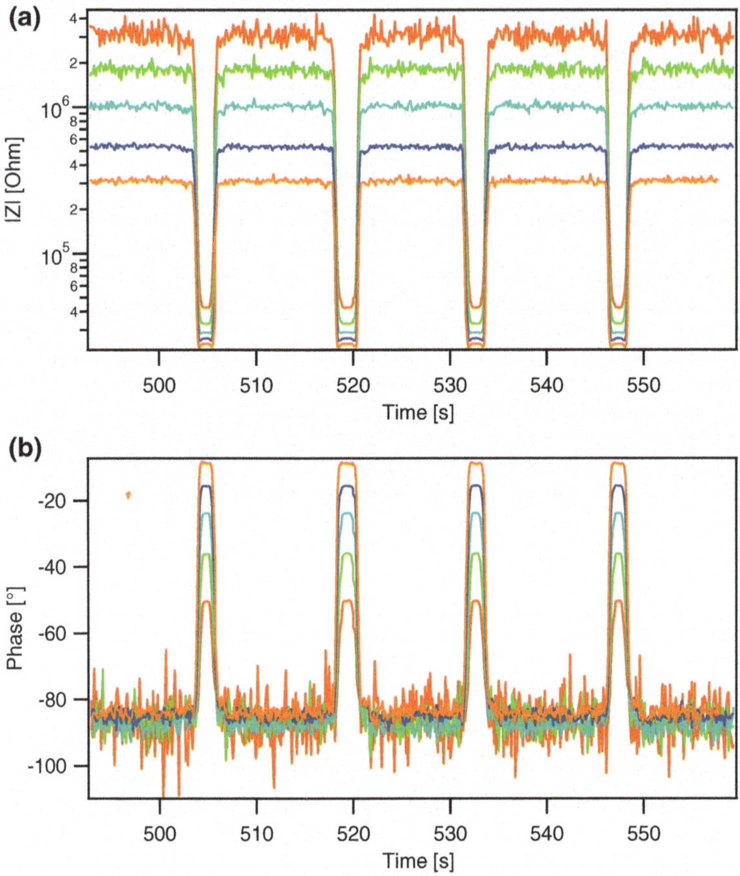

Fig. 5.13 **a** The variation of the impedance modulus during the passage of fluid segments between the electrodes. The individual frequencies of the applied multi-sine wave are indicated by the legend. Measurements acquired using the IMPSPEC system **b** The variation of the phase angle during the passage of 100-mM saline solution fluid segments between the electrodes

for quick and precise biological and chemical analysis. In biosystems, it is of great importance to monitor parameters, such as, cell density, cell growth and cell viability. Cell metabolism is determined by biophysical and biochemical variables that depend directly on cell growth and other metabolic processes. Microwave measurement technology has opened up a broad range of applications for measuring dielectric material properties that are suitable for application as contactless nondestructive sensors in microsystems [41].

The previous sections explained how contactless conductivity detection is of high interest for many applications in microfluidics and it is particularly attractive for measuring the concentration of inorganic and organic ions and matter that are not directly accessible by optical means. Microwave measurements offer many of the advantages of impedimetric measurement at RF frequencies, but also possess several advantages, particularly with regard to miniaturization and their relative insensitivity to insulation layer thickness. The sensor described in this chapter was developed by Nacke et al. [42, 43]. Wessel et al. [44, 45] have also developed microwave sensors for monitoring cell cultivation directly in PTFE tubes operating at 7 and 200 GHz.

Methodology

The rapid development of communications technology in the last two decades has led to the development of components and techniques that can extend dielectric spectroscopy to microwave frequencies, so that dielectric properties of solid matter and liquids in the frequency range from 100 MHz to 100 GHz can be measured. That is, relaxation phenomena with characteristic time constants between fs and some μs are accessible to practical applications. A further advantage of microwave measurement is evident: microwave measurement is much quicker than impedimetric measurement at lower frequencies and is more suited to high-throughput.

Dielectric spectroscopy is a measurement technology that depends on molecular interactions between electrical charges and matter. Figure 5.14 shows the basic set-up of a microwave reflection sensor. The analyser generates a signal that is transmitted through an applicator antenna so that it interacts with the material under test. The antenna collects part of the signal reflected from the material under test and the analyser measures this reflected signal and references against the input signal.

In comparison with measurement at lower frequencies, an additional advantage of microwave sensing is that it is possible to clearly separate conductivity changes from permittivity changes. The dielectric spectrum can be well described by a Debye-type relaxation function with an additional term for the saline dependence (simplified) [46]:

$$\varepsilon = \varepsilon_{(\infty)} + \frac{\varepsilon_{(0)} - \varepsilon_{(\infty)}}{1 + j\omega\tau} - \frac{j\sigma}{\varepsilon_0\,\omega} \tag{5.8}$$

where τ is the relaxation time of the water dipole, angular frequency is defined by $\omega = 2\pi f$, f is the signal frequency, $\varepsilon_{(\infty)}$ the permittivity at high frequency, $\varepsilon_{(0)}$ the permittivity at low frequency and ε_0 the permittivity of free space.

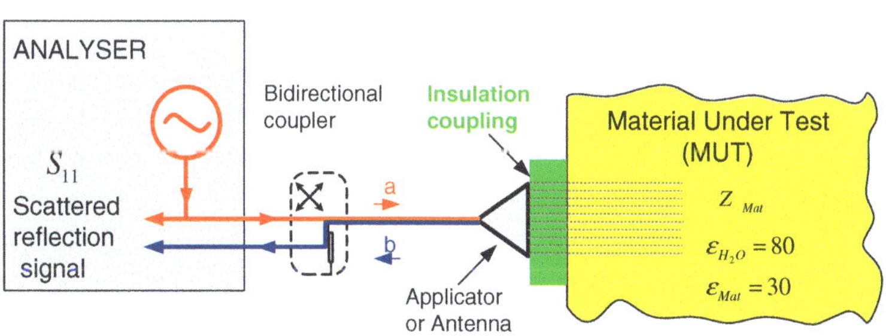

Fig. 5.14 Schematic drawing of the basic setup for microwave reflection measurements

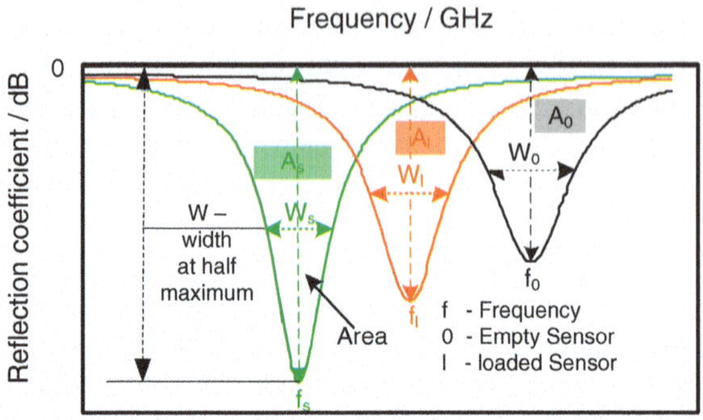

Fig. 5.15 Frequency response of a resonator, with measured S_{11} of a loaded and unloaded sensor and the basics of parameter estimation

Various frequency-dependent measurement parameters are affected by the interaction of microwaves with dielectric materials [47–49], for instance, the reflection coefficient (S_{11}), the magnitude, the partition of magnitude and phase of the signal. The change in resonant frequency, the quality factor (Q-factor) and the attenuation can also be used to characterize the system depending on the load of the resonator sample volume with the material under test.

Microwave sensing is based on how the electrophysical parameters of polar liquids (water and aqueous solutions) depend on frequency in the microwave range. Variations of the water content in a mixture can be related to a shift of the resonant frequency. The width at half maximum and the area between the area between the S_{11} and $S_{11} = 0$ are typically used as indicators of the frequency response of a resonator and are shown schematically in Fig. 5.15. Figure 5.15 plots the reflection coefficient (S_{11} expressed in dB) as a function of frequency for mixtures of varying water content.

The geometry of the measurement cell and sensor are taken into account by the factors k_1 and k_2 [47–49]:

$$\varepsilon' = 1 + \frac{f_0 - f_1}{f_0} \cdot k_1 \tag{5.9}$$

$$\varepsilon'' = \left(\frac{1}{Q_1} - \frac{1}{Q_0}\right) \cdot k_2 \tag{5.10}$$

Calibration of solutions of known values for the real and imaginary parts of the complex dielectric constant, ε' and ε'' respectively, allows the factors k_1 and k_2 to be experimentally determined, so that measurement system artifacts that cannot be modeled by simulation can be accounted for.

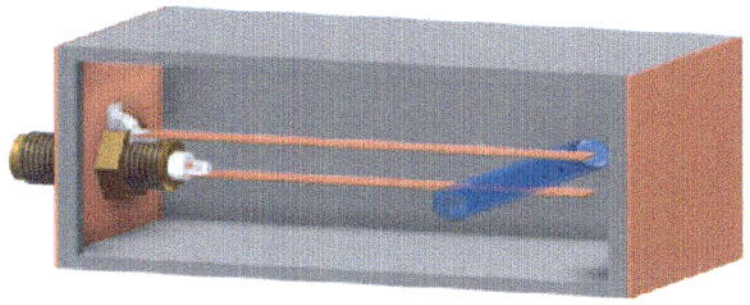

Fig. 5.16 Schematic view of the microstripline resonator

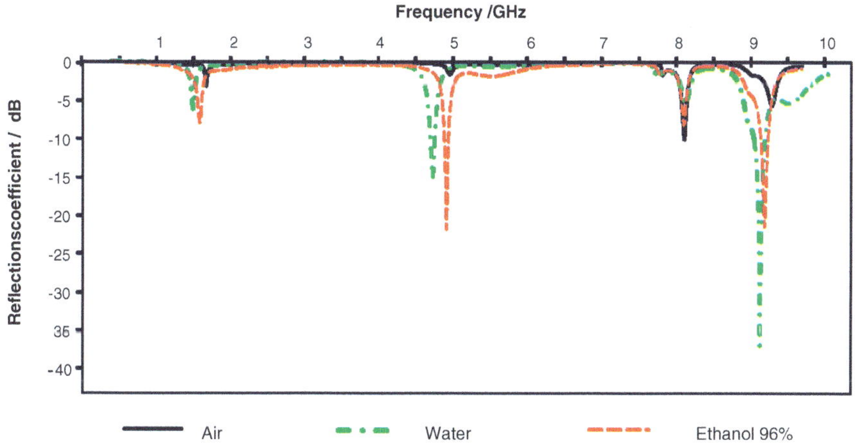

Fig. 5.17 Frequency response of the microstrip resonator sensor

5.5.2 Example of Experimental Realization if Microwave Sensing in Microsegmented Flow

A stripline waveguide was integrated in a resonator (stripline in a cavity) and operated in reflection measurement mode (see Fig. 5.16). The stripline waveguide consists of two copper cantilevers of 36 mm length and 3 mm width separated from each other by 3 mm. The stripline waveguide was housed in an aluminium box (51 × 26 mm) with direct connection of a coaxial cable to an attached SMA connector. Four single resonant peaks in the frequency range up to 10 GHz were observed in measurements. The variation of these single resonant peaks allowed the wideband determination of the material properties of the fluids under test. A PTFE tube was positioned between the stripline, through which the sample fluids could be pumped. Silicone tubing of 3 mm outside diameter and 2.5 mm inner diameter was also used for the measurements.

An Anritsu MS MS4644A vector analyzer was used to measure and record the results. All measurements were carried out at constant room temperature.

Figure 5.17 shows the frequency response of the sensor with regard to air, water and ethanol being present in the tubing.

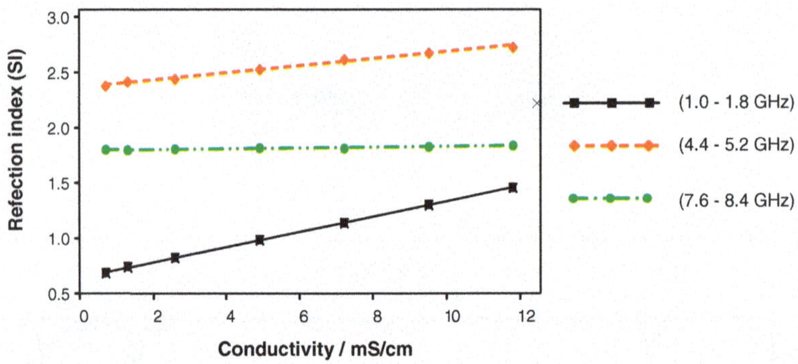

Fig. 5.18 Measurement of KCl concentration at three single resonance peaks (see Fig. 5.5) in the frequency range from 1 to 9 GHz

Equation 5.11 was developed to evaluate the area underneath the resonant curve. The reflection index $SI_{(S11)}$ is given by:

$$SI_{(S11)} = \frac{A_1}{f_0 - f_1} \tag{5.11}$$

In Fig. 5.18, $SI_{(S11)}$ is plotted against conductivity for three separate resonance peaks.

Figure 5.19 shows the dependence of $SI_{(S11)}$ on the concentration of an aqueous solution of ethanol.

The sensing of air bubbles or of phase boundaries is of importance for many applications in microfluidics. For example, in applications of segmented flow, aqueous solutions are embedded in an immiscible carrier liquid. For the analysis of droplet position, speed and volume, the sensing of the presence or absence of the droplet

Fig. 5.19 Reflection index $SI_{(S11)}$ as a function of the Ethanol and Water dilution

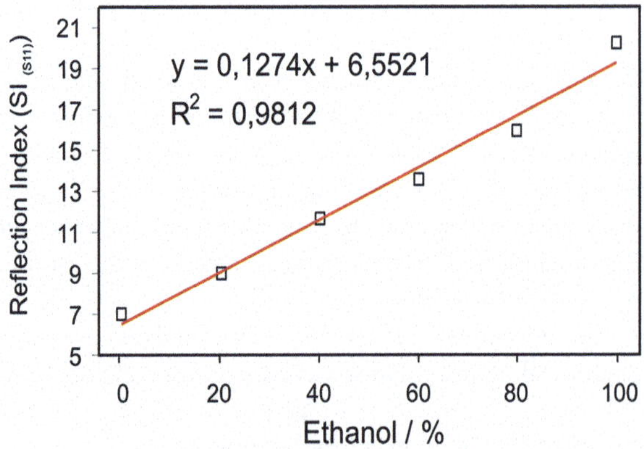

is of great importance. Figure 5.20 shows an example of the sensing of air bubble detection in an aqueous carrier liquid.

5.6 Summarizing Conclusions for Electrical Characterization in Microsegmented Flow

Digital microfluidics is a technique that uses droplets as single experiments for high-throughput screening in biology and chemistry. Digital microfluidics requires the development and optimization of microfluidic hardware for generating, handling, and sensing droplets. Increasing the throughput of sensing for digital microfluidics is necessary for many applications. A typical cell cytometer can measure 10,000 cells per second. If this speed is to be replicated by droplet–based microfluidic systems, it requires sensor and actuator systems that can sense at microsecond intervals and switch droplets in hundreds of microseconds. In the last few decades, high-speed electronics has become commonplace in the telecommunications industry as part of the race to increase the bandwidth of data transmission. This has led to the development of many interesting and relatively inexpensive electronic components that can be used for high-throughput sensing in digital microfluidics: (i) fiber-optic telecommunications has driven the development of optoelectronics and optical sensing based on this technology is the most direct solution to many sensing problems in droplet-based microfluidics, (ii) microwave electronics is the basis of transmission in modern wireless telecommunications and operates at much higher frequencies than

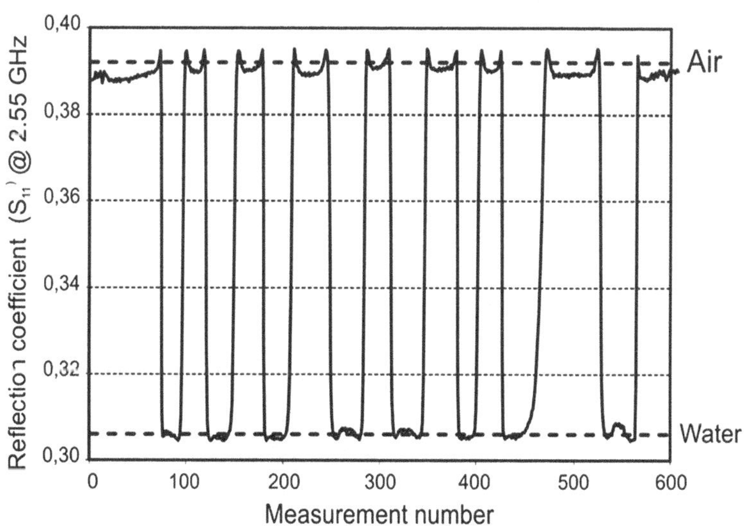

Fig. 5.20 Injected air bubbles of various volumes in a small water filled tube

typical RF electronics and (iii) telecommunications technology is based on high-speed digital switching. Programmed microprocessors are necessary to deal with the acquired data in real time and translate this into high-speed physical digital microfluidic switching see (Chap. 3) of this book for more information). For the promise of digital microfluidics to be fulfilled, it is necessary to develop optoelectronic and mechatronic hardware and corresponding control systems.

Acknowledgments Thomas Nacke would like to thank the Thuringian Ministry of Education, Science and Culture for financial support. B. Cahill would like to thank the European Union for financially supporting the Marie Curie ERG project EWETDYNAM under reference number PERG05-GA-2009-247784.

References

1. L.K. Baxter, *Capacitive Sensors: Design and Applications*, 1st edn. (IEEE Press, New York, 1997)
2. W.C. Heerens, Application of capacitance techniques in sensor design. J. Phys. E Sci. Instrum. **19**, 897–906 (1986)
3. J.P. van der Linden, Liquid-liquid separation in disc-stack centrifuges. Ph.D. Thesis, TU Delft, (1987)
4. J.M. Köhler, A. Funfak, J. Cao, D. Kürsten, S. Schneider, P.M. Günther, in *Addressing of Concentration Spaces for Bioscreenings by Micro Segmented Flow with Microphotometric and Microfluorimetric Detection*. In Optical Nano-and Microsystems for Bioanalytics, (Springer Berlin Heidelberg, 2012), pp. 47–81
5. A. Funfak, J. Cao, O.S. Wolfbeis, K. Martin, J.M. Köhler, Monitoring cell cultivation in microfluidic segments by optical pH sensing with a micro flow-through fluorometer using dye-doped polymer particles. Microchim. Acta **164**, 279–286 (2009)
6. P.M. Günther, F. Möller, T. Henkel, J.M. Köhler, G.A. Groß, Formation of monomeric and novolak azo dyes in nanofluid segments by use of a double injector chip reactor. Chem. Eng. Technol. **28**, 520–527 (2005)
7. J. Schemberg, A. Grodrian, R. Römer, G. Gastrock, K. Lemke, Online optical detection of food contaminants in microdroplets. Eng. Life Sci. **9**, 391–397 (2009)
8. J.Z. Chen, A.A. Darhuber, S.M. Troian, S. Wagner, Capacitive sensing of droplets for microfluidic devices based on thermocapillary actuation. Lab Chip **4**, 473–480 (2004)
9. N. Srivastava, M.A. Burns, Electronic drop sensing in microfluidic devices: automated operation of a nanoliter viscometer. Lab Chip **6**, 744–751 (2006)
10. X. Niu, M. Zhang, S. Peng, W. Wen, P. Sheng, Real-time detection, control, and sorting of microfluidic droplets. Biomicrofluidics **1**, 044101–044112 (2007)
11. C. Elbuken, T. Glawdel, D. Chan, C.L. Ren, Detection of microdroplet size and speed using capacitive sensors. Sens. Actuators A **171**, 55–62 (2011)
12. J. Nichols, A. Ahmadi, M. Hoorfar, H. Najjaran, J.F. Holzman, In situ digital microfluidic conductance sampling. Sens. Actuators A **152**, 13–20 (2009)
13. M.J. Schertzer, R. Ben-Mrad, P.E. Sullivan, Using capacitance measurements in EWOD devices to identify fluid composition and control droplet mixing. Sens. Actuators B **145**, 340–347 (2010)
14. M.J. Schertzer, R. Ben-Mrad, P.E. Sullivan, Automated detection of particle concentration and chemical reactions in EWOD devices. Sens. Actuators B **164**, 1–6 (2012)
15. S.C.C. Shih, R. Fobel, P. Kumar, A.R. Wheeler, A feedback control system for high-fidelity digital microfluidics. Lab Chip **11**, 535–540 (2011)
16. M.A. Murran, H. Najjaran, Capacitance-based droplet position estimator for digital microfluidic devices. Lab Chip **12**, 2053–2059 (2012)

17. A. Ernst, W. Streule, N. Schmitt, R. Zengerle, P. Koltay, A capacitive sensor for non-contact nanoliter droplet detection. Sens. Actuators A **153**, 57–63 (2009)
18. A. Ernst, W. Streule, R. Zengerle, P. Koltay, Quantitative volume determination of dispensed nanoliter droplets on the fly, in *Solid-State Sensors, Actuators and Microsystems Conference, 2009. TRANSDUCERS 2009. International (IEEE, 2009)*, pp. 1750–1753
19. A. Ernst, L. Ju, B. Vondenbusch, R. Zengerle, P. Koltay, Noncontact determination of velocity and volume of nanoliter droplets on the fly. Sens. J **11**, 1736–1742 (2011)
20. A. Ernst, K. Mutschler, L. Tanguy, N. Paust, R. Zengerle, P. Koltay, Numerical investigations on electric field characteristics with respect to capacitive detection of free-flying droplets. Sensors **12**, 10550–10565 (2012)
21. M. Takeuchi, Q. Li, B. Yang, P.K. Dasgupta, V.E. Wilde, Use of a capacitance measurement device for surrogate noncontact conductance measurement. Talanta **76**, 617–620 (2008)
22. K. Arshak, D. Morris, A. Arshak, O. Korostynska, K. Kaneswaran, Development of oxide thick film capacitors for a real time pressure monitoring system. Mater. Sci. Eng. C **27**, 1406–1410 (2007)
23. Y. Hotta, Y. Zhang, N. Miki, A flexible capacitive sensor with encapsulated liquids as dielectrics. Micromachines **3**, 137–149 (2012)
24. I.K. Kiplagat, P. Kubán, P. Pelcová, V. Kubán, Portable, lightweight, low power, ion chromatographic system with open tubular capillary columns. J. Chromatogr. A **1217**, 5116–5123 (2010)
25. A.J. Zemann, E. Schnell, D. Volgger, G.K. Bonn, Contactless conductivity detection for capillary electrophoresis. Anal. Chem. **70**, 563–567 (1998)
26. R.M. Guijt, C.J. Evenhuis, M. Macka, P.R. Haddad, Conductivity detection for conventional and miniaturised capillary electrophoresis systems. Electrophoresis **25**, 4032–4057 (2004)
27. P. Kuban, P.C. Hauser, Capacitively coupled contactless conductivity detection for microseparation techniques—recent developments. Electrophoresis **32**, 30–42 (2011)
28. B.P. Cahill, R. Land, T. Nacke, M. Min, D. Beckmann, Contactless sensing of the conductivity of aqueous droplets in segmented flow. Sens. Actuators B **159**, 286–293 (2011)
29. B.P. Cahill, R. Land, J. Metze, Simulation of the measurement of the impedance of aqueous droplets in segmented flow. Chem. Eng. Trans. **24**, 535–540 (2011)
30. B.P. Cahill, Optimization of an impedance sensor for droplet-based microfluidic systems. Proc. SPIE **8066**, 80660F–80668 (2011)
31. T. Lederer, S. Clara, B. Jakoby, W. Hilber, Integration of impedance spectroscopy sensors in a digital microfluidic platform. Microsyst. Technol. **18**, 1163–1180 (2012)
32. T. Lederer, B.P. Stehrer, S. Bauer, B. Jakoby, W. Hilber, Utilizing a high fundamental frequency quartz crystal resonator as a biosensor in a digital microfluidic platform. Sens. Actuators A **172**, 161–168 (2011)
33. S. Sadeghi, H. Ding, G.J. Shah, S. Chen, P.Y. Keng, C.J.C. Kim, R.M. van Dam, On chip droplet characterization: a practical, high-sensitivity measurement of droplet impedance in digital microfluidics. Anal. Chem. **84**, 1915–1923 (2012)
34. M.C. Hofmann, M. Funke, J. Büchs, W. Mokwa, U. Schnakenberg, Development of a four electrode sensor array for impedance spectroscopy in high content screenings of fermentation processes. Sens. Actuators B **147**, 93–99 (2010)
35. H.E. Ayliffe, A.B. Frazier, R.D. Rabbitt, Electric impedance spectroscopy using microchannels with integrated metal electrodes. J. Microelectromech. Syst. **8**, 50–57 (1999)
36. U. Pliquett, D. Frense, M. Schönfeldt, C. Frätzer, Y. Zhang, B. Cahill, M. Metzen, A. Barthel, T. Nacke, D. Beckmann, Testing miniaturized electrodes for impedance measurements within the β-dispersion—a practical approach. J. Electr. Bioimpedance **1**, 41–55 (2010)
37. S.C.C. Shih, I. Barbulovic-Nad, X. Yang, R. Fobel, A.R. Wheeler, Digital microfluidics with impedance sensing for integrated cell culture and analysis. Biosens. Bioelectron. **42**, 314–320 (2013)
38. A. Barthel, D. Beckmann, J. Friedrich, I. Mock, T. Nacke, P. Peyerl, IMPSPEC - Ein hard- und softwarekonzept für die (bio)-impedanzspektroskopie. Chem. Ing Tech. **77**, 1948 (2005)

39. T. Nacke, A. Barthel, J. Friedrich, M. Helbig, J. Sachs, M. Schäfer, P. Peyerl, U. Pliquett, A new hard and software concept for impedance spectroscopy analysers for broadband process measurements, in *13th International Conference on Electrical Bioimpedance and the 8th Conference on Electrical Impedance Tomography*, vol. 17, pp. 194–197. doi:10.1007/978-3-540-73841-1_52 (2007)
40. R. Land, B.P. Cahill, T. Parve, P. Annus, M. Min, Improvements in design of spectra of multisine and binary excitation signals for multi-frequency bioimpedance measurement, in *2011 Annual International Conference of the IEEE Engineering in Medicine and Biology Society*, pp. 4038–4041. doi:10.1109/IEMBS.2011.6091003 (2011)
41. C. Song, P. Wang, A radio frequency device for measurement of minute dielectric property changes in microfluidic channels. Appl. Phys. Lett. **94**, 023901 (2009)
42. T. Nacke, A. Barthel, M. Meister, B.P. Cahill, Application of high frequency sensors for contactless monitoring in disposable bioreactors. Chem. Ing. Tech. **85**, 179–185 (2013)
43. T. Nacke, A. Barthel, B.P. Cahill, R. Klukas, M. Meister, C. Pflieger, D. Beckmann, Contactless wide-band nearfield microwave sensing techniques in microfluidic applications, in *14th International Meeting on Chemical Sensors IMCS 2012*, pp. 411–414. doi:10.5162/IMCS2012/4.5.4. (2012)
44. J. Wessel, J.C. Scheytt, K. Schmalz, B. Cahill, G. Gastrock, Microwave biosensor for characterization of compartments in teflon capillaries, in *42nd European Microwave Conference EuMC 2012*, pp. 534–537. (2012)
45. J. Wessel, K. Schmalz, B. Cahill, G. Gastrock, C. Meliani, Contactless characterization of yeast cell cultivation at 7 GHz and 240 GHz, in 2013 IEEE 13th Topical Meeting on Silicon Monolithic Integrated Circuits in RF Systems (SiRF), (Austin, TX, USA, 2013), pp. 147–149. doi:10.1109/SiRF.2013.6489461
46. A. Stogryn, G. Desargant, The dielectric properties of brine in sea ice at microwave frequencies. IEEE Trans. Antennas Propag. **33**, 523–532 (1985)
47. M. Venkatesh, G. Raghavan, An overview of dielectric properties measuring techniques. Can. Biosyst. Eng. **47**, 15–30 (2005)
48. K.K. Joshi, I. Allika, C. Wadke, Nondestructive microstrip resonator technique for the measurement of moisture / permittivity in crude oil, in *Proceedings of the XXVIIIth URSI General Assembly*, (New Delhi, India, 2005), pp. 324–330
49. R. Knöchel, R. Jahns, W. Taute, C. Döscher, A resonator-based moisture meter for high moisture levels, in *Proceedings of the First European Conference on Moisture Measurement*, (Weimar, Germany, 2010), pp. 53–62

Part II
Chemical Application in Micro Continuous-Flow Synthesis of Nanoparticles

Chapter 6
Solid Particle Handling in Microreaction Technology: Practical Challenges and Application of Microfluid Segments for Particle-Based Processes

Frederik Scheiff and David William Agar

Abstract This chapter reviews the current applications, challenges and trends of solid particle utilization in microfluidic devices. The physics of particle behavior are discussed and serve as a fundament to engineer a suspension dosage and phase separation, as well as for clogging countermeasures. Special attention is finally drawn on the competitiveness with alternative concepts for heterogeneously catalyzed processes.

6.1 Application of Solids in Microfluidics

A reflection of history illustrates that the field of microfluidics has been initialized with research on gas chromatographs at Stanford University and ink jet nozzles at IBM in the late 1970s [1]. In the 1990s, research on microfluidics lifted off and soaked many fields. Microreaction technology has been one of the first [2].

Since the 1990s biotechnological or medical "lab on a chip devices" and since about 2000 nanoparticle synthesis have driven the development of microfluidics. Particle handling in microfluidic devices has become a relevant issue throughout these fields. If one considers vascular transport of cells as a natural case of treating particulate matter in microchannels, it appears remarkable that solid processing in microfluidics has been considered with reservations for many years. Research on particulate matter in microreactors has been cut down to particle immobilization in micro-fixed bed and wall-coated reactors or to particle-induced clogging resulting from salt formation, degradation and side reactions. Interestingly, the idea of suspension catalysis has not been mentioned at all in reviews on microreactors for

F. Scheiff (✉) and D. W. Agar
Technical University of Dortmund, Department of Biochemical and Chemical Engineering,Emil-Figge-Straße 66,044227 Dortmund, Germany
e-mail: frederik.scheiff@bci.uni-dortmund.de

D.W.Agar
e-mail: agar@bci.tudortmund.de

J. M. Köhler and B. P. Cahill (eds.), *Micro-Segmented Flow*, Biological and Medical Physics, Biomedical Engineering,
DOI: 10.1007/978-3-642-38780-7_6,

catalytic reactions from 2005 and 2007 [3, 4]. However, a paradigm shift has taken place since 2009 [5]. The progress in understanding the clogging mechanism and successful particle handling in nanoparticle synthesis and biotechnology opened the door for microreactions of suspended particles—heterogeneous-catalyzed reactions and solid phase extraction for instance [6]. Special attention has been drawn on suspension catalysis, which would extend microreaction technology to a whole class of reactions unaddressed so far [7]. Three types of microreactor suspension catalysis are in the main focus of current research projects:

- Liquid–solid reactors:

Vankayala et al. investigate the hydrogenation of aromatic nitro compound catalyzed by suspended Pd/activated carbon particles in a suspension microreactor [8]; other heterogeneous catalyzed liquid reactions have been investigated in micro-packed bed or wall-coated catalytic reactor and could be transferred to suspension catalysis [3, 4].

- Gas–liquid–solid segmented flow:

Gas–liquid–solid reactions are widespread in chemical industry. Hydrogenation of nitroaromatic or other compounds are prominent examples and well investigated in micro-fixed bed and wall coated catalytic reactors [9, 10]. The literature on gas–liquid-suspension microreactors is rather decent, instead. Huang et al. [11] trapped particles sterically to hydrogenate 3-nitrotoluene in a "catalyst trap microreactor", Enache et al. [12] used about 50 μm-sized Rh/Al_2O_3 particles for the hydrogenation of resorcinol in milli-capillary slurry reactor, until Buisson et al. [13] reported the successful microreactor slurry hydrogenation in Corning® Advanced Flow^{TM} reactors. However, systematic studies within gas–liquid segmented flow have been addressed just recently with the hydrogenation of 3-methyl-1-pentyn-3-ol with a Pd/SiO_2 catalyst by Liedtke et al. [14].

- Liquid-liquid–solid segmented flow microreactors:

Solid catalyzed liquid–liquid reactions or solid catalyzed reactive extraction have been rarely reported so far. Olivon [15] applied liquid–liquid droplet and slug flow in microcapillaries for catalyst screening within the confined volume. Ufer et al. provided a proof-of-concept for suspension catalysis with liquid–liquid mass transfer in segmented flow. A catalytic transfer hydrogenation has been investigated for this purpose in 1.6 mm ID plastic capillary reactors at TU Dortmund University [5, 16]. A current project at TU Dortmund University is committed to resolving fluid dynamics and mass transfer in suspension slug flow microreactors [17].

After tracing back the history of solid handling in microfluidics to the status quo, current challenges are to be reviewed within this chapter. Although rather remote, microreaction technology (this chapter) nanoparticle synthesis (Part II) and biological applications (Part III) share a common ground of technological challenges (i–iv), which will be discussed according the structure of Fig. 6.1.

The feed of the suspensions (i) is not trivial, as soon as particle are prone to sedimentation and agglomeration. Most established microfluidic concepts fall short

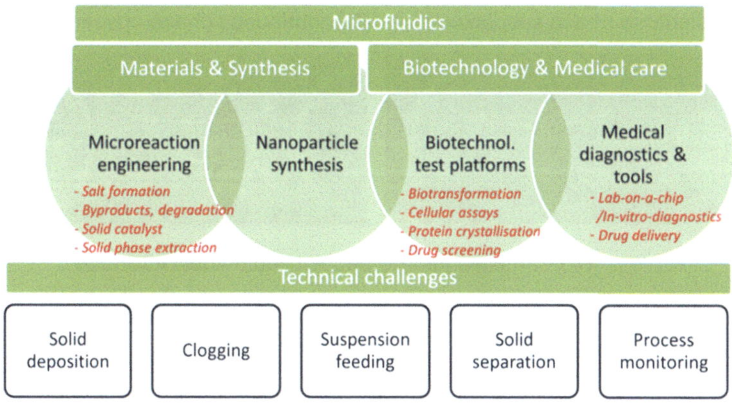

Fig. 6.1 Application of suspended solids in microfluidics

with respect to that issue, so that prototypes, designed recently and accounting for sedimentation, are presented. The deposition of non-neutrally buoyant particles in microfluidic devices impedes a robust reactor operation and is closely related to the mechanisms of clogging (ii). Guidelines and countermeasures will be presented. And unlike conventional scale liquid–solid equipment, liquid–liquid–solid separation (iii) must be re-invented for the micro scale. A final part of this chapter is committed to the assessment of alternative microreactor concepts (iv).

6.2 Particle Transport Behavior in Micro Segmented Flow

A basis will be laid by introducing particle behavior in liquid–liquid segmented flow (see sections one and two). Micro- or nanoparticles in liquid–liquid slug flow are suspended either in disperse or continuous phase or aggregate at the liquid–liquid interface (Fig. 6.2).

Suspending particles in the disperse phase (case a) provides precisely defined particle residence times and is inherently safe from clogging. By contrast, particles in the continuous segment (case b) may deposit on the microchannel surface and have a non-uniform residence time distribution. Before the cases a and b are discussed in detail, particle aggregation at the liquid–liquid interface (case c) shall be mentioned. Liquid–Liquid–solid reactions have been reported to profit notably from the good access of reactants to the catalyst at the interface and this concept is termed "phase boundary catalysis" [5, 18, 19]. Particle adsorption at the interface is determined by interfacial tensions γ_i and has been successfully promoted by surface modifications [19]. According to thermodynamic considerations derived by Pieranski [20], particles P are trapped at the interface of fluids A and B, if they meet the following condition:

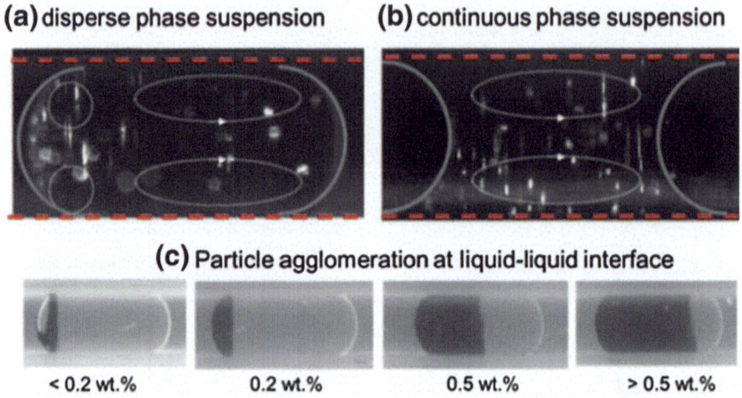

Fig. 6.2 Types of particle suspensions in liquid–liquid slug flow; **a** disperse suspension, e.g. fluorescent Al_2O_3 in aqueous slug and continuous paraffin oil phase; **b** continuous segment suspension, fluorescent TMS-coated SiO_2 particles in continuous segment of paraffin oil in and disperse water phase; **c** activated carbon in toluene-water slug flow, initially suspended in continuous toluene phase [17]

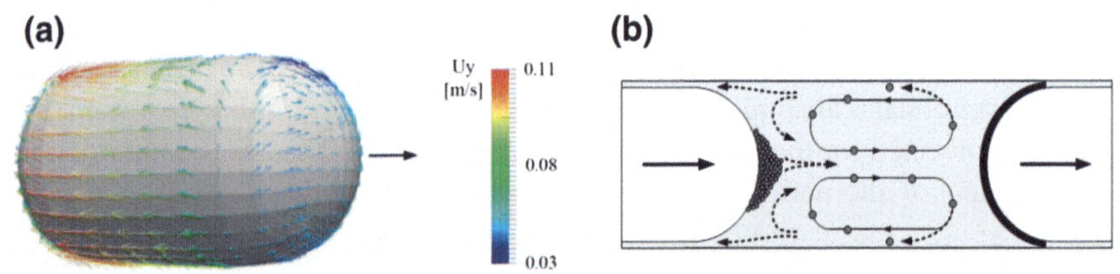

Fig. 6.3 **a** CFD simulation of surface velocity (OpenFoam® simulation of water-toluene slug flow in circular 1.6 mm ID capillary at $u_{\mathrm{Slug}} = 0.02$ m/s [17]); **b** schematic of cap formation [5]

$$|\gamma_{p-A} - \gamma_{p-B}| < \gamma_{A-B} \tag{6.1}$$

In this case, the total interfacial energy of the three phase system has a minimum at a certain depth of immersion. An energy well prevents particle detachment from the interface. This activation energy is large for microparticles, "10^7 times the thermal energy for polystyrene microparticles adsorbed at the water-air interface" [21]. So particle adsorption at the interface is rather rigid. Certainly, nanoparticle accumulation is significantly less stable with activation energies estimated at $5k_BT$ of typical oil-water-interfaces [22]. As reviewed by Bresme and Oettel, the dense concentration of particles at the interface and the contact to both fluid media induces attractive and/or repulsive interparticle forces—electrostatic interactions, van-der-Waals forces, capillary forces, flocculation and solvation forces among others [21]. However, the fluid-dynamic motion of the interface outweights the other forces. Particles once trapped at the interface are carried partly to the nose, but dominantly to the rear cap. Recent CFD simulations of water-toluene slug flow in circular capillaries support the suggestion that this phenomenon is an immediate consequence of interfacial motion (Fig. 6.3a).

According to Fig. 6.2c and the results of Ufer et al. [5], particles form a ring at the rear cap when the solid concentration is low. Higher concentrations result in a closed cap and the accumulation stretches out over the cylindrical part of the slug.

The detailed discussion of case a and b with a suspension encapsulated in a liquid slug or the continuous phase must consider numerous forces on the particles—exerted by the fluid, external fields, interfaces or particle–particle interaction [23, 24]. The ratio of these forces results in versatile particle motion pattern and three types of particles have been identified according to their behavior.

- metal/-oxide nanoparticles (10–100 nm, high density)
- metal/-oxide microparticle (1–100 μm, high density)
- cells (1–10 μm, almost neutrally buoyant).

In the Langrangian reference frame, the equation of motion of a single particle is written as [23, 25]:

$$\vec{F}_{inertia} = \frac{\pi}{6} x_P^3 \rho_P \frac{d\vec{u}_p}{dt} = \frac{\pi}{6} x_P^3 \rho_P \frac{d^2 s_p}{dt^2} = \sum \vec{F}_i \tag{6.2}$$

Table 6.1 summarizes the relevant forces F_i and provides state-of-the-art correlations. In practice, aqueous-organic slug flow with hydrophilic metal/-oxide particles suspended in the aqueous phase is the predominant case and encapsulating the aqueous suspension in the disperse segment rules out particles deposition at the channel wall. Thus, forces acting on nano- and microparticles are scaled for dispersed suspensions (case a). It can be seen from Fig. 6.4a that drag and pressure gradient force push particles along the internal circulation streamlines. Inertia and virtual mass force act against these forces parallel to the fluid streamline. Basset's history force is not relevant, as it only comes into play for highly instationary flow conditions that are beyond the characteristics of slug flow. Additionally, the force on a particle has a vectorial component perpendicular to the fluid streamline. Gravity is directed downwards, Saffman and Magnus force move the particles to regions of higher relative fluid velocity, i.e. away from the center of the Taylor vortex. The inertial lift force is opposite, but is only relevant at higher Re_P numbers. Hence, the motion normal to the fluid is governed by gravity, Saffman and Magnus force. This force balance must be modified at the front or rear end of the Taylor vortex as shown in Fig. 6.4a.

Due to drag and the pressure gradient force, particles follow the streamline, as long as inertia, virtual mass, centrifugal, Saffman or Magnus force are overridden. It should be mentioned that only a vectorial fraction of gravity compensates or enhances the effect of Saffman or drag force, if particles do not circulate on the vertical plane. This concept remains abstract as long as the ratio of these numerous forces is uncertain. Thus, previous correlations were used to estimate these forces, assuming point particles, one-way-coupling with the fluid and stationary flow. This estimate reveals that only a handful of forces determine the particle motion. Although certain sim-

Table 6.1 Forces on particles in liquid–liquid slug flow

Type	Force	Equation		Validity	Reference
Particle-fluid	Drag	$\vec{F}_{drag} = \frac{\pi}{4} x_P^2 \frac{\rho_F}{2} (\vec{u}_F - \vec{u}_p)^2 C_D$		$Re_P \ll 1$ & atoms per particle >200	[24]
		$F_{drag} = 3\pi \mu x_p (u_F - u_p)$	(6.3)		[26]
	Virtual mass	$\vec{F}_{VM} = \frac{\pi}{6} x_P^3 \frac{\rho_F}{2} (\vec{a}_F - \vec{a}_p)$	(6.4)	Spherical particles	[23]
	Basset history	$\vec{F}_{Basset} = \frac{3}{2} \sqrt{\pi \mu \rho_F} \int_{to}^{t} \frac{\vec{a}_F - \vec{a}_p}{\sqrt{t-t'}} dt'$	(6.5)	Highly instationary flow	[23]
	Saffman[a]	$\frac{F_{Saff}}{F_{Saff,0}} = \begin{cases} -140\epsilon^5 \text{In}(\epsilon^{-2}) \ for \ \epsilon \ll 1 \\ 0.3[1 + \tanh(2.5\log(\epsilon + 0.191))][0.667 + \tan(6(\epsilon - 0.32)] \\ for \ 0.1 \le \epsilon \le 20 \\ 1 - 0.287\epsilon^2 \quad for \ \epsilon \gg 1 \end{cases}$	(6.6) with	(Non-) rotating particle	[27–33]
		$\vec{F}_{Saff,0} = 1.615\mu(\lvert u_F - u_p \rvert) x_p^2 \sqrt{\frac{1}{v} \vec{\nabla}(u_F - u_P)}$	(6.7)		
		and $\epsilon = \frac{\sqrt{Re \vec{\nabla} u_F}}{Re_s} = \sqrt{\frac{2v}{d_{cap} u_{slug}}}$	(6.8)		
	Inertial lift	$\vec{F}_{Inertial\,lift} = -\rho_P \vec{\nabla} u_F C_L x_P^4$	(6.9)	$Re_P \gg 1$	[28]
	Magnus[a]	$\vec{F}_{Magnus} = \frac{\pi}{8} x_P^3 \rho_F \vec{\omega} x \Delta \vec{u}_{rel}$	(6.10)	Spherical particles $Re_P \ll 1$	[34]
	Dynamic pressure	$\vec{F}_P = \frac{\pi}{6} x_P^3 \rho_f \frac{d\vec{u}_F}{dt}$	(6.11)	Spherical particles, instationary flow	[35]
	Dean	$F_{Dean} = 5.4 \cdot 10^{-4} \pi \mu \, De^{1.63} x_P$	(6.12)	Spherical particles Curved micro-channel, $Re_P \ll 1$	[36]

(continued)

Table 6.1 (continued)

Type	Force	Equation		Validity	Reference
	Centrifugal = particle inertia – dynamic pressure	$F_{centri} = \frac{\pi}{6} x_P^3 (\rho_P u_{\varphi,P}^2 - \rho_F u_{\varphi,F}^2) \frac{1}{r}$	(6.13)	Spherical particles, constant radius of curvature	[37]
	Brownian motion	$\vec{F}_B = \xi_i \sqrt{\frac{\pi S_o}{\Delta t}}\ with\ S_o = \frac{216 \upsilon k_B T}{\pi^2 \rho_F x_P^5 (\frac{\rho_P}{\rho_F})^2 C_c}$	(6.14)	Applicability of Stokes-Einstein-equation	[25, 38]
		And $\varepsilon = \frac{diffusion}{convection} = \frac{1}{\bar{u}} \sqrt{\frac{2D}{\Delta t}}$ (6.15) with $D = \frac{k_B T}{3 \pi \mu x_P}$	(6.16)		
External fields	Gravity-Buoyancy	$\vec{F}_g = \frac{\pi}{6} x_P^3 (\rho_P - \rho_F) \vec{g}$	(6.17)		
External forces	Electrophoresis [39], dielectrophoresis [40], magnetophoresis [41], acustic [42], thermophoresis, radiation, optophoresis [43]				
Particle–particle	Diffusion [44], in-/elastic collision [45], van-der-Waals force [46], electrostatic/steric/magnetic repulsion, Coulomb force [47]				
Particle-interface	In-/elastic collision, electrostatic interaction, capillary forces, van-der-Waals forces, fluctuation forces, solvation [21, 48]				

[a] Saffman force is a special case of Magnus force, valid when the flow is governed by slip-shear, rather than slip-spin as assumed in Magnus force and it includes particle rotation. It is strictly valid at zero relative Reynolds number in creeping shear flow [49, 50]. Hence, either Saffman or Magnus force should be considered. Herein, both are calculated to show that lift in shear flow exceeds rotational Magnus lift

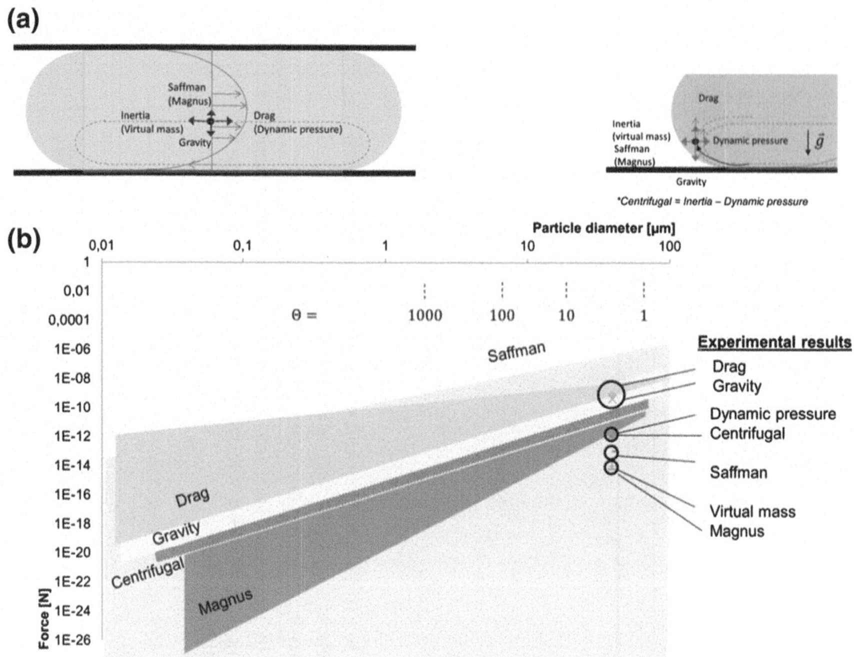

Fig. 6.4 Particle force balance in segmented flow: **a** schematic of force balance (*side view*, i.e. gravity shows vertically downward), **b** Estimate of forces and experimental results (estimates are valid for the boundary conditions given in subsequent table; experimental results are obtained by extracting particle trajectories from fluorescence microscopy video sequences that track three individual in-house fabricated, fluorescent Rhodamin B-labelled Al_2O_3 particles with $x_P = 40\,\mu m$ or PIV microparticles (SiO_2, $10\,\mu m$, RITC-labelled, sicastar® redF (micromod GmbH (Germany))), suspended in H_2O—paraffin oil slug flow within a 1.0 mm FEP capillary at $u_{Slug} = 0.0123$ m/s [17]

plifying assumptions are inevitable, the results shed light on the particle behavior observed experimentally.

The fundamentals of these theoretical estimations are summarized in Table 6.2. The a priori calculation of the governing forces has been validated experimentally with the equations and simplification described before, but using experimental results of particle and fluid flow rather than assumption on the flow structure (Fig. 6.4b). This provides insight into the particle circulation and the segregation in the posterior section of a slug as reported by several authors [5, 15, 56]. The ratio of drag and gravity, known as Shields parameter Θ from sedimentation theory [57], has also been included in this diagram:

$$\Theta = \frac{drag}{gravity} = \frac{9\mu}{x_p^2 \Delta\rho g} u_{slug} \tag{6.18}$$

Despite some deviations, a priori estimations match the experimental results and the forces can be classified into four groups, according to their magnitude:

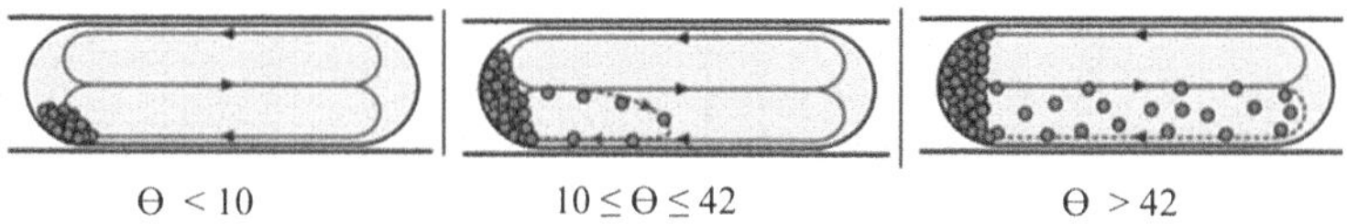

Fig. 6.5 Schematic particle concentration profiles as a function of Shields parameter Θ [56]

i. Drag and gravity are strongest. Gravity is not negligible in this kind of microfluidic application.
ii. Dynamic pressure force and the centrifugal acceleration gain a notable magnitude, evoked by the internal circulation pattern in slug flow.
iii. Saffman force is small, because the shear gradient vanishes on the center line of the slug. But it attains large values closer to the vortex center and Saffman force is to be considered in this region.
iv. Virtual mass and Magnus force are proven to be negligible under the assumptions proposed, e.g. for circular particles.

The particle motion is in the viscous regime. Particle Reynolds numbers Re_P indicates that inertial effects come only into play for large particles and high velocities ($Re_P = 1 \cdot 10^{-4}(10\,\text{nm}), 1 \cdot 10^{-2}(1\,\mu\text{m}), 1 \cdot 10^{0}(100\,\mu\text{m})$ at $u_{\text{Slug}} = 0.01\,\text{m/s}$).The ratio of drag and gravity is close to unity in the range of 1–100 μm particles. According to Shields parameter $\Theta < 100$, sedimentation plays a crucial role. Based on Θ, Kurup and Basu [56] found that particles are completely settled and aggregated in the rear end of a slug at $\Theta < 10$. Partial segregation and suspension is found at intermediate values of $10 < \Theta < 42$, and a maximum fraction of suspended particles is attained above a threshold of $\Theta > 42$ (Fig. 6.5).

But this explanation is not exhaustive and extrapolation to other particle diameters is intricate. For instance, drag force may not scale with x_p, because the relative velocity depends on particle size size as well, so that F_{drag} scales $x_p \cdot u_P$. Moreover, Kurup and Basu do not explain why particles accumulate in the co-rotating vortex at the rear cap under conditions, where sedimentation is ruled out ($\Theta > 42$).

To find evidence for the mechanism of particle segregation, local concentration profiles of 15 and 40 μm sized Rhodamin B-labelled Al_2O_3 particles at u_{Slug} = 2–32 mm/s have been obtained with fluorescence microscopy at 4x magnification (Fig. 6.6). One must take into account that the co-rotating vortex in the posterior section (at large values of z) is exceedingly large in water- paraffin oil slug flow and this leads to bimodal axial profiles at intermediate velocities. At lower velocities, particles are segregated at the bottom in the rear vortex and the radial profiles prove gravitational settling. In the opposite case of higher velocities, particles spread symmetrically in vertical direction. The fraction of particles suspended in the anterior vortex increases from $\Theta = 10$ to $\Theta = 30$, which is in good agreement with the postulate of Kurup and Basu. But larger particles are suspended at lower values of Θ, which refers to previous considerations. Gravity is entirely overridden at $\Theta \geq 180$ in this experiment.

Table 6.2 Boundary conditions for theoretical estimation of forces on particles in suspension slug flow

General Validity	- Conventional hydrophilic particles, completely suspended in an aqueous dispersed phase, and solid concentrations of ≤ 0.1 vol.%.	- Straight microchannels are assumed, so that Dean vortices do not occur	
		- Circular microchannel cross section, $d_{cap} = 1.0\,\text{mm}$	
	- Particle-interface- and particle-particle-interactions are absent [51]	- Superficial flow velocity = $u_{Slug} = 1.23\,\text{cm/s}$	
		- Volumetric ratio of aqueous:organic phase = 1:1	
	- Application-specific external forces are not considered	- Aqueous carrier fluid ($\rho_F = 997\text{kg/m}^3$, $\mu_F = 1\,\text{mPa s}$)	
	- Basset's history force is neglected	- Particle diameter $x_P = 0.01\,\mu\text{m}$ (10 nm) – $100\,\mu\text{m}$	
Force	**Eqn.**	**Boundary conditions/assumptions**	**Additional parameters**
Drag	(6.3)	1. Maximum relative velocity: $(u_F - u_P)_{\max} = u_F = u_{slug}$; *max. fluid velocity and particle static relative to slug reference frame* 2. Minimum relative velocity: $(u_F - u_P)_{\min} = u_{Stokes} = \frac{x_p^2 g \Delta\rho}{18\mu}$ ($\rho_P = 3920\,\text{kg/m}^3$); *min. relative velocity between fluid and particle is Stokes sedimentation velocity* [28, 52]	Maximum drag increases with u_F
Saffman/Inertial lift	(6.6)	1. Maximum relative velocity: $(u_F - u_P)_{\max}$ cf. "drag"; 2. Minimum relative velocity: $(u_F - u_P)_{\min}$ cf. "drag"; 3. $\vec{\nabla}(u_F - u_P) \cong \vec{\nabla}_{u_F} \cong \frac{4u_{slug}}{d_{cap}}$;	Saffman force is weakened due to inertial effects at higher $u_{F\&}$

(continued)

Table 6.2 (continued)

Force	Eqn.	Boundary conditions/assumptions	Additional parameters
		4. Minimum shear on symmetry axis of slug results in zero Saffman force, 5. Wall effect neglected	narrower d_{cap}
		1. Maximum relative velocity: $(u_F - u_P)_{\max}$ cf. "drag",	
Magnus	(6.10)	2. Minimum relative velocity: cf. "drag"	Magnus force increases with u_F and smaller d_{cap}
		3. $\omega = 0.5\Delta u_F$ at $\mathrm{Re}_P = 0$ [53], Δ_{u_f} cf. Saffman force, Rotation only induced by flow, hence ω is perpendicular to u	
Centrifugal		1. $r = 1/4 d_{cap}$, 2. $\rho_P = 3920\,\mathrm{kg/m^3}$, 3. Maximum $u_\varphi = u_{slug}$ [54]	Centrifugal force increases with u_F and smaller d_{cap}
	(6.13)	4. Minimum $u_\varphi = 0.47 u_{slug}$ [55]	
Brownian motion	(6.15)	1. Stokes-Einstein equation applied for calculation of particle diffusion coefficient D,	Effect of Brownian motion amplified by elevating temperature and narrow capillaries
		2. $T = 298K$; 3.$\Delta t = \frac{d_{cap}}{2u_{slug}}$, $\overline{u} = u_{slug}$	
Gravity - Buoyancy	(6.17)	1. Maximum gravity: $\rho_P = 1050\mathrm{kg/m^3}$ (Polystyrene particles);	Gravitational influence increases with decreasing fluid density
		2. Minimum gravity: $\rho_P = 3920\mathrm{kg/m^3}$ (Al_2O_3 particles)	
Virtual mass	No reasonable assumption for the magnitude of $\vec{a}_F - \vec{a}_P$ accessible		
Dynamic pressure	No reasonable assumption for the magnitude of $\vec{a}_F$ accessible		

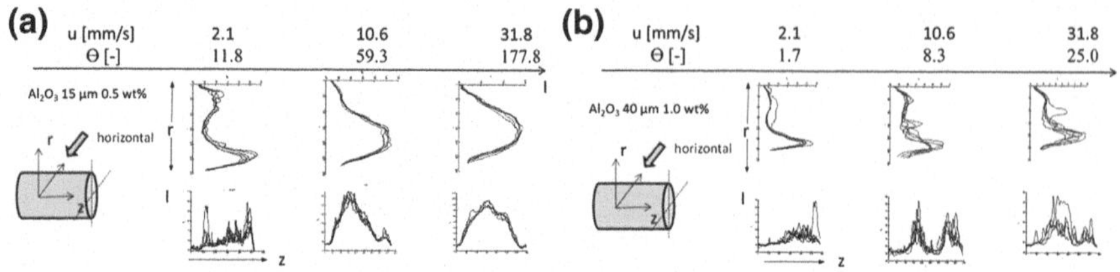

Fig. 6.6 Radial (r) and axial (z) fluorescence intensity profiles of Rhodamin B-labelled Al_2O_3 particles in H_2O—paraffin oil slug flow within a circular 1.0 mm ID FEP capillary 20 cm downstream from the T-junction; 5 particles per flow rate, horizontal view at 3 velocities $u = u_{Slug}$; **a** $x_P = 15\,\mu m$ 0.5 wt.%; **b** $x_P = 40\,\mu m$ 1.0 wt.% [17]

So conclusions valid for particles of 10–100 µm and $\rho_P \gg \rho_F$ are:

- $\Theta \ll 10$: Particles settle and accumulate at the rear end completely,
- $10 < \Theta < 30$: The amount of particles suspended in the main vortex increases, but gravity shifts all particles to the lower vortex,
- $30 < \Theta < 180$: Particles are suspended in the main vortex predominantly and the degree of uniform vertical distribution increases,
- $\Theta > 180$: Gravity is overridden and particles are distributed uniformly in upper and lower main vortex.

It has been argued so far that particle segregation at the rear end is due to gravity. But a second mechanism appeared when tracking the local particle concentration profile of 10 µm SiO_2 particles at axial positions of 20 and 50 cm downstream from the T-junction (Fig. 6.7). Particles were expected to be suspended homogeneously, since Θ was as large as 487 and 1460. Instead, particles, suspended in the main vortex at 20 cm, migrate entirely to the rear vortex subsequently. This effect is triggered by flow velocity.

Fang et al. as well as Kurup and Basu have reported similar observations with DNA or 10 µm fluorescent particles concentrated at the rear end [58, 59]. Up to now, Saffman force, centrifugal action or particle–particle interaction could serve as reasoning. And deeper insight as well as a confirmation must be provided by further investigations. However, it is now certain that particle segregation can be due to gravity, which would call for a high velocity in slug flow, or by fluid-dynamic aspects, favoring slow slug flow. Hence, process conditions and particle properties must be well optimized to prevent particle segregation. So, what can be concluded from Fig. 6.7 for particle $\leq$ 10 µm is:

- Drag is the dominant force on particles $\leq$10 µm; Submicron- and nanoparticles should be uncoupled from gravitational effects,
- Centrifugal, inertial or lift forces are suggested to cause segregation other than by gravity; as conjectured by recent publications [60, 61], nanoparticle may be not suspended homogeneously within liquid slugs,
- Submicron particles are particularly prone to agglomeration or repulsion.

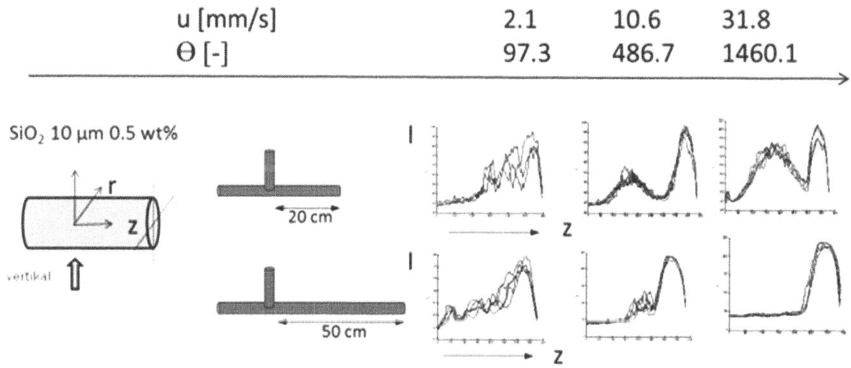

Fig. 6.7 Axial (z) fluorescence intensity profiles of silicate microparticles (SiO_2, 10 μm, RITC-labeled, sicastar® redF (micromod GmbH (Germany))), suspended in H_2O—paraffin oil slug flow within a 1.0 mm FEP capillary at a distance of 20 and 50 cm downstream from slug generation (*vertical view*) [17]

Furthermore, additional forces can arise:

- Van-der-Waals forces between particles or between particle-wall can achieve a significant intensity, according to Zoetjeweij [46]:

$$F_{vdW} = \frac{\Lambda x_p}{12z_0^2}\left(1 + \frac{2a^3}{z_o x_p}\right) \cong \frac{\Lambda x_p}{12z_0^2} \tag{6.19}$$

  The maximum of F_{vdw}, given by assuming non-deformable particles, a contact area of $a = 0$, $z_0 = 0.4$ nm [46] for a Hamaker constant $\Lambda = 5.3 \cdot 10^{-20}$ J for $Al_2O_3{-}H_2O{-}Al_2O_3$ interactions [62], ranges from $2.8 \cdot 10^{-13}$ N for 10 nm particles to $2.8 \cdot 10^{-9}$ N for 100 μm particles.
- Dean force can arise in spiral microchannels [36, 63]. F_{De} can be as large as $5.0 \cdot 10^{-14}$N for 10 nm particles to $5.0 \cdot 10^{-10}$ N for 100 μm particles at the flow rates assumed above and radii of 0.01 m.
- Other forces may be exerted by external fields, magnetic or electric for instance. Even purely electrostatic forces can achieve $F_{el} = 10^{-14}$ N [64], dielectrophoretic force may amount to $F_{diel} = 10^{-12}$ N [65] magnetic forces can be in the same order of magnitude $F_{mag} = 10^{-12}$ N [66].

And some additional remarks on the effect of particle-particle and particle-interface-interaction are necessary for a comprehensive review:

i. Particle-fluid interaction and rheology: "two-way-coupling" The particle-fluid becomes bidirectional or "two-way-coupled", when the solid concentration exceeds 0.001 vol.% [51]. Kurup and Basu [56] explain the rear end accumulation with an increase in viscosity that dampens the circulation and thus traps particles in the co-rotating vortex. Even the circulation in the main vortex could be curbed at high solid contents.

ii. Particle-particle-interaction: "four-way-coupling" At solid concentrations >0.1 vol.% particles experience the presence of other particles by flow disturbance, in-/elastic collisions, diffusion or agglomeration. Most metal, metal-oxide or ceramic spheres in aqueous media possess a zeta potential larger than 30 mV (absolute), because of electrostatic stabilization [48], so that they are not prone to agglomeration. A deeper discussion of interparticle forces can be found in literature [48, 51].
iii. Brownian motion: This adds random noise to the particle motion. Brownian motion is relevant for nano-, at low flow rates within narrow channels. But convective transport masks Brownian motion in usual slug flow applications so that diffusion can be neglected.

6.3 Feeding of Particles and Suspensions in Microsegmented Flow

Suspension dosage is the first issue to address in practical applications of liquid–liquid suspension flow. An appropriate particle feeding device must consider sedimentation of particles and the stability of suspensions according to previous considerations. Further specifications like constant, pulsation-free delivery, abrasion resistance, maximum shear stress exerted on solid and connectivity to microchannels curb the range of applicable technologies. These requirements can be refined in terms of particle types with the help of previous force considerations:

- Nanoparticles (e.g. Gold-nanoparticles, 1–100 nm, 19.32 g/cm^3 [67]):

Sedimentation is negligible as long as the colloid suspension is stable; otherwise, commercially available devices that stabilize the suspension are advisable—ultrasound, for instance. The time to settle down a distance of 1 cm ranges from days up to years; Conventional feeding techniques like syringe pumps, gear pumps and micropumps are applicable, if the material resists abrasion [68].

But, the high density difference between the nanoparticles and the surrounding liquid causes a significant concentration gradient in the perpendicular direction which results from the antagonistic effect of gravitational sedimentation and Brownian motion.

- Cells (e.g. 1–100 μm, 1.05 g/cm^3 [69]):

The time to sediment a distance of 1 cm is as large as days to minutes. This timescale is sufficient for most analytic applications of cell suspensions. The power input to maintain a well-suspended state can be low; however, the maximum shear stress tolerated by cells limits the range of applicable techniques. Two technical solutions are commercially available (by cetoni® NeMIX, Germany) and make use of syringe pumps, combined with low stress stirrers.

- Microparticles (e.g. 1–100 μm, 2–4 g/cm^3):

Particles settle within seconds to hours. This calls for external stirring power input. And the delivery connection, prior to the slug generation must be designed in such way that particle deposition is prevented. Moreover, most micropumps are not compatible with respect to abrasion.

While the technical solutions for suspension dosage of nanoparticles and biological matter are available, microparticle feeding is far from being managed satisfactorily. All concepts for microfeeding of suspensions can be condensed to four principles (Table 6.3).

Feeding the powder inline into the fluid flow (a)) is available only at mass flow rates that exceed the microfluidic range of 10 mL/min and solid concentrations of 20 wt.%. A review of conventional techniques for metering of solids is provided by Yang and Evans [70]. But this lack of concepts for microfeeding of dry powder has evoked research on piezoelectric or ultrasound-induced vibration of feeding capillaries that had internal diameters of several millimeters down to tens of microns [71–75]. Dispensing a few grams per second down to 10 μg/s would match the requirements of suspension segmented flow, but sealing and connection to microchannels has not been handled yet. At the current stage, inline powder feeding is not readily applicable for microfeeding. Withdrawing a suspension from an unstirred feed tank (b)) requires a stable suspension that is not prone to sedimentation. This limits the concept to neutrally buoyant spheres and nanoparticles. But stirring the reservoir (c)) would allow the feed of non-neutrally buoyant spheres, microparticles in particular. A technical solution for suspending microparticles uniformly, has been patented recently [76]. However, concept (c) relies on micropumps that do not wear. A review on micropumps is given by Laser and Santiago [68]. Piezoelectric-driven pumps have been proven capable for suspension delivery by Jang et al. [77] with flow-rates up to 1 mL/min and solid concentrations of 0.001 g/mL of 20 μm polystyrene microspheres. Johnston et al. [78] have developed a micro-throttle pump that could supply 5 μm polystyrene beads at concentrations of 0.3 vol.% and flow-rates of 0.1 mL/min and allowed conveying of cells [79]. Still, these micropumps are designed for lab-on-a-chip applications in very fine microstructures and small flow rates, accordingly. Hence, solid-resistant micropumps are the bottleneck for supplying suspensions to microchannels. At the present stage, stirring the reservoir and feeding the capillary by displacement of a solid or fluid boundary is the preferred approach (d)). Either commercial devices or prototypes have been designed, but the way of displacing the suspension, the feeding volume and mixing properties impede a universal recommendation. Olivon and Sarrazin use a 25 mL magnetically stirred syringe for feeding carbon-based or zeolite catalysts of ~10 μm diameter at concentrations up to 5 g/L for the catalytic acylation of anisole [15]. Commercially available, the *cetoni®neMIX stirred syringe* (*cetoni®GmbH*, Germany) works similarly with a permanent magnet fixed on a flexible tip in 2.5 or 5 mL syringe that is actuated contactless by a rotating external magnet, but cannot feed larger quantities. It has been observed that aluminium oxide particles sediment in >50 mL syringes in zones close to the plunger or at the tip, which are unaffected from the stirrer bar rotation. For this reason, stirred syringes can be applied in analytic research with small solid concentrations, small

Table 6.3 Principles for microfeeding of suspensions

Principle	(a) Inline powder feeding	(b) Unstirred reservoir and inline pump	(c) Stirred reservoir and inline pump	(d) Stirred reservoir and delivery by displacement
Examples	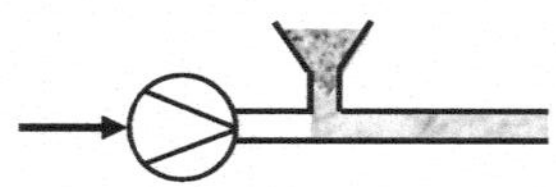 Ultrasound or piezoelectric-induced vibration of powder-filled capillary	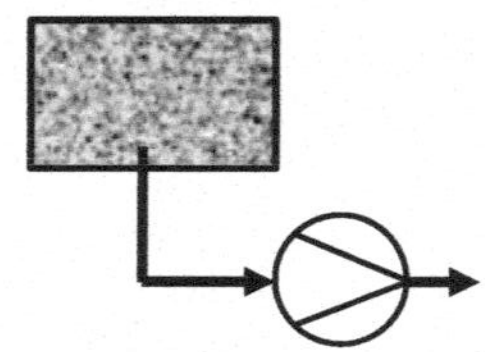 Fixed-valve or micro throttle pump	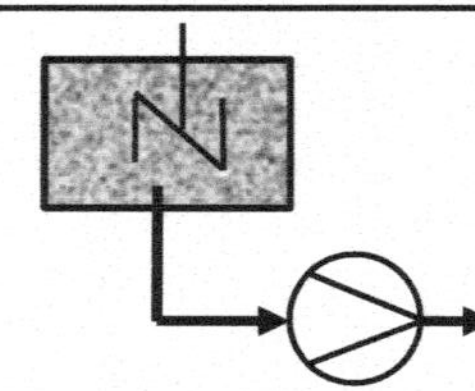 Fixed-valve or micro throttle pump	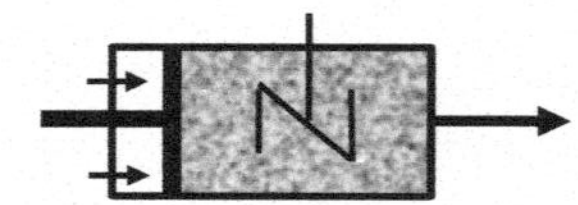 Stirred syringe pump, *cetoni®NeMIX syringe stirrer*, *cetoni®NeMIX stirring bottle*, gas blanket displacement from stirred reservoir membrane-assisted stirred feed tanks
Applicability	Not applicable at current state	Nanoparticles, neutrally buoyant spheres for ultra-low flow rates	Nanoparticles, cells, microparticles for ultra-low flow rates	Nanoparticles, cells, microparticles

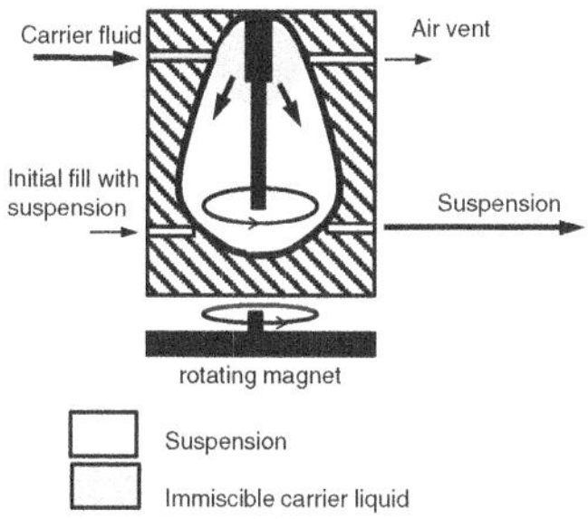

Characteristics	value
Volume/dosable volume	1.9/1.5 mL
Rotation speed min/max	14.2/426.6 rpm
Suspendable particles	$SiO_2 \leq$ 20 μm, $Al_2O_3 \leq$ 3 μm, yeast-cells

Fig. 6.8 *cetoni® NeMIX stirring bottle*—principle and features

volumes and shear-stress sensitive materials. A different approach is the displacement of the stirred suspension via a fluid interface. The *cetoni® neMIX stirring bottle* (cf. Fig. 6.8) utilizes a movable interface between suspension and an immiscible carrier liquid in a pear-shaped volume with a magnetically driven mixing tip. This device has been designed for cell suspensions. Its applicability to catalytic microparticle suspensions has been investigated at TU Dortmund University in cooperation with *cetoni®GmbH* (Germany) [17]. Tests with 10, 20 and 55 μm SiO_2 and 3 μm Al_2O_3 particles at concentrations of 0.3–5.0 wt.% in aqueous solutions and cyclohexane as carrier liquid have been conducted at flow rates of 0.05–0.25 mL/min. The mild mixing, the small dosable volume and the discontinuous handling impede conveying of microparticle suspensions at constant solid concentrations. Only slurries of particles $\leq$ 20 μm (SiO_2) or $\leq$ 3 μm (Al_2O_3) are well-suspended, whereas larger particles settle down. An influence of concentration or flow rate has not been observed. A further drawback is the manual refilling and the experimenter is kept in the dark about how to do so without sedimentation in the syringe and tubings connected. In our tests, particles deposited during the initial filling. After all, bioanalytics requirements may be fulfilled, but feeding micro particulate catalyst slurries is not resolved with the stirring bottle.

It is not surprising that research groups at CPE/CNRS Lyon and TU Dortmund University have prototyped suspension dosage devices for feeding microparticle slurries from the scratch. At TU Dortmund University, Ufer et al. have published the successful delivery of magnetically stirred, aqueous Al_2O_3 and SiO_2(<100 μm) suspensions and a mixture of activated carbon catalyst (20–50 μm) in toluene with concentrations up to 20 wt.% via a pressurized gas blanket in a flask [5]. This concept refined as shown in Table 6.4 and a similar device has been designed by the laboratory of Mechanical Engineering of TU Dortmund University at the same time.

The discharge of suspension from the mixed chamber is aligned downwards, because sedimentation in horizontal capillaries has been verified with separate experiments of 5 wt.% SiO_2(< 44 μm) suspensions in 1.6 mm ID FEP capillaries.

The film thickness has been measured in terms of relative fraction of the capillary height as a function of time at three axial distances from the bend (Fig. 6.9). The film grows nonlinearly and attains a terminal thickness, which scales inversely with flow rate. Once the film has built up to the maximum height, the initial suspension

Table 6.4 Membrane-assisted suspension feeding device and concentration of suspension after discharge [17]

Membrane-assisted discharge from stirred suspension chamber	Suspension concentration after discharge of aqueous Al_2O_3 suspensions (3, 40, 110 μm; 0.1, 0.2 vol.%) from the membrane-assisted feeding device in a 1.6 and 1.0 mm FEP capillary; measured gravimetrically at 22 °C
Schematic	1.0 mm capillary
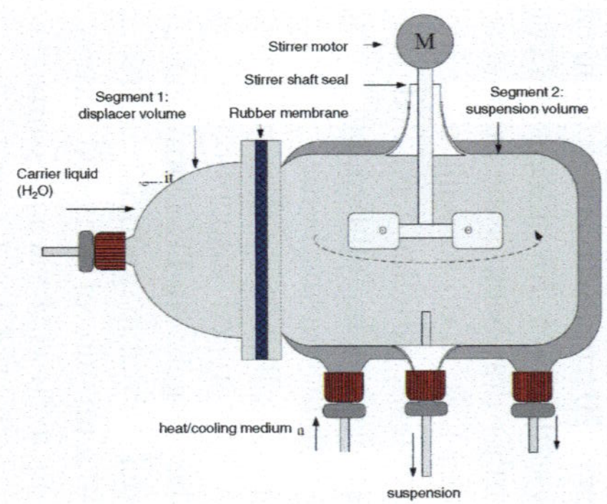	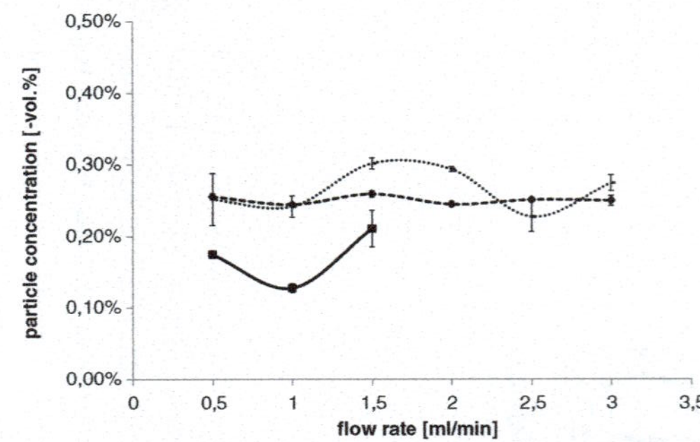

(continued)

Table 6.4 (continued)

	Membrane-assisted discharge from stirred suspension chamber	Suspension concentration after discharge of aqueous Al_2O_3 suspensions (3, 40, 110 µm; 0.1, 0.2 vol.%) from the membrane-assisted feeding device in a 1.6 and 1 0 mm FEP capillary; measured gravimetrically at 22 °C
Operating behavior	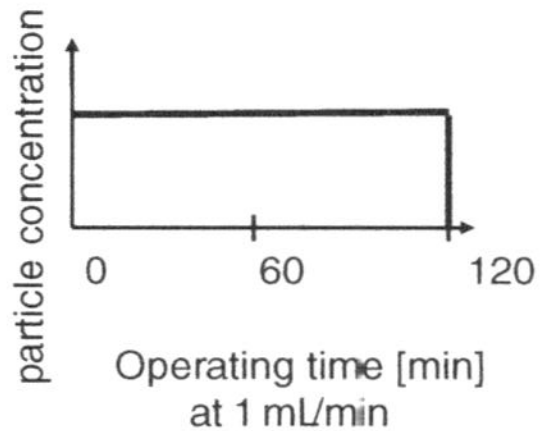	
$\dot{V}_{max}$	120 mL	
$t_{90\,\%}$	<5 s	
Suspension power	Al_2O_3, $\leq$ 110 µm, 5 wt% ($\rho_{Al2O3} = 3920\ kg/m^3$)	1.6 mm capillary 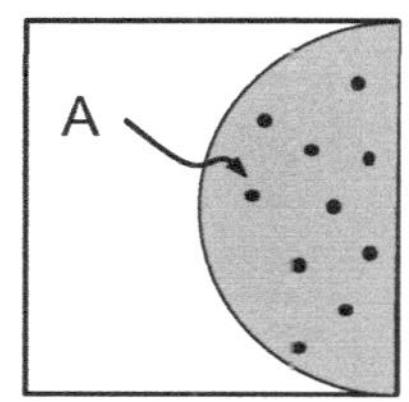

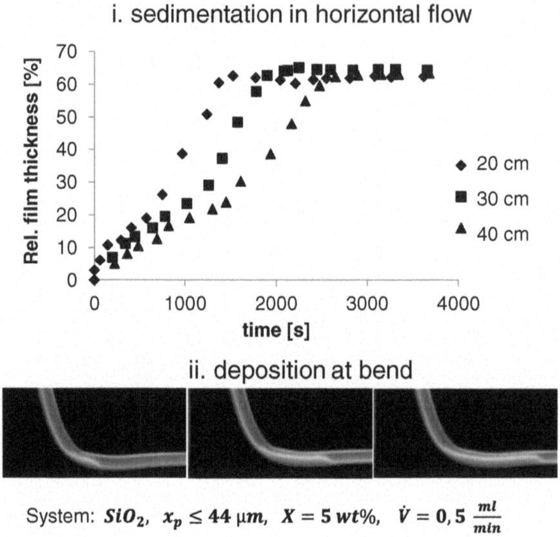

Fig. 6.9 Sedimentation in horizontal capillary and deposition in bend [17]

concentration remains constant on its way through the horizontal path. But this time is far too long compared to the operating time, which is limited by the suspension volume in the feeding devices. This volume is separated by an impermeable Nitril membrane from a second chamber, the displacer volume, which is necessary for the feeding principle: Water is injected at the desired flow rate into that separate chamber, which expands and displaces a flexible membrane. The suspension is then displaced into the vertical capillary. Thanks to resigning the gas blankets in Ufers device, the response to flow rate adjustments is almost instantaneous ($t_{90\,\%} < 5$ s) and a mechanical stirrer bar allows for mixing of "difficult" suspensions—suspensions that are prone to sedimentation or suspensions of magnetic particles, for example. The operating time has been extended compared to previous apparatus by designing a larger suspension chamber of 120 mL volume and even jacket for heating/cooling has been included. The discharge capillary can be of 1/16"–1/8" OD and is oriented downwards for reason explained before. But particle sedimentation into the discharge capillary and the particle transport velocity is the sum of fluid flow and sedimentation velocity. According to our experimental results (Table 6.4), the concentration can exceed the default value, if flow rates are below a certain threshold. This threshold increases with sedimentation tendency—that is to say with particle size. Furthermore, this effect has been found to vanish at low default concentrations and within smaller capillaries. Nevertheless, the experimental errors at each data point are surprisingly low, proving that reliable control of suspension concentration in segmented flow is possible. The latest device fulfills all requirements for suspension segmented flow applications in microreaction technology, for this reason. It should be mentioned that Liedtke et al. have developed an almost identical suspension feeding device for gas–liquid suspension slug flow independently, at the same time [14]. The idea of a vertical downward-oriented supply has been transferred from this work to the project at CPE/CNRS Lyon.

To summarize the discussion of particle dosage for liquid–liquid segmented flow, nanoparticle supply is relatively simple, because of negligible sedimentation tendency and applicable micropumps. The feed of particulate, biological matter calls for shear-stress-free techniques and two commercial concepts have been proposed. The feed of microparticle suspension turned out to be most challenging, but a technical solution can look like the latest development of a membrane-assisted prototype. In the near future, the development of direct powder feeding into microchannels or the design of microparticle-resistant and low shear-stress micropumps would supersede complex devices with a stirred reservoir and delivery by displacement. Stirred reservoir and inline micropumps would simplify the suspension dosage in microfluidics.

6.4 Clogging Risk and Clogging Prevention

The second practical issue of solid processing in microchannels is clogging. Putting microreactors into jeopardy of channel blockage has been carefully avoided in the past by ruling solids out of microfluidic operations. But this excludes a broad variety of chemical reactions and processes from microfluidic implementation [80]. Unintentional solid formation by precipitation has paved the way for research on the mechanism of channel blockage, since the year 2000. The mechanisms are too complex for an all-encompassing discussion, but the main aspects are described to provide a basis for discussing appropriate countermeasures:

Three mechanisms of microchannel blocking can be distinguished according to the ratio of channel dimension d_{cap} and particle size x_p:

- Size exlusion ($d_{cap}/x_P \leq 1$) [81, 82]: Particles larger than the microchannel cannot pass the inlet and obstruct the orifice. Due to manufacturing tolerances, size exclusion is also relevant once particles have entered the microchannel and lead to clogging at channel constrictions.
- Bridging ($d_{cap}/x_P \leq 9$) [82–88]: Particles or clusters that are up to nine times smaller than the channel can form bridges and arches across the cross-section and plug the channel.
- Fouling ($d_{cap}/x_P \geq 1$) [85, 87]: Solid deposition on the wall constricts the channel and can block the cross section in the long run. This mechanism can also trigger size exclusion and bridging. Synonyms for fouling are solid deposition or internal plugging.

While clogging due to size exclusion can be prevented effectively by matching channel dimensions and particles appropriately, more attention must be paid to bridging and fouling. Early results of Ramachandran and Fogler from 1999 demonstrate bridging under conditions of electrostatic repulsion between polystyrene particles (188–249 μm) and the cylindrical wall [82]. The hydrodynamic force must exceed the repulsion so that a critical velocity is required for bridging. Plugging is promoted by higher flow rates. While bridging occurred at ratios $d_{cap}/x_P \leq 3$–4 in this study, Stoeber and Mustin [86] noticed in 2008 that a small fraction of larger particles in polydisperse suspensions of 0.47–1.5 μm polystyrene particles within 20 × 27 μm channels is enough to induce bridging. And in 2010, Hartmann et al. [83] observed bridging at ratios of $d_{cap}/x_P \leq 9$ for particles not electrostatically stabilized. So it is certain at this stage that the risk of bridging should be minimized by choosing a ratio d_{cap}/x_P larger than about nine with respect to the maximum particle or cluster size. But even suspensions of particles that are small compared to the channel cross section can block microchannels [82, 89]. Referring to Wyss et al. [85], particle retention on the wall is governed by short-range attractive van-der-Waals, once particles reach a minimum distance to the wall that is less than the Debye length [90] by overriding Saffman and the wall-induced lift force [64] and the interaction potential well given by electrostatic repulsion. Since attractive and repulsive forces scale with flow velocity likewise, Wyss suggests that fouling is independent from flow velocity and solid concentration, as long as one blinds out interception and gravitational deposition. Stoeber and Liepmann [87] as well as Wyss et al. [85] gave evidence that the number of particles passing through the channel and the ratio d_{cap}/x_P determine the time until fouling results in ultimate blocking of the channel, i.e.

$$t_{Clogging} \sim \frac{1}{V\Phi} \cdot \left(\frac{d_{cap}}{x_p}\right)^4 \tag{6.20}$$

The ratio d_{cap}/x_P is particularly important. The chance for clogging scales to the power of four [85, 87, 88]. So from an engineer's position, materials that do not feature adhesive forces and a ratio $d_{cap}/x_P > 100$ are advisable [84]. Further complexity of clogging in microchannels arises from two ways of interaction between these three mechanisms:

i. Particle deposition/fouling constrict the channel cross section. Before the channel cross section narrows to zero, bridging or size exclusion can occur at $d_{cap}/x_P \leq 9$.
ii. Clusters of coagulating nano- and microparticles affect each of the three clogging mechanisms, e.g. fouling due to gravitational deposition or bridging and size exclusion by undershooting the critical channel-to-particle ratio. Therefore, cluster formation of suspended particles in flow is highly important, as even most accurate precautions with respect to d_{cap}/x_P will fail, if particles coagulate. Gudipaty et al. [90] have reported that shear-induced interparticle collision followed by particle aggregation due to van-der-Waals forces competes with shear-induced break-up of clusters. The research work of Georgieva et al. [89] from 2010 highlights that coagulation causes clogging even in accurately designed applications. Particle deposition and fouling have been ruled out by appropriate choice of materials and the d_{cap}/x_P exceeds 100, so that bridging was not expected. Channel blocking has been surprising to this end. But cluster formation in the converging flow field of a slit and hetero-coagulation between

Table 6.5 Approaches and guidelines for prevention of clogging in microchannels

Approach	Practical guideline
Prevent coagulation and cluster formation	- Remove exceptionally large particles from polydisperse probes [89] - Stabilize particles sterically or electrostatically - Promote cluster break up by high shear stress [83, 91] - Avoid curved geometries and inertial deposition [83, 92]
Microchannel design	- Use aspect ratios $d_{cap}/x_P > 100$ - Select wall materials that inhibit attraction of particles [83, 92]
Process procedures	- Flush microchannel or dissolve deposits occasionally [83, 92] - Use flow structures that prevent particle-wall contact
External forces	- Apply external forces to maintain uniform dispersions [83, 92]

nanoparticles and micron-sized objects in the raw material (clustered primary particles or impurities) in particular lead to particle retention at a slit—even though micron-sized objects have been concentrated < 100 ppm and they were at least ten times smaller than the channel size. Moreover, clogging occurs faster at low flow rates, because of larger interparticle contact time and less cluster break up.

That demonstrates that microfluidic systems set up carefully with regard to materials and the aspect ratio are not inherently safe from clogging. More guidelines must be considered for a robust design:

This compendium of guidelines (Table 6.5) results in a broad spectrum of technical measures against clogging. Some are inherently safe, whereas others only reduce the risk of clogging.

- Remove plugs:

Flush the microchannel or dissolve solid deposits. It does not prevent clogging proactively, but reinstalls the operational readiness of the microfluidic system [83].

- Reduce the risk of clogging by external forces:

Keeping the solid well-suspended reduces the risk of clogging effectively. The same holds true for preventing coagulation. One technical approach is the agitated cell reactor, presented by Browne et al. [93]. Mechanical vibration [94] of the microreactor and magnetically-coupled mixers [92] in the micro flow path are further examples. An effective way to keep solids suspended and to prevent agglomeration are ultrasound bath [94–96], acoustic irradiation, piezoelectric-induced ultrasound [97] or particle focusing in standing acoustic waves (cf. further references in [83])

- Prevention of clogging by the choice of flow structure:

The most convenient approach is establishing a flow regime that is inherently safe from blockage, if possible. Inertial focusing of particles on a certain streamline guarantees that particles are not deposited at the wall. Another example is the annular flow, generated in concentric nozzles of a microchannel designed by Takagi et al. for TiO_2 nanoparticle synthesis [98]. The far most prominent way to handle solids in microchannels inherently free from clogging is gas–liquid- or liquid–liquid slug flow [5, 94, 98, 99]. Particles are either trapped in the disperse phase or swept away by an adjacent slug, when suspended in the continuous phase. However, whether particles in the continuous phase touch the wall and are transported through the wall film is yet unresolved. Two experimental observations in Figs. 6.2 and 6.10 are ambiguous. Surface-modified silicon dioxide particles remain trapped in a certain continuous segment and are not transported through the wall film between the FEP capillary wall and water slug (Fig. 6.2). Aluminium oxide particle (suspended in water) may enter the gap between glass wall and disperse phase (Fig. 6.10). So dispersed phase suspensions are recommended for ruling out clogging and particle retention, leading to uniform solid phase residence times without accumulation in the microchannel.

After all, solid handling in the liquid–liquid slug flow itself can be well managed. Only the part of purely liquid–solid flow between suspension feed reservoir and mixer for slug generation remains prone to clogging. So this part is crucial for such operations and must be well-designed with previous guidelines and some further design aspects. One is the vertical alignment of the supply capillary from the suspension dosage device to the slug generating mixer below. This prevents gravitational deposition (Fig. 6.9) and reduces the risk of clogging by bridging or size exclusion, which would come when deposits constrict the cross section. A second additional rule is to avoid any sharp reduction of the channel cross section. Any reduction is exposed to two ways of failure. Firstly, particles are prone to deposit on the front edge of the channel reduction and can form clusters due to the intimate contact and long contact time. These agglomerates can slide into the reduction and block the cross section by bridging or size exclusion. The second mode of failure is the coagulation in the converging flow field and subsequent bridging or size exclusion, as reported by Georgieva et al. [89]. The clogging of microchannels at constrictions has been verified experimentally (cf. Fig. 6.11). The experimental results match previous publications and Eq. (6.20). The aspect ratio d_{cap}/x_P amounts to 40 in the large channel and drops to 20 in the narrow tube. Bridging in the 1.6 mm capillary is ruled out, according to the threshold of $d_{cap}/x_P = 9$ from Hartmann et al. [83]. The time for clogging is inversely proportional to the number of particles that pass the reduction, as given by Eq. (6.20), which indicates the validity of this scaling law for channel constrictions.

A final remark on clogging in the suspension supply capillary is committed to the choice of wall materials. Aqueous suspensions of Al_2O_3 and SiO_2 as well as activated carbon dispersed in toluene did not show any retention on the surface of hydrophobic FEP capillaries. But general advice cannot be drawn from these findings. One should rather choose the wall material based on experiments with the suspension of interest.

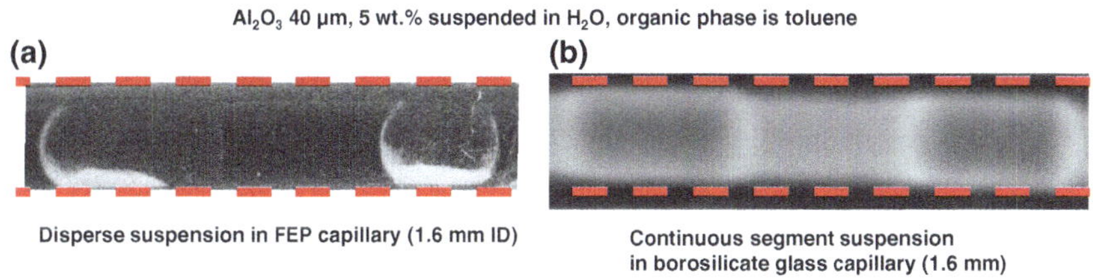

Fig. 6.10 Dispersed phase and continuous phase suspension of Al_2O_3 particles; the opaque appearance of particle-unladen slugs in **b** indicate particles in the wall film [17]. **a** Dispersed suspension in FEP capillary (1.6 mm ID). **b** Continuous segment suspension in borosilicate glass capillary (1.6 mm)

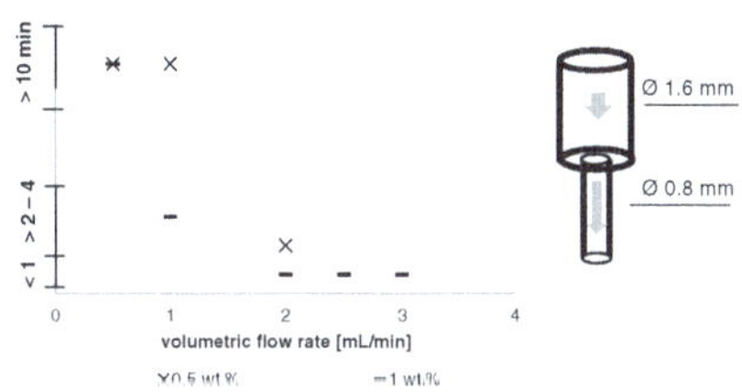

An aqueous suspension of Al_2O_3 40 µm particle with a concentration of 0.5 or 1.0 wt.% has been led through a vertical joint of circular 1.6 mm FEP and 0.8 mm Teflon channels at flow rates from 0.5 to 3.0 mL/min. If clogging did not occur within a time of 10 minutes, the operating conditions were defined as being "safe". Clogging within 2-4 minutes indicates that the system is at "high risk" and clogging within less than one minute characterizes conditions that are "not operable". Optical observation of the transparent capillaries verifies that particle deposition and clogging starts at the junction.

Fig. 6.11 Time for clogging at a sharp channel constriction of 1.6 mm FEP and 0.8 mm Teflon microchannels as a function of flow rate of Al_2O_3 suspension (40 µm, 0.5 and 1.0 wt.%) [17]

After all, fundamental knowledge of clogging mechanism and the guidelines for a robust design demonstrate that slug flow of suspension is inherently safe from plugging. Only the part of the assembly, which is upstream to the slug generation, needs some additional measures. A vertical alignment without reductions or enlargements of the cross section and an appropriate material selection are easy to implement and sufficient. Thus, it is incontestable that suspension slug flow microreactors can be run reliably and general reservations on solids in microreactors are unsubstantiated.

6.5 Downstream Phase Separation

6.5.1 General Aspects of Separation in Micro Segmented Flow

Downstream separation is the final unit of solid handling in suspension slug flow. Splitting up the three-phase flow requires two sequential separation steps—liquid–liquid and liquid–solid separation (Table 6.6). A route *A* is liquid–liquid separation in a first step and treating the suspension with liquid–solid separation methods, afterwards.

The second route (B) is the separation of particles from suspension slug flow at first and liquid–liquid separation, secondly. Which of those routes is preferred,

Table 6.6 Downstream separation routes for suspension slug flow applications

Application	Separation step 1	Separation step 2
Route A	**Liquid–liquid:**	**Liquid–solid:**
	Interruption of liquid–liquid mass	*Separation of non-reactive/*
	transfer/reaction	*mass transfer-less suspension*
Nanoparticle synthesis;	- Recovery/make up/recycle of particle	- Isolation of solid product
Solid phase extraction (SPE);	free reactant phase	- Recovery/make up/recycle of liquid phase
Protein crystallization	- Inert phase recovery/recycle	
Biotechnological analysis	- Inert phase recovery/recycle	- Isolation of solid product
		- Recovery/make up/recycle of liquid phase
Solid catalyzed liquid–liquid reaction;	- Isolation of product and/or catalyst	Catalyst recovery/ regeneration/recycle
Reactive extraction;	phase for further downstream	
Two-phase biotransformation	purification and catalyst recovery	
Route B	**Liquid–solid:**	**Liquid–liquid:**
	Interruption of liquid–solid mass-transfer/	*Separation of non-reactive/*
	reaction	*mass transfer-less liquid–liquid mixture*
Screening of catalysts with dispersed liquid–solid flow and inert continuous phase	- Catalyst recovery/ regeneration/recycle	- Inert phase recovery/recycle

depends on whether mass-transfer/reaction must be first interrupted between both liquids (route A) or between liquid and solid (route B) as shown in Table 6.6. Most applications call for an immediate liquid–liquid separation interrupting mass-transfer or reaction at a distinct moment, prior to liquid–solid separation, hence route A.

Microfluidic liquid–liquid separation has been strongly investigated in numerous publications. Special techniques utilizing micro scale physics have been developed, since gravitational separation using density differences drops out for microfluidic applications [100, 101]. A large variety of liquid–liquid microseparators use stratified flow structures , i.e. laminar flow, contacting of immiscible liquids via membranes [102, 103], stabilization of liquid–liquid interface by surface-wettability [104–107] or guided flow structures [108]. Reviews of these devices can be found elsewhere [100, 109, 110] and more attention should be drawn on separation methods that are applicable to slug flow:

6.5.2 Micro Settlers

Micro-settlers have been investigated in particle-free liquid–liquid mixer-settler extraction devices. In 2001, BASF (Ludwigshafen, Germany) and IMM (Mainz, Germany) investigated fine emulsions of water/acetone/butyl acetate and water/ succinic acid/butanol at flow rates of 3.3–6.6 mL/min in cylindrical glass settlers of 15 and 150 mL volume. Due to settler residence time of 22.5—45 min., mass-transfer in the settler contributed significantly to the total extraction efficiency [111]. Similar observations have been made by Wojik et al. [110], who performed extraction of acetone from n-heptane with water. The contribution of mass transfer was almost as large as in the mixer element [110]. In a recent publication, Kumar et al. describe a micro-settler that reduced the settling time down to 30–40 s in a much smaller settler of 1.2 mL volume [112]. However, micro-settlers have not been adopted to liquid-suspension flow. But it is obvious they should be applicable as long as $\rho_P > \rho_F$ (for particles suspended in the denser liquid phase) leads to sedimentation of particles or $\rho_P < \rho_F$ (for particles suspended in the lighter liquid phase) causes flotation and a long contact time during separation is acceptable. If the first condition is not fulfilled, particle aggregate at the liquid–liquid interface and impede a sharp cut-off of the phases. The second requirement rules out the use of micro-settlers for most applications mentioned in Table 6.6. Hence, micro-settlers are recommended with reservations only.

6.5.3 Micro-Hydrocyclones and Curved Branches

These limitations apply for liquid-suspension separation with micro-hydrocyclones and curved branches as well. These apparatus replace gravitational impact by centrifugal and lift forces. Micro-hydrocyclones have not been deeply studied yet. Numerical simulations are available for mini hydrocyclones with 5 mm diameter [113, 114]. But the capacity of these hydrocyclones of 1000 mL/min exceeds the regime of slug flow in microchannels. Hydrocyclones that are truly micro have only been reported by Wengeler et al. [115] and Bhardwaj et al. [116]. Bhardwaj et al. farbricated a device with an internal bore of 350 μm, a capacity of 0.15 mL/min and a cut-off sizes down to 1 μm. Curved microchannels facilitate liquid–liquid separation as well. Hydrodynamic lift forces move the lighter phase to outer radii, while the heavier phase flows close to the inner wall of the curved microchannel. Ookawara et al. [117] fabricated a curved microchannel with a 112×160 μm cross section and an arc radius of 10 mm at angles of 30°. This device was proven capable for complete separation of water/oil emulsion at droplet size down to 8 μm. However, whether this technique is applicable to slug flow, i.e. with break-up or stratifying slug flow, is yet unclear.

6.5.4 Wettability and Capillarity Separators: Membranes, Pore Combs, Branches

The majority of miniaturized liquid–liquid separation techniques exploits the selective wettability of organic or aqueous phase to hydrophobic/hydrophilic materials and capillary pressure within slugs. That is why this concept matches specifically the nature of slug flow. Only the transfer from liquid–liquid to liquid-suspension slug flow is almost unaddressed. Early reports of this micro-separation method for liquid–liquid slug flow have been given by Kashid et al. in 2005 [118]. They used a symmetric Y branch to separate kerosene-water and n-butanol-water slug flow with total flow rates of 0.16–2.0 mL/h, shown in Fig. 6.12. However, the underlying driving force of this separation process remained uncertain until 2007, when Kralj et al. [101] fabricated the membrane separator (cf. Fig. 6.12) comprising a PTFE membrane with 0.1–1.0 μm pore size for splitting DMF-water flow at flow rates of 0.01–2.0 mL/min. They observed that the organic fluid was withdrawn selectively through the hydrophobic membrane, whereas break-through of water was impeded as long as the capillary pressure of water exceeds the difference in hydrodynamic pressure across the membrane and the organic side fluidic resistance is much lower than the aqueous. In the same year, Kolehmainen and Turunen [119] observed successful separation of water from different Shellsol-TEHP mixtures, indifferent from using steel or glass plates within the plate-type coalesce (cf. Fig. 6.12). This proves that rather than a specific value of wettability, the discrimination between preferential wetting of aqueous and organic phase is sufficient. Angelescu et al. [120] and Castell et al. [121] added a further approach for the geometric design to this separation technique. Angelescu designed a pore comb structure with 15 × 15 μm pores in a PDMS device. This was compared to 30 nm PTFE membrane for separating various oils from water. Castell et al. [121] fabricated a device comprising 140 pores of 36 × 130 μm size made of PTFE for separation of chloroform and water. They derived the operating limits in terms of total flow rates and provided an experimental verification. The theory on the wettability & capillary pressure-induced separation has been extended beyond the break through regime by Scheiff et al. [122] in 2011. The degree of contamination was found to be a function of fluidic resistance in both outlets. From that, simple equations that predict the operating quality as well as an approach for the separation control were provided. The experimental verification was conducted with the simplest design of the wettability & capillary separation devices so far. 0.82 or 0.51 ID, flat-tipped stainless steel needles were introduced into a 0.63 mm ID plastic microcapillary and the separation efficiency fairly insensitive to the geometric details (cf. Fig. 6.12). The latest research results on this separation method were published by Gaaker et al. [123] in 2012. They proposed a slit-shaped separator. However, only Castell et al. [121] reported a successful proof-of-concept for separating suspensions (1 wt.% neutrally-buoyant, fluorescent polystyrene microparticle ($x_p = 2.0\,\mu$m) suspended in aqueous slugs segmented by chloroform within their device of 36 × 130 μm). So, withdrawing a continuous segment suspension was proven successful. And the same holds true for the separation of

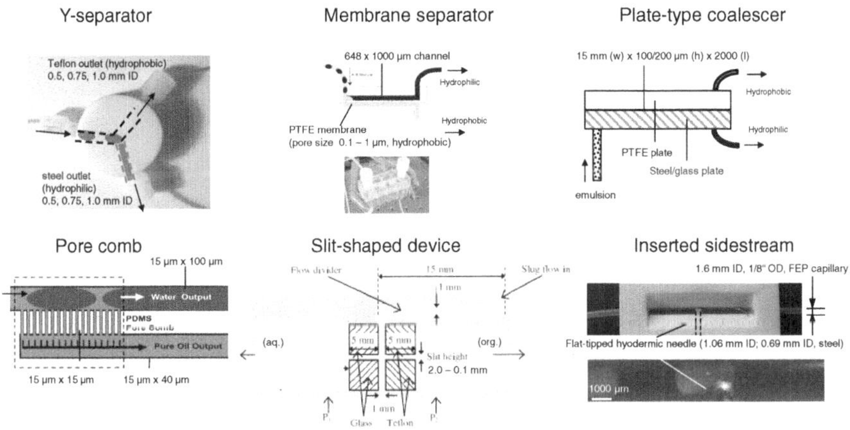

Fig. 6.12 Wettability & capillarity separators—Y-separator for slug flow separation (Kashid et al. [118]), Membrane separator (Kralj et al. [101]), plate-type coalesce (Kolehmainen and Turunen [119]), Pore comb structure (Angelescu et al. [120]), lit-shaped device (Gaaker et al. [123]), Inserted sidestream branch (Scheiff et al. [122])

suspensions in discontinuous segments, as presented here for the first time [17]. The device of Scheiff et al. [122] was modified in such way that a steel needle of 0.7 mm ID was inserted into 1.6 mm FEP capillaries at right angles. Disperse aqueous suspension of SiO_2(37 μm, 0–5 wt.%) and Al_2O_3 (40 μm, 0–5 wt.%) were completely recovered from slug flow with toluene at flow rates from 0–1 mL/min (Fig. 6.13). But the separation was incomplete in a range of 1.0–2.0 mL/min, and vanished at >2.0 mL/min.

A more intricate situation is the liquid–liquid separation with surface-active particles aggregated on the slug surface. Activated carbon (0.4–12 μm, suspended at 1 wt.% in toluene) was withdrawn with the phase in the sidestream by 70 %, whereas 30 % remained in the straight microchannel (Fig. 6.13). Hence, this separation is not applicable to surface active particle at this state.

By virtue of wettability and capillary pressure, slug flow is split into one stream of clear fluid and one of suspension. That raises design considerations for discharging the suspension, which are similar to the upstream section:

i. This channel must be aligned vertically downwards and channel constrictions in the suspension outlet must be avoided to prevent clogging. Membrane-, pore comb- or slit-devices require modifications to this end.
ii. A second important aspect is the choice of materials. For instance, particles withdrawn with the aqueous phase must not attach to the steel or glass surfaces in this discharge (Al_2O_3 and SiO_2 particles fulfill this requirement, as tested experimentally).
iii. The capillary pressure is inversely proportional to the channel diameter. Hence, small capillary diameters result in larger capillary pressures and larger operating

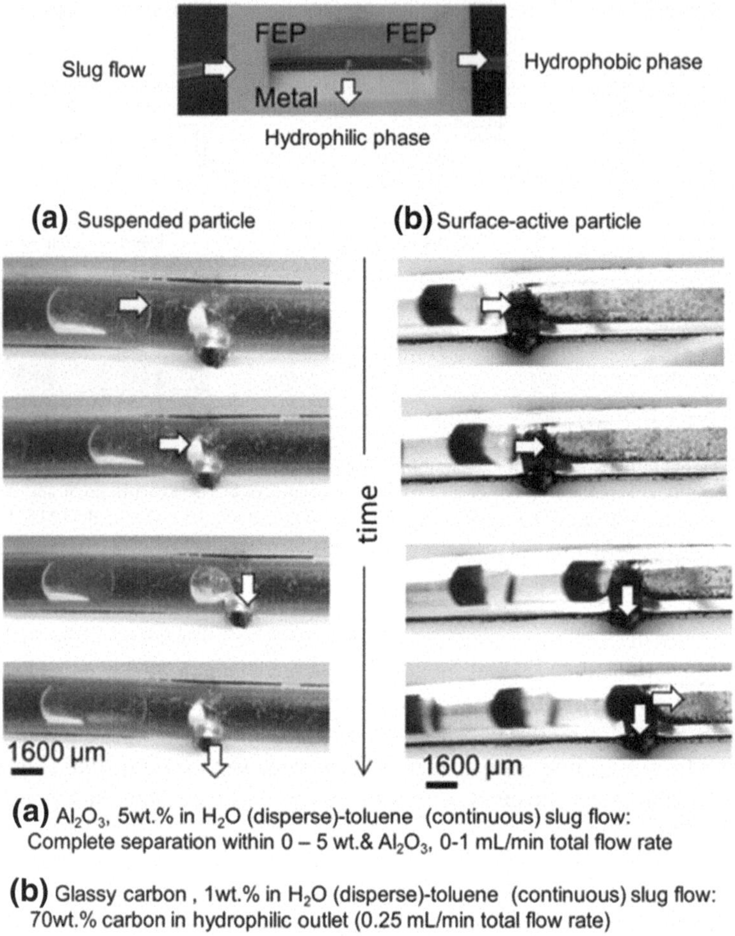

Fig. 6.13 Results of suspension slug flow separation with inserted sidestream separator [17]

regimes, accordingly. However, the aspect ratio d_{cap}/x_P in the suspension discharge must exceed the threshold discussed before.

iv. Complete phase separation is achieved by adjusting the fluid-dynamic resistance, i.e. the pressure drops. Any kind of backpressure regulation must be clogging-proof. Valves in the suspension discharge are inappropriate; but valves in the clean fluid outlet and regulation of hydrostatic pressure or ambient pressure at the outlet are eligible options.

What it comes down to is that current membrane- and pore comb-devices as well as the slit-shaped separator fall short according to the aspect ratio or vertical alignment. The only technique applicable with modifications is the inserted sidestream (Fig. 6.13). In summing up, the best technique available to split a suspension from the second liquid in slug flow is wettability & capillary separation. Micro-settler may work well for applications that are not sensitive to the contact time and with appropriate density ratios; micro-hydrocyclones and curved branches need further development, but appear promising. The second step of isolating the three phase system is liquid–solid separation. It depends on the application, whether the suspension must be concentrated or whether solid and fluid matter must be isolated completely. Several techniques for continuous particle concentration in microfluidic suspension flow are available—pinched flow fractionation, inertial focusing, filtration, sedimentation, electrophoresis & magnetophoresis, just to name a few. Exhaustive reviews can be found in the literature [124–126]. Some techniques mentioned before have been adapted to liquid–solid systems as well. Bhardwaj et al. [116] used the micro-hydrocyclone and Oozeki et al. [127] the curved branch for liquid-solid separation. For complete isolation of fluid and solid matter, techniques particularly designed for microfluidic purposes are not available at present. Hence, operations like filtration and drying are conducted with conventional laboratory tools. After all, approaches for liquid–liquid–solid separation on micro scale are available on an academic basis up to now. Further prototyping and commercial launch is desired. Otherwise, downstream separation may become a bottleneck in implementing solid-based microfluidic processes.

6.6 Heterogeneously Catalyzed Reactions in Microfluidic Processes

6.6.1 Application of Suspension Slug Flow for Heterogeneously Catalyzed Reactions

The final section of this chapter is committed to the performance of suspension slug flow in heterogeneous reactions, i.e. with solid particles serving as catalyst. Micro-packed bed reactors, wall-coated microreactors and membrane or mesh microreactors are alternative principles for integrating catalysts into microreactors. Each type is eligible for specific reactions. The number and kind of phases involved in a solid-catalyzed reaction is the most essential aspect in that matter. Gas-solid, liquid–solid and gas–liquid–solid multiphase reactions have been covered in numerous publications, text-books and review articles [3, 4, 10]. But liquid–liquid–solid (l–l–s) reactions have not been discussed from the author's knowledge. Hence, Table 6.7 provides a comprehensive guideline for the choice of reactor type for the first time. Heat and mass transfer parameters as well as engineering characteristics are taken into consideration. Types of reactions that are preferential for each microreactor are proposed with the help of Table 6.8, assuming that all reactor types are available. However, preparation of a catalytic wall film or mesh may be infeasible with some catalysts or slug flow not accessible for viscous liquids with small interfacial tensions in some situations. Hence, one is urged to check the availability of a microreactor type

Table 6.7 Qualitative comparison of reactor performance and applicability for different types of reactions

Reactor type	Reactor performance criteria													Type of reactions		
	Schematic	X	k_L	a_L	k_s	a_s	h	Δp	RTD	Fabrication, reactor life time	Robustness	Flexibility	Catalyst recovery	Kinetic	Heat	Selectivity
Micro-packed bed	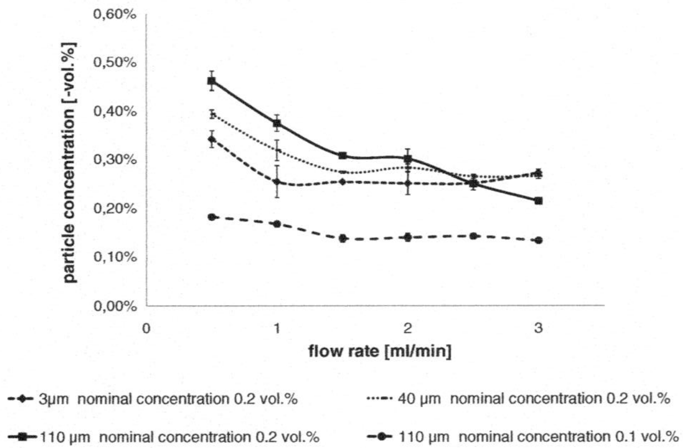	+	+	+	–	–	–	–	–	o	+	–	o	Slow, chemically controlled	Isothermal, slightly endo-/exothermic	Inselective chemical reaction
Suspension slug flow	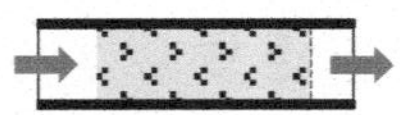	o	+	+	+	+	o	+	+	+	o	+	+	Fast, mass-transfer-controlled	Moderate heat of reaction	Highly selective reactions
Wall-coated micro-reactor	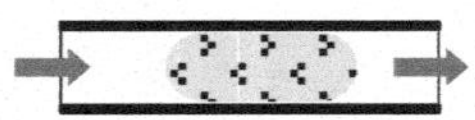	–	+	+	–	–	+	+	o	–	o	–	–	Fast, mass-transfer-controlled	Large heat of reaction	Moderate selective reactions
Micro-membrane/mesh		–	o	o	o	o	+	+	+	–	+	–	o	Very fast, mass transfer-controlled	Large heat of reaction	Highly selective reaction

Reaction sensitive towards X, a, k, RTD, T-profile

Table 6.8 Types of liquid–liquid–solid reactions and corresponding microreactor design criteria

Type of reaction	Schematic	Examples	Design priority
Supported catalyst-philic phase	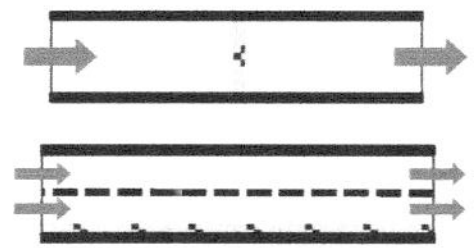	Oxidation of alcohols in $scCO_2$ with PEG-stabilised Pd-nanoparticles [142]	$k_L a_L$
Solid-catalyzed liquid–liquid reaction	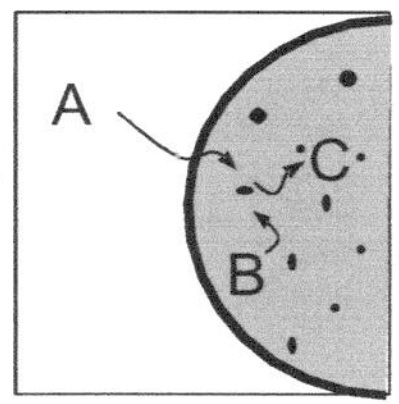	Catalytic transfer hydrogenation (CTH) [5], Catalytics hydrogenation of fatty acids, Epoxidation of alkenes [19], Hydroxylation of benzene [143] Transesterfication of biodiesel, Hydratisation of alkenes, chemo-enzymatic epoxidation [144], immobilized PTC [145]	k_S a_S k_{total} h
Reactive extraction	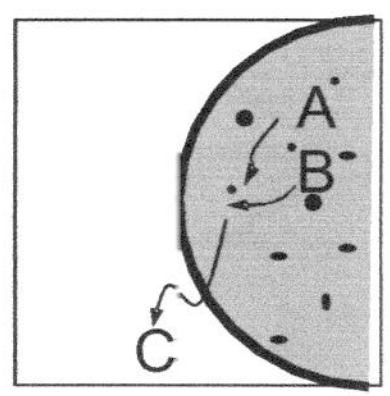	Hydration of olefins, Enzyme-catalyzed hydrolysis of fatty acids, Synthesis of antibiotics in aqueous two-phase systems [146], further bioconversions in aqueous two-phase systems [147]	Catalyst load

(continued)

Table 6.8 (continued)

Type of reaction	Schematic	Examples	Design priority
Solid-catalyzed liquid reaction with additional, inert liquid	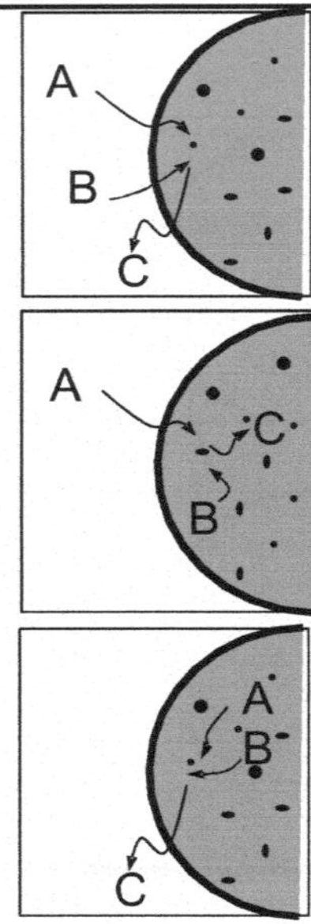	Fischer-Tropsch-Synthesis [15]	$k_L a_L$, RTD, precise control of X and T
Extraction and affinity chromatography	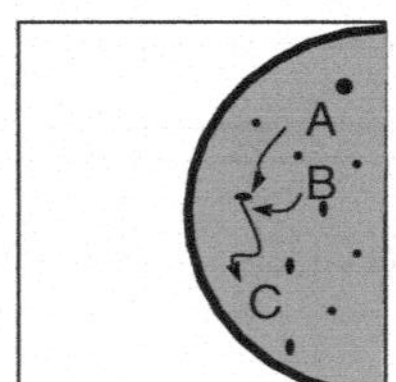	Extraction of (*R*,*S*)-propranolol from organic phase and selective binding to molecular imprinted polymer beads in aqueous phase [6]	k_L, a_L, k_S, a_S, k_{total}, h Catalyst load

for the intended reaction first. The second step in reactor choice refers to the nature of reaction. Slow, chemically controlled reactions require a large catalytic activity X, while mass-transfer is less crucial. If the reaction is fast, mass transfer will be the bottleneck. Liquid–liquid and liquid–solid mass-transfer must be accelerated by large interfacial or surface areas a_L and a_S as well as convective transport in the liquid phase and the hydrodynamic film surrounding the catalyst. And it is important that mass transfer enhancement focusses on the limiting step, either liquid–liquid or liquid–solid mass transfer. The heat of reaction is a second criterion for reactor selection and the ability to control of process conditions precisely relevant for kinetic studies or selectivity-sensitive reactions. As said before, micro-packed bed, micro membrane/mesh and wall-coated microreactors have not been evaluated for liquid–liquid, solid-catalyzed reactions to the author's knowledge, yet. But the reactor performance—presented in Table 6.7 and discussed subsequently—is deduced from well-characterized, gas–solid, liquid–solid and gas–liquid–solid reactions and compared with experimental studies on the suspension slug flow microreactor, for the first time.

6.6.2 Micro-Packed Bed

High catalytic loads ranging from 13 to 60 vol.% have been reported in micro-packed bed reactors [128, 129]. Such large catalytic activities are not achieved in any other microreactor and makes micro-packed beds attractive for slow reactions. But flow structure and wetting of the catalyst bed are rather random. Bubbly, slug flow or stratified flow as well as liquid bridging are just a few types of flow in gas–liquid flow through micro-packed beds [128, 130]. Flow maldistribution and broad RTDs must be taken into account, as indicated by low Peclet number $Pe = 0.1 - 0.3$ (Marquez et al. [131]). Kiwi-Minsker and Renken [132] give $Pe_{ax} = 2 - 3$ within $Re \cdot Sc = 100 - 10$ for a packed bed, compared to $Pe_{ax} = 2 - 7$ within wall-coated microreactor for homogeneous flow. The wetting of the catalyst is important for efficient catalyst contact. But it is conjectured that the hydrodynamic layer on the particle surface leads to poor liquid–solid mass transfer. In micro-packed beds, the liquid–solid mass transfer is slower ($k_S = 0.01 - 1\,\mathrm{m}^{-1}$) compared to wall-coated microreactors ($k_S = 0.09 - 4\,\mathrm{m}^{-1}$) (values are given for identical Sc numbers and void fractions [132]). However, thanks to intensive mixing and large interfacial areas ranging from $1660\,\mathrm{m}^{-1}$ [10, 128] to $16{,}000\,\mathrm{m}^{-1}$ [10, 129] liquid–liquid mass transfer in slug flow through packed beds is fast, although the energy dissipation is rather large (2–5 kW m^{-3} [129]). The energy is dissipated by creation of large interfacial areas and by the detrimental pressure drop in packed beds [129, 130]. Heat transfer may be insufficient for highly endo-/exothermic reactions. Further engineering aspects are that packed beds can be fabricated with virtually all conventional catalyst microparticles and the immobilization prevents clogging. However, exchange of catalyst, catalyst recovery and preparation of the bed is time-consuming. After all, micro-packed bed reactors are well suitable for kinetically-controlled, nearly isothermal, non-selective reactions and could be envisaged for fine chemical production scale, i.e. with special attention drawn to process robustness rather than flexibility.

6.6.3 Suspension Slug Flow Microreactor

The suspension slug flow microreactor has not been considered in previous reviews and text books. But Ufer et al., Olivon and Sarrazin as well as the authors' research work at TU Dortmund University suggest that this concept is promising [5, 15–17]. Olivon and Sarrazin conducted a proof-of-concept for catalyst screening within slug flow of microparticle suspensions (8–10 μm, 1 wt.%). The kinetics of the Friedel-Crafts acylation of anisole was monitored in slug flow either with spatial time resolution in slug flow or as a function of reaction time in batch experiments. Both approaches resulted in identical conversion-time-profiles.

Ufer et al. [5] have demonstrated the effective performance of solid catalyzed liquid–liquid reactions in slug flow microcapillary reactors. They have conducted the catalytic transfer hydrogenation of *m*-nitrotoluene to *m*-nitrotoluidine, catalyzed by Pd/C (10 wt.% Pd on 20–50 μm carbon particles), as shown in Fig. 6.14. The carbon catalyst is surface-active and accumulates at the liquid–liquid interface. The amount of catalyst and the flow velocity have been varied and benchmarked with a stirred tank reactor. These results demonstrate the good performance of the suspension slug flow reactor in mass-transfer-limited reactions. The conversion is comparable to the stirred tank benchmark, especially at high velocities and intensive internal circulation, accordingly (Fig. 6.14). The catalytic activity is not annihilated by accumulation of particles on the rear surface of the slug. This has been rationalized with two experiments: the increase in catalyst concentration results in higher conversions at equal residence time (Fig. 6.14) and introducing gas bubbles at the front or rear end of a slug reduces the conversion. However, it does not drop to zero, indicating that serial liquid–liquid mass transfer through the film and subsequent liquid–solid mass transfer from the slug to the catalyst play a role as well [5].

The reactor can be assessed referring to these experimental findings and the aspects of suspension slug flow discussed throughout this chapter: catalyst concentration up to 20 vol.% are accessible without clogging. That is lower than the micro-packed bed, but outmatches the catalytic wall or mesh microreactors. The liquid–liquid mass transfer in slug flow is excellent, $k_L a_L$ amounts to 0.02–1.5 s^{-1}. And the wall film gives rise to large interfacial areas of 3,000–4,000 m^{-1} [16]. Similar holds true for the heat and liquid–solid mass transfer. The microparticles are either accumulated at the interface, which reduces the diffusive transport path length. Or they are intensively stirred within the internal vortices, which enhances the particle film transport. However, more quantitative investigation of the efficiency of liquid–solid mass transfer is needed. Other relevant aspects are the uniform residence time distribution and precise control of catalyst load. That is in contrast to fine dispersed suspension droplet flow, which might be envisaged for overcoming liquid–liquid mass transfer limitation. Another advantage is the low pressure drop in liquid–liquid

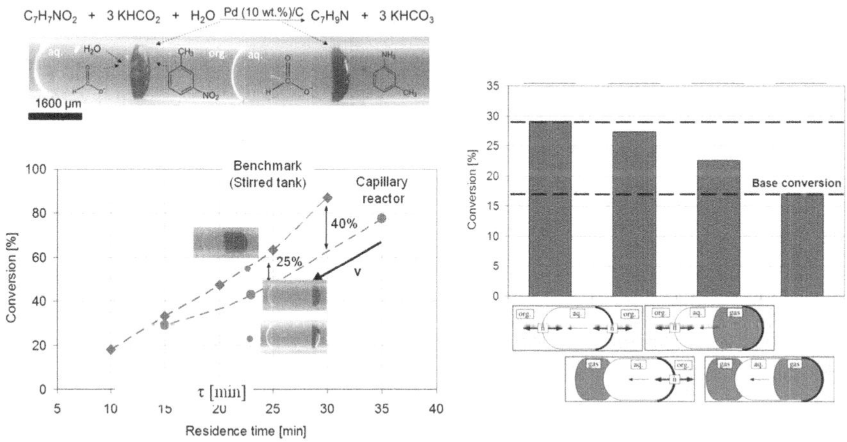

Fig. 6.14 Schematic, conversion versus residence time and effect of mass transfer on catalytic transfer hydrogenation [16]

slug flow. Even suspensions of 10 wt.% should not increase viscosity and pressure drop greatly. Also engineering considerations make the suspension slug flow reactor attractive. The reactor design is absolutely simple. And most striking, catalyst recovery or rapid tests of various catalyst materials are easy issues, because one resigns from catalyst immobilization. Hence, suspension slug flow microreactors may be a good choice for mass-transfer-limited and/or selectivity-sensitive reactions that do not have a harsh enthalpy of reaction. A further application may be kinetic studies and catalyst screening, because residence time and reaction conditions can be control precisely and exchange of catalytic material is simple (Tables 6.7, 6.8).

6.6.4 Wall-Coated Microreactor

Again, solid-catalyzed liquid–liquid reactions have not been tested in this type of microreactor until now. But the number of gas–liquid reactions in wall-coated microreactor is vast and the reader referred to reviews [3, 10, 132, 133] and milestone publications [134, 135]. Only small amounts of catalytic material are deposited on the wall and the liquid–solid surface area a_S is roughly that of the channel wall so that the catalytic activity is low. Assuming that liquid–liquid fluid dynamics in the slug flow regime, diffusion through the hydrodynamic layer close to the wall drives the liquid solid mass transfer. Kreutzer et al. have demonstrated that mass transfer from the disperse phase to the wall follows two paths—i. transfer from the disperse slug into the film, ii. transfer from the continuous film to the wall or, iii. transport from the slug cap into the continuous liquid, followed by step ii. [134] interestingly and against intuitive expectations, mass transfer through the film increases with reducing velocities, because of decreasing film thickness. Liquid-liquid mass transfer k_L is same as in suspension slug flow, attaining large values of up $1.4\,s^{-1}$

[118]. Similar applies for a_L. And k_S outmatches micro-packed bed values as outlined above. Important for large enthalpies of reaction is the excellent heat transfer, because of the short distance between heat release and cooling at the wall. The energy dissipation equals the suspension slug flow reactor and is small compared to micro-packed beds. However, the residence time distribution of the catalyst-contacting phase may not be as narrow as in disperse suspension slug flow. A significant dispersion arises from the wall film transport, according to experimental findings of Günther et al. [136]. What it comes down to is that heat and liquid–liquid transfer properties are excellent, liquid–solid transport is reasonable, but the catalytic load rather low. Hence, wall-coated microreactors match the requirements of fast, mass-transfer-limited reactions—especially of those, which are highly exo- or endothermic. But the preparation of the catalyst deposit puts notable costs on that concept. Details for the laborious preparation are given elsewhere [4, 132, 137]. And one cannot change, recover or regenerate the immobilized catalyst in most cases. To this end, wall-coated microreactors can open access to novel chemistry of endo-/exothermic reactions or provide insight into mechanistic studies on catalysis [138].

6.6.5 Membrane/Mesh Microreactor

Two kinds of membrane/mesh microreactors can be distinguished. Abdallah et al. [139] have used a utilized a mesh with 5 μm bores to stabilize the gas–liquid interface with an area of $a_L = 2000\,\text{m}^{-1}$, while the catalyst was deposited on the liquid-side wall. Shaw has used Pt, Pd or Rh-coated meshes for catalysis of liquid–liquid reduction of nitrobenzene and compared this reactor with Abdallahs concepts [140]. Zero conversion was observed, when the catalyst was deposited on the wall of one phase exclusively, but 12 % conversion were obtained with the Pt-coated mesh. This indicates that "in-series"-transfer through phases that are separated by a mesh requires short diffusive pathways. Anyway, membrane and mesh microreactors provide a low amount of catalyst in the reactor. That is applicable for very fast, highly endo-/exothermic or selectivity-sensitive reactions. But the incipient studies provide no clear insight into the mass transfer performance. Good control of temperature and residence time as well as low pressure drop and inherent phase separation are further beneficial features. Fabrication and catalyst recovery can be laborious or impossible. On the other hand, the reactor might be robust in terms of fluid-dynamic conditions, as long as wetting/contact hysteresis effects do not promote break-through [10]. Kulkarni et al. indicate narrow residence time distribution in gas–liquid mesh microreactors, but Peclet number varies from 0.5–1 and 0.5–2 due to contact angle hysteresis [141]. In conclusion, membrane/mesh microreactors might serve as an alternative for wall-coated microreactors, as these types are largely comparable, except for the inherent phase separation provided by the membrane/mesh microreactor.

The global picture on alternative microreactor concepts reveals that the suspension slug flow microreactor takes an intermediate position in terms of catalytic load, heat and mass transfer. It offers a good compromise of numerous design criteria - making it attractive for many reactions. Simple fabrication, low installation costs and flexible adaption to multiple purposes support this rating.

6.7 Conclusion on Particle Handling and Synthesis in Micro Segmented Flow

In summing up, little doubt has been left that the field of solid handling in segmented flow is vast. That arises from the variety of scientific areas *referring* to the topic. Second are the complex underlying physics, which are far from being fully explained. More fundamental research is desired to gain full knowledge of this matter. Third are the engineering issues: the mechanical engineering part of the microfluidic community is addressed to *developing* apparatus for suspension dosage or micropumps, capable of conveying suspensions. Further complexity arises from the unresolved physics of clogging and particle-particle interaction. Neither experimental nor numerical approaches have been able to regard the interaction of numerous clogging mechanisms. However, much progress has been made on downstream separation. This gives hope that the remaining challenges can be resolved in near future. All these considerations culminate in the comparative assessment of the slug flow microreactor. Demonstrating its capability relative to alternative concepts leads to the outlook that segmented flow of solids will gain more relevance in the framework of microfluidic *applications*.

Acknowledgments The author appreciates the support in experimental work of A. Ufer, F. Thygs, D. Stümmler, M. Goldbeck, J. Nietfeld, M. Schlüter & J.Gies. Special thanks to J. Franzke and N. Ahlmann (Leibniz institute ISAS. e.V. (Dortmund,Germany)) for vital discussions on the fluorescence microscopy and to N. Antweiler & F. Kaske for their stimulating contribution to this chapter. The author's research at TU Dortmund University, Laboratory for Technical Chemistry B (Dortmund, Germany) has been funded by: Deutsche Forschungsgemeinschaft DFG (German Research Foundation), GZ AG26/13-2.

References

1. P. Gravesen, J. Branebjerg, O.S. Jensen, Microfluidics-a review. J. Micromech. Microeng. **3**(4), 168 (1993)
2. V. Hessel, Mirroring the evolution of microreaction technology: the vast contributions of IMRET. Green Process. Synth. **1**(4), 313–314 (2012)
3. L. Kiwi-Minsker, A. Renken, Microstructured reactors for catalytic reactions. Catal. Today **110**(1–2), 2–14 (2005)
4. E. Klemm, H. Döring, A. Geisselmann, S. Schirrmeister, Microstructured reactors in heterogenous catalysis. Chem. Eng. Technol. **30**(12), 1615–1621 (2007)

5. A. Ufer, D. Sudhoff, A. Mescher, D.W. Agar, Suspension catalysis in a liquid-liquid capillary microreactor. Chem. Eng. J. **167**(2–3), 468–474 (2010)
6. O.K. Castell, C.J. Allender, D.A. Barrow, Novel biphasic separations utilising highly selective molecularly imprinted polymers as biorecognition solvent extraction agents. Biosens. Bioelectron. **22**(4), 526–533 (2006)
7. R. Dittmeyer, N. Kockmann, Alle Mikro. CITPlus **3**, 19–21 (2011)
8. B.K.H.S. Vankayala, R. Dittmeyer, Hydrogenation of aromatic nitro compunds in a slurry micro reactor, 8th European Congress of Chemical Engineering, Berlin, 2011
9. M. Ueno, T. Suzuki, T. Naito, H. Oyamada, S. Kobayashi, Development of microchannel reactors using polysilane-supported palladium catalytic systems in capillaries. Chem. Commun. **14**, 1647–1649 (2008)
10. V. Hessel, P. Angeli, A. Gavriilidis, H. Loewe, Gas-liquid and gas-liquid-solid microstructured reactors: contacting principles and applications. Ind. Eng. Chem. Res. **44**(25), 9750–9769 (2005)
11. J. Huang, J. Weinstein, R.S. Besser, Particle loading in a catalyst-trap microreactor: experiment vs. simulation. Chem. Eng. J. **155**(1–2), 388–395 (2009)
12. D.I. Enache, G.J. Hutchings, S.H. Taylor, R. Natividad, S. Raymahasay, J.M. Winterbottom, E.H. Stitt, Experimental evaluation of a three-phase downflow capillary reactor. Ind. Eng. Chem. Res. **44**(16), 6295–6303 (2005)
13. B. Buisson, S. Donegan, D. Wray, A. Parracho, J. Gamble, P. Caze, J. Jorde, C. Guermmeur, Slurry hydrogenation in a continuous flow reactor for pharmaceutical application, Teknoscienze, Milano, ITALIE (2009)
14. A. Liedtke, F. Bornette, R. Philippe, C. De Bellefon, Hydrodynamics, mass transfer and reaction in a Gas-Liquid-Solid Taylor flow slurry micro-reactor: First Results, IMRET 12, Lyon (France) (2012)
15. K. Olivon, F. Sarrazin, Heterogeneous reaction with solid catalyst in droplet-flow millifluidic device. Chem. Eng. J. **277**, 97–102 (2013)
16. A. Ufer, M. Mendorf, A. Ghaini, D.W. Agar, Liquid/liquid slug flow capillary microreactor. Chem. Eng. Technol. **34**(3), 353–360 (2010)
17. F. Scheiff, Experimental investigation of mass-transfer-limited, heterogeneous-catalyzed liquid-liquid reactions in capillary microreactors. TU Dortmund University, Department of Biochemical and Chemical Engineering, Laboratory for Technical Chemistry B, Dortmund (2012)
18. H. Nur, S. Ikeda, B. Ohtani, Phase-boundary catalysis: a new approach in alkene epoxidation with hydrogen peroxide by zeolite loaded with alkylsilane-covered titanium oxide. Chem. Commun. **22**, 2235–2236 (2000)
19. H. Nur, S. Ikeda, B. Ohtani, Phase-boundary catalysis of alkene epoxidation with aqueous hydrogen peroxide using amphiphilic zeolite particles loaded with titanium oxide. J. Catal. **204**(2), 402–408 (2001)
20. P. Pieranski, Two-dimensional interfacial colloidal crystals. Phys. Rev. Lett. **45**(7), 569–572 (1980)
21. F. Bresme, M. Oettel, Nanoparticles at fluid interfaces. J. Phys.: Condens. Matter **19**(41), 413101 (2007)
22. Y. Lin, A. Boeker, H. Skaff, D. Cookson, A.D. Dinsmore, T. Emrick, T.P. Russell, Nanoparticle assembly at fluid interfaces: structure and dynamics. Langmuir **21**(1), 191–194 (2004)
23. S. Schütz, M. Piesche, G. Gorbach, M. Schilling, C. Seyfert, P. Kopf, T. Deuschle, N. Sautter, E. Popp, T. Warth, CFD in der mechanischen Trenntechnik. Chem. Ing. Tech. **79**(11), 1777–1796 (2007)
24. R.W. Johnson, *Handbook of Fluid Dynamics* (Taylor & Francis, London 1998)
25. D. Wen, L. Zhang, Y. He, Flow and migration of nanoparticle in a single channel. Heat Mass Transf. **45**(8), 1061–1067 (2009)
26. V. Rudyak, A. Belkin, E. Tomilina, Force acting on a nanoparticle in a fluid. Tech. Phys. Lett. **34**(1), 76–78 (2008)

27. P.G. Saffman, The lift on a small sphere in a slow shear flow. J. Fluid Mech. **22**(02), 385–400 (1965)
28. E.S. Asmolov, The inertial lift on a spherical particle in a plane Poiseuille flow at large channel Reynolds number. J. Fluid Mech. **381**, 63–87 (1999)
29. J.B. McLaughlin, Inertial migration of a small sphere in linear shear flows. J. Fluid Mech. **224**, 261–274 (1991)
30. J.B. McLaughlin, The lift on a small sphere in wall-bounded linear shear flows. J. Fluid Mech. **246**, 249–265 (1993)
31. P. Cherukat, J.B. McLaughin, D.S. Dandy, *A computational study of the inertial lift on a sphere in a linear shear flow field* (Elsevier, Kidlington, Royaume-Uni, 1999)
32. R. Mei, *An approximate expression for the shear lift force on a spherical particle at finite Reynolds number* (Elsevier, Kidlington, Royaume-Uni, 1992)
33. D.S. Dandy, H.A. Dwyer, A sphere in shear flow at finite Reynolds number: effect of shear on particle lift, drag, and heat transfer. J. Fluid Mech. **216**, 381–410 (1990)
34. S.I. Rubinow, J.B. Keller, The transverse force on a spinning sphere moving in a viscous fluid. J. Fluid Mech. **11**(03), 447–459 (1961)
35. T. Zhu, T. Nozaki, Y.H. Xie, J. Han, J. Jiang, Numerical simulation and visualization experiment of solid particle motion affected by parameters of flow in tapered drum rotating type separator, 7th International Conference on System Simulation and Scientific Computing. ICSC 2008. Asia Simulation Conference, 2008. pp. 867–871
36. S.S. Kuntaegowdanahalli, Inertial microfluidics for continuous particle separation in spiral microchannels, University of Cincinnati (2009)
37. Z.I. Abarbanel, I.Y. Viner, Sedimentation of particles in a centrifugal force field. Russ. Phys. J. **12**(5), 566–570 (1969)
38. R. Lindken, M. Rossi, S. Grosse, J. Westerweel, Micro-particle image velocimetry ([small micro]PIV): recent developments, applications, and guidelines. Lab Chip **9**(17), 2551–2567 (2009)
39. P. Dietrich, K. Lerche, J. Reusch, R. Nitzsche, The characterization of silica microparticles by electrophoretic mobility measurements. Chromatographia **44**(7), 362–366 (1997)
40. M. Dürr, J. Kentsch, T. Müller, T. Schnelle, M. Stelzle, Microdevices for manipulation and accumulation of micro- and nanoparticles by dielectrophoresis. Electrophoresis **24**(4), 722–731 (2003)
41. M. Kawano, H. Watarai, Two-dimensional flow magnetophoresis of microparticles. Anal. Bioanal. Chem. **403**(9), 2645–2653 (2012)
42. P.B. Muller, R. Barnkob, M.J.H. Jensen, H. Bruus, A numerical study of microparticle acoustophoresis driven by acoustic radiation forces and streaming-induced drag forces. Lab Chip **12**(22), 4617–4627 (2012)
43. W. Li, E. James Davis, Measurement of the thermophoretic force by electrodynamic levitation: microspheres in air. J. Aerosol Sci. **26**(7), 1063–1083 (1995)
44. P. Hinderliter, K. Minard, G. Orr, W. Chrisler, B. Thrall, J. Pounds, J. Teeguarden, ISDD: a computational model of particle sedimentation, diffusion and target cell dosimetry for in vitro toxicity studies. Part. Fibre Toxicol. **7**(1) 36 (2012)
45. P. Fede, O. Simonin, Effect of particle-particle collisions on the spatial distribution of inertial particles suspended in homogeneous isotropic turbulent flows, ed. by M. Deville, T.-H. Lê, P. Sagaut,*Turbulence and Interactions* (Springer, Heidelberg, 2009) pp. 119–125
46. M.L. Zoeteweij, J.C.J. van der Donck, R. Versluis, Particle removal in linear shear flow: model prediction and experimental validation. J. Adhes. Sci. Technol. **23**(6), 899–911 (2009)
47. Z. Adamczyk, P. Weronski, Application of the DLVO theory for particle deposition problems. Adv. Colloid Interface Sci. **83**(1–3), 137–226 (1999)
48. Y. Min, M. Akbulut, K. Kristiansen, Y. Golan, J. Israelachvili, The role of interparticle and external forces in nanoparticle assembly. Nat. Mater. **7**(7), 527–538 (2008)
49. C.T. Crowe, *Multiphase Flow Handbook* (CRC Press, Boca Raton, 2006)
50. T.B. Drew, J.W. Hoopes, T. Vermeulen, *Advances in Chemical Engineering* (Academic Press, New York, 1966)

51. S. Elghobashi, On predicting particle-laden turbulent flows. Appl. Sci. Res. **52**(4), 309–329 (1994)
52. R.C. Jeffrey, J.R.A. Pearson, Particle motion in laminar vertical tube flow. J. Fluid Mech. **22**(04), 721–735 (1965)
53. G.B. Jeffery, The motion of ellipsoidal particles immersed in a viscous fluid. Proc. R. Soc. Lond. A **102**(715), 161–179 (1922)
54. I. Dittmar, Numerical study of a liquid/liquid slug flow in a capillary microreactor, PAMM, 82th Annual Meeting of the International Association of Applied Mathematics and Mechanics (GAMM), Graz, 2012
55. U. Miessner, Lindken, R., Westerweel, J., 3D—Velocity measurements in microscopic two-phase flows by means of micro PIV, 14th International Symposium on Applications of Laser Techniques to Fluid Mechanics, Lisbon, Portugal, 2008
56. G.K. Kurup, A.S. Basu, Field-free particle focusing in microfluidic plugs. Biomicrofluidics **6**(2), 022008–022010 (2012)
57. M.C. Miller, I.N. McCave, P.D. Komar, Threshold of sediment motion under unidirectional currents. Sedimentology **24**(4), 507–527 (1977)
58. W.-F. Fang, S.-C. Ting, C.-W. Hsu, Y.-T. Chen, J.-T. Yang, Locally enhanced concentration and detection of oligonucleotides in a plug-based microfluidic device. Lab Chip **12**(5), 923–931 (2012)
59. G.K. Kurup, A.S. Basu, Shape dependent Laplace vortices in deformed liquid-liquid slug flow, Engineering in Medicine and Biology Society, EMBC, 2011 Annual International Conference of the IEEE, pp. 4034–4037
60. V. Sebastian Cabeza, S. Kuhn, A.A. Kulkarni, K.F. Jensen, Size-controlled flow synthesis of gold nanoparticles using a segmented flow microfluidic platform. Langmuir **28**(17), 7007–7013 (2012)
61. D.V. Ravi Kumar, B.L.V. Prasad, A.A. Kulkarni, Segmented flow synthesis of Ag nanoparticles in spiral microreactor: role of continuous and dispersed phase. Chem. Eng. Technol. **192**(0), 357–368 (2012)
62. C.J. van Oss, *Interfacial Forces in Aqueous Media*, 2nd edn. (Taylor & Francis, London, 2006)
63. D. Di Carlo, Inertial microfluidics. Lab Chip **9**(21), 3038–3046 (2009)
64. X. Zheng, Z. Silber-Li, The influence of Saffman lift force on nanoparticle concentration distribution near a wall. Appl. Phys. Lett. **95**(12), 124105-124105-124103 (2009)
65. T. Kulrattanarak, R.G.M. van der Sman, C.G.P.H. SchroÃn, R.M. Boom, Classification and evaluation of microfluidic devices for continuous suspension fractionation. Adv. Colloid Interface Sci. **142** (1–2), 53–66 (2008)
66. A. Shina, Characterizing magnetic particle transport for microfluidic applications. Faculty of Virginia Polytechic Institute Faculty of Virginia Polytechic Institute and State University Blacksburg (2008)
67. I. US Research Nanomaterials, The advanced nanomaterial provider, Housten (2012)
68. D.J. Laser, J.G. Santiago, A review of micropumps. J. Micromech. Microeng. **14**(6), R35 (2004)
69. W.H. Grover, A.K. Bryan, M. Diez-Silva, S. Suresh, J.M. Higgins, S.R. Manalis, Measuring single-cell density. Proc. Natl. Acad. Sci. **108**(27), 10992–10996 (2011)
70. S. Yang, J.R.G. Evans, Metering and dispensing of powder; the quest for new solid freeforming techniques. Powder Technol. **178**(1), 56–72 (2007)
71. X. Lu, S. Yang, J.R.G. Evans, Microfeeding with different ultrasonic nozzle designs. Ultrasonics **49**(6–7), 514–521 (2009)
72. M. Mracek, J. Wallaschek, A system for powder transport based on piezoelectrically excited ultrasonic progressive waves. Mater. Chem. Phys. **90**(2–3), 378–380 (2005)
73. S. Matsusaka, M. Urakawa, H. Masuda, Micro-feeding of fine powders using a capillary tube with ultrasonic vibration. Adv. Powder Technol. **6**(4), 283–293 (1995)
74. Y. Yong, L. Xiaochun, Experimental and analytical study of ultrasonic micro powder feeding. J. Phys. D: Appl. Phys. **36**(11), 1349 (2003)

75. S.C. Tsai, Y.L. Song, T.K. Tseng, Y.F. Chou, B.J. Chen, C.S. Tsai, High-frequency, silicon-based ultrasonic nozzles using multiple Fourier horns. IEEE Trans. Ultrason. Ferroelectr. Freq. Control **51**(3), 277–285 (2004)
76. J. Feldhusen, H. Schlüter, J. Thiemann, J. Lemburg, *Apparatus and mothd for the uniform ditribution of microparticles in a liquid* (Charitee - Universitätsmedizin Berlin, Germany, 2011)
77. L.-S. Jang, C.J. Morris, N.R. Sharma, R.L. Bardell, F.K. Forster, Transport pf particle-laden fluids through fixed-valve micropumps. J. Microelectromech. Syst. **1**, 503–509 (1999)
78. I.D. Johnston, M.C. Tracey, J.B. Davis, C.K.L. Tan, Microfluidic solid phase suspension transport with an elastomer-based, single piezo-actuator, micro throttle pump. Lab Chip **5**(3), 318–325 (2005)
79. M.J. Davies, I.D. Johnston, C.K.L. Tan, M.C. Tracey, Whole blood pumping with a microthrottle pump. Biomicrofluidics **4**(4), 044112–044110 (2010)
80. D.M. Roberge, L. Ducry, N. Bieler, P. Cretton, B. Zimmermann, Microreactor technology: a revolution for the fine chemical and pharmaceutical industries? Chem. Eng. Technol. **28**(3), 318–323 (2005)
81. G.H. Goldsztein, J.C. Santamarina, Suspension extraction through an opening before clogging. Appl. Phys. Lett. **85**(19), 4535–4537 (2004)
82. V. Ramachandran, Fogler, Plugging by hydrodynamic bridging during flow of stable colloidal particles within cylindrical pores. J. Fluid Mech. **385**(-1), 129–156 (2000)
83. R.L. Hartman, J.R. Naber, N. Zaborenko, S.L. Buchwald, K.F. Jensen, Overcoming the challenges of solid bridging and constriction during Pd-Catalyzed C-N bond formation in microreactors. Org. Process Res. Dev. **14**(6) 1347–1357 (2010)
84. K. Sharp, R. Adrian, On flow-blocking particle structures in microtubes. Microfluid. Nanofluid. **1**(4), 376–380 (2005)
85. H.M. Wyss, D.L. Blair, J.F. Morris, H.A. Stone, D.A. Weitz, Mechanism for clogging of microchannels. Phys. Rev. E **74**(6), 061402 (2006)
86. B. Mustin, B. Stoeber, Design considerations for robust suspension-based microfluidic systems for biomedical applications, IMS3TW 2008: IEEE 14th International Mixed-Signals, Sensors, and Systems Test Workshop, 2008, pp. 1–6
87. B.L.D. Stoeber, Design, Fabrication and testing of a MEMS syringe, Technical Digest of the, Solid-State Sensor and Actuator Workshop, Hilton Head Island, South Carolina, USA, 2002, pp. 77–80
88. U. Häfeli, A. Mokhtari, D. Liepmann, B. Stoeber, In vivo evaluation of a microneedle-based miniature syringe for intradermal drug delivery. Biomed. Microdevices **11**(5), 943–950 (2009)
89. K. Georgieva, D.J. Dijkstra, H. Fricke, N. Willenbacher, Clogging of microchannels by nanoparticles due to hetero-coagulation in elongational flow. J. Colloid Interface Sci. **352**(2), 265–277 (2010)
90. T. Gudipaty, M. Stamm, L. Cheung, L. Jiang, Y. Zohar, Cluster formation and growth in microchannel flow of dilute particle suspensions. Microfluid. Nanofluid. **10**(3), 661–669 (2011)
91. T. Gothsch, J.H. Finke, S. Beinert, C. Lesche, J. Schur, S. Büttgenbach, C. Müller-Goymann, A. Kwade, Effect of microchannel geometry on high-pressure dispersion and emulsification. Chem. Eng. Technol. **34**(3) 335–343 (2011)
92. P. Koos, L. Browne Duncan, V. Ley Steven, Continuous stream processing: a prototype magnetic field induced flow mixer. Green Process. Synth. **1**, 11 (2012)
93. D.L. Browne, B.J. Deadman, R. Ashe, I.R. Baxendale, S.V. Ley, Continuous flow processing of slurries: evaluation of an agitated cell reactor. Org. Process Res. Develop. **15**(3) 693–697 (2011)
94. T. Horie, M. Sumino, T. Tanaka, Y. Matsushita, T. Ichimura, J.-i. Yoshida, Photodimerization of maleic anhydride in a microreactor without clogging. Org. Process Res. Develop. **14**(2), 405–410 (2010)
95. J.r. Sedelmeier, S.V. Ley, I.R. Baxendale, M. Baumann, KMnO4-Mediated Oxidation as a Continuous Flow Process. Org. Lett. **12**(16), 3618–3621 (2010)

96. T. Noel, J.R. Naber, R.L. Hartman, J.P. McMullen, K.F. Jensen, S.L. Buchwald, Palladium-catalyzed amination reactions in flow: overcoming the challenges of clogging via acoustic irradiation. Chem. Sci. **2**(2), 287–290 (2011)
97. S. Kuhn, T. Noel, L. Gu, P.L. Heider, K.F. Jensen, A Teflon microreactor with integrated piezoelectric actuator to handle solid forming reactions. Lab Chip **11**(15), 2488–2492 (2011)
98. M. Takagi, T. Maki, M. Miyahara, K. Mae, Production of titania nanoparticles by using a new microreactor assembled with same axle dual pipe. Chem. Eng. J. **101**(1–3), 269–276 (2004)
99. S.L. Poe, M.A. Cummings, M.P. Haaf, D.T. McQuade, Solving the clogging problem: precipitate-forming reactions in flow. Angew. Chem. Int. Ed. **45**(10), 1544–1548 (2006)
100. R.L. Hartman, K.F. Jensen, Microchemical systems for continuous-flow synthesis. Lab Chip **9**(17), 2495–2507 (2009)
101. J.G. Kralj, H.R. Sahoo, K.F. Jensen, Integrated continuous microfluidic liquid-liquid extraction. Lab Chip **7**(2), 256–263 (2007)
102. D.A. Wenn, J.E.A. Shaw, B. Mackenzie, A mesh microcontactor for 2-phase reactions. Lab Chip **3**(3), 180–186 (2003)
103. W.E. TeGrotenhuis, R.J. Cameron, M.G. Butcher, P.M. Martin, R.S. Wegeng, Microchannel devices for efficient contacting of liquids in solvent extraction. Sep. Sci. Technol. **34**(6–7), 951–974 (1999)
104. A. Aota, M. Nonaka, A. Hibara, T. Kitamori, Countercurrent Laminar Microflow for Highly Efficient Solvent Extraction. Angew. Chem. Int. Ed. **46**(6), 878–880 (2007)
105. A. Aota, K. Mawatari, S. Takahashi, T. Matsumoto, K. Kanda, R. Anraku, A. Hibara, M. Tokeshi, T. Kitamori, Phase separation of gas-liquid and liquid-liquid microflows in microchips. Microchim. Acta **164**(3), 249–255 (2009)
106. B. Zhao, J.S. Moore, D.J. Beebe, Surface-directed liquid flow inside microchannels. Science **291**(5506), 1023–1026 (2001)
107. J.T. Tesfai, R.N. Perry, E.L. Jablonski, Water-in-oil emulsion separation within a milli-fluidic device. J. Colloid Interface Sci. **354**(2), 895–899 (2011)
108. M. Tokeshi, T. Minagawa, K. Uchiyama, A. Hibara, K. Sato, H. Hisamoto, T. Kitamori, Continuous-flow chemical processing on a microchip by combining microunit operations and a multiphase flow network. Anal. Chem. **74**(7), 1565–1571 (2002)
109. W.H. Ehrfeld, Volker; Löwe, *Holger Microreactors* (Wiley-VCH, Weinheim, 2005)
110. A. Wojik, R. Marr, Mikroverfahrenstechnische Prinzipien in der Flüssig/Flüssig-Extraktion. Chem. Ing. Tech. **77**(6), 653–668 (2005)
111. K. Benz, K.P. Jäckel, K.J. Regenauer, J. Schiewe, K. Drese, W. Ehrfeld, V. Hessel, H. Löwe, Utilization of micromixers for extraction processes. Chem. Eng. Technol. **24**(1), 11–17 (2001)
112. S. Kumar, B. Kumar, M. Sampath, D. Sivakumar, U. Kamachi Mudali, R. Natarajan, Development of a micro-mixer-settler for nuclear solvent extraction. J. Radioanal. Nucl. Chem. **291**(3), 797–800
113. C.A. Petty, S.M. Parks, Flow structures within miniature hydrocyclones. Miner. Eng. **17**(5), 615–624 (2004)
114. G. Zhu, J.-L. Liow, A. Neely, Computational study of the flow characteristics and separation efficiency in a mini-hydrocyclone. Chem. Eng. Res. Des. **90**(12), 2135–2147 (2012)
115. R. Wengeler, H. Nirschl, F. Herbstritt, W. Ehrfeld, Studies on a micro hydrocyclone for liquid-liquid separation. Filtration **6**(1), 21–26 (2006)
116. P. Bhardwaj, P. Bagdi, A.K. Sen, Microfluidic device based on a micro-hydrocyclone for particle-liquid separation. Lab Chip **11**(23), 4012–4021 (2011)
117. S. Ookawara, T. Ishikawa, K. Ogawa, Applicability of a miniaturized micro-separator/classifier to oil-water separation. Chem. Eng. Technol. **30**(3), 316–321 (2007)
118. M.N. Kashid, Y.M. Harshe, D.W. Agar, Liquid-liquid slug flow in a capillary: an alternative to suspended drop or film contactors. Ind. Eng. Chem. Res. **46**(25), 8420–8430 (2007)
119. E. Kolehmainen, I. Turunen, Micro-scale liquid-liquid separation in a plate-type coalescer. Chem. Eng. Process. **46**(9), 834–839 (2007)
120. D.E. Angelescu, B. Mercier, D. Siess, R. Schroeder, Microfluidic capillary separation and real-time spectroscopic analysis of specific components from multiphase mixtures. Anal. Chem. **82**(6), 2412–2420 (2010)

121. O.K. Castell, C.J. Allender, D.A. Barrow, Liquid-liquid phase separation: characterisation of a novel device capable of separating particle carrying multiphase flows. Lab Chip **9**(3), 388–396 (2009)
122. F. Scheiff, Mendorf, M., Reis, N., Mackley, M., Agar, D. W., The Separation of immiscible liquid slugs within plastic microchannels using a hydrophilic sidestream, 8th European Congress of Chemical Engineering, Berlin, Deutschland (2011)
123. W.A. Gaakeer, M.H.J.M. de Croon, J. van der Schaaf, J.C. Schouten, Liquid-liquid slug flow separation in a slit shaped micro device. Chem. Eng. J. **207–208**(0), 440–444
124. N. Pamme, Continuous flow separations in microfluidic devices. Lab Chip **7**(12), 1644–1659 (2007)
125. D. Gossett, W. Weaver, A. Mach, S. Hur, H. Tse, W. Lee, H. Amini, D. Di Carlo, Label-free cell separation and sorting in microfluidic systems. Anal. Bioanal. Chem. **397**(8), 3249–3267 (2010)
126. A. Lenshof, T. Laurell, Continuous separation of cells and particles in microfluidic systems. Chem. Soc. Rev. **39**(3), 1203–1217 (2010)
127. N. Oozeki, S. Ookawara, K. Ogawa, P. Löb, V. Hessel, Characterization of microseparator/classifier with a simple arc microchannel. AIChE J. **55**(1), 24–34 (2009)
128. S. McGovern, G. Harish, C.S. Pai, W. Mansfield, J.A. Taylor, S. Pau, R.S. Besser, Multiphase flow regimes for hydrogenation in a catalyst-trap microreactor. Chem. Eng. J. **135**(Suppl 1(0)), S229–S236 (2008)
129. M.W. Losey, M.A. Schmidt, K.F. Jensen, Microfabricated multiphase packed-bed reactors: characterization of mass transfer and reactions. Ind. Eng. Chem. Res. **40**(12), 2555 2562 (2001)
130. S. Krishnamurthy, Y. Peles, Gas-liquid two-phase flow across a bank of micropillars. Phys. Fluids **19**(4), 043302–043314 (2007)
131. N. Márquez, P. Castaño, M. Makkee, J.A. Moulijn, M.T. Kreutzer, Dispersion and holdup in multiphase packed bed microreactors. Chem. Eng. Technol. **31**(8), 1130–1139 (2008)
132. L. Kiwi-Minsker, A. Renken, Microstructured reactors, in *Handbook of Heterogeneous Catalysis*, ed. by G. Ertl, H. Knözinger, F. Schüth, J. Weitkamp (Wiley-VCH, Weinheim, 2008)
133. M.N. Kashid, A. Renken, L. Kiwi-Minsker, Gas-liquid and liquid-liquid mass transfer in microstructured reactors. Chem. Eng. Sci. **66**(17), 3876–3897 (2011)
134. M.T. Kreutzer, F. Kapteijn, J.A. Moulijn, J.J. Heiszwolf, Multiphase monolith reactors: Chemical reaction engineering of segmented flow in microchannels. Chem. Eng. Sci. **60**(22), 5895–5916 (2005)
135. J. Kobayashi, Y. Mori, K. Okamoto, R. Akiyama, M. Ueno, T. Kitamori, S. Kobayashi, A microfluidic device for conducting gas-liquid-solid hydrogenation reactions. Science **304**(5675), 1305–1308 (2004)
136. A. Gunther, S.A. Khan, M. Thalmann, F. Trachsel, K.F. Jensen, Transport and reaction in microscale segmented gas-liquid flow. Lab Chip **4**(4), 278–286 (2004)
137. M.N. Kashid, L. Kiwi-Minsker, Microstructured reactors for multiphase reactions: state of the art. Ind. Eng. Chem. Res. **48**(14), 6465–6485 (2009)
138. J. Yue, J.C. Schouten, T.A. Nijhuis, Integration of microreactors with spectroscopic detection for online reaction monitoring and catalyst characterization. Ind. Eng. Chem. Res. **51**(45), 14583–14609 (2012)
139. R. Abdallah, V. Meille, J. Shaw, D. Wenn, C. de Bellefon, Gas-liquid and gas-liquid-solid catalysis in a mesh microreactor. Chem. Commun. **4**, 372–373 (2004)
140. J. Shaw, Micromesh contactors for gas/liquid and liquid/liquid reactions with preliminary tests of reductions at catalytic surfaces, 9th PIN meeting, London (2003)
141. A.A. Kulkarni, A.K. Gorasia, V.V. Ranade, Hydrodynamics and liquid phase residence time distribution in mesh microreactor. Chem. Eng. Sci. **62**(24), 7484–7493 (2007)
142. P. Tundo, A. Perosa, Multiphasic heterogeneous catalysis mediated by catalyst-philic liquid phases. Chem. Soc. Rev. **36**(3), 532–550 (2007)
143. A. Bhaumik, P. Mukherjee, R. Kumar, Triphase catalysis over titanium silicate molecular sieves under solvent-free conditions: I. Direct hydroxylation of benzene. J. Catal. **178**(1), 101–107 (1998)

144. S. Bhattacharya, A. Drews, E. Lyagin, M. Kraume, M.B. Ansorge-Schumacher, Efficient chemo-enzymatic epoxidation using silcoat-Novozym®435: characterizing the multiphase system. Chem. Eng. Technol. **35**(8) 1448–1455 (2012)
145. H.J. Glatzer, L.K. Doraiswamy, Rate enhancements due to autocatalysis and heterogenization in phase transfer catalysis: a comparative study. Chem. Eng. Sci. **55**(21), 5149–5160 (2000)
146. O. Hernandez-Justiz, R. Fernandez-Lafuente, M. Terreni, J.M. Guisan, Use of aqueous two-phase systems for in situ extraction of water soluble antibiotics during their synthesis by enzymes immobilized on porous supports. Biotechnol. Bioeng. **59**(1), 73–79 (1998)
147. E. Andersson, B. Hahn-Haegerdal, Bioconversions in aqueous two-phase systems. Enzyme Microb. Technol. **12**(4), 242–254 (1990)

Chapter 7
Micro Continuous-Flow Synthesis of Metal Nanoparticles Using Micro Fluid Segment Technology

Andrea Knauer and J. Michael Köhler

Abstract The micro segmented flow technique is very promising for the synthesis of metal nanoparticles, in particular for plasmonic nanoparticles and is very useful for combinatorial syntheses and screenings of new types of nanomaterials. In this chapter, the specific properties and technical as well as scientific challenges related to metal nanoparticles, the advantages of micro segmented flow and draw-backs of conventional synthesis for metal nanoparticles as well as the general applicability and the potential for the application of micro segmented flow for the preparation of metal and semiconductor nanoparticles are discussed. The specific conditions of micro segmented flow are described relating to the critical steps of reactant mixing, nucleation, and particle growth. It is shown that the intensification of local transport in the microfluidic system causes a significant improvement in particle homogeneity. In the formation and handling of metal particles, aspects of redox reactions, electrochemical parameters, and aspects of coordination chemistry have to be reconsidered. Ligands, which are able to interact with the metal ions in solution or with the forming nanoparticles, have a strong effect on the particle formation, their transport behavior, and interaction. The effect of fast reactant mixing supported by intensive segment-internal convection due to high flow rates is used in order to obtain uniform conditions for nucleation as well as for the particle growth. It is explained why non-spherical particles are of particular interest for different applications and how their quality can be improved by the application of microfluidic synthesis techniques, too. The formation of silver prisms by a micro continuous-flow synthesis in micro fluid segments will be given as a typical example allowing the tuning of the optical properties of the colloidal solutions. Finally, it is demonstrated that the micro segmented flow technique is well suited for an automated variation of composition of reactant mixtures. Thus, it is possible to screen a large quantity of different compositions in one single experimental run, combined with a minimum of consumed chemicals.

A. Knauer (✉) · J. M. Köhler
Institute of Chemistry and Biotechnology, Ilmenau University of Technology, PF 10 05 65, 98684 Ilmenau, Germany
e-mail: andrea.knauer@tu-ilmenau.de

J. M. Köhler and B. P. Cahill (eds.), *Micro-Segmented Flow*, Biological and Medical Physics, Biomedical Engineering,
DOI: 10.1007/978-3-642-38780-7_7,

The integration of miniaturized optical devices allows an online monitoring and the real-time detection of the effect of changing transport and reaction conditions on the properties of synthesized nanomaterials.

7.1 Introduction in Metal Nanoparticle Synthesis by Micro Fluid Segment Technique

The synthesis of nanoparticles attracts more and more attention in science and technology. The fascination of this class of materials can be explained by two main aspects: Firstly, there are unique combinations of properties of such materials coming from the classical bulk materials on one side and secondly, quantum effects and the specific properties of molecules and clusters on the other side. The dominance and the merging of both traditional material properties and special properties arising from the small size and the limited number of atoms and electrons in a single nanoparticle as well as the high effect of the surface states for the appearance of a particle lead to the possibility of realization of a large spectrum of new properties and new functions and give the potential of a variation of physical properties to a large extend. [1–7] The second is the enormous potential of variation of particle composition and the internal particle architecture [8]. Nanoparticles can be formed by all types of classical materials, spanning over the spectrum from metal over semiconductors, inorganic dielectric materials, and diverse inorganic molecular solids to low-molecular weight organic materials, polymers and biopolymers [2, 9–13]. The arrangement of atoms and molecular units, the topological connection between molecular building units and the spatial distribution of different types of chemical bonds open a huge range of variation possibilities in nanoparticle architectures (Fig. 7.1). In addition, different elements, different molecules, and completely different material classes can be combined [11, 14]. Surfaces and intraparticle interfaces can be chemically modified in order to enhance the substantial multiplicity of nanoparticles [15, 16]. Taking these potentials in mind, the variability of nanoparticle materials is unlimited. Thus, it can be expected that new types and new useful functions of nanoparticles can be detected or developed now and in the distant future.

Metal nanoparticles are a comparatively simple and long-known class of nanomaterials [17]. Noble metal nanoparticles as gold or silver nanoparticles are used for

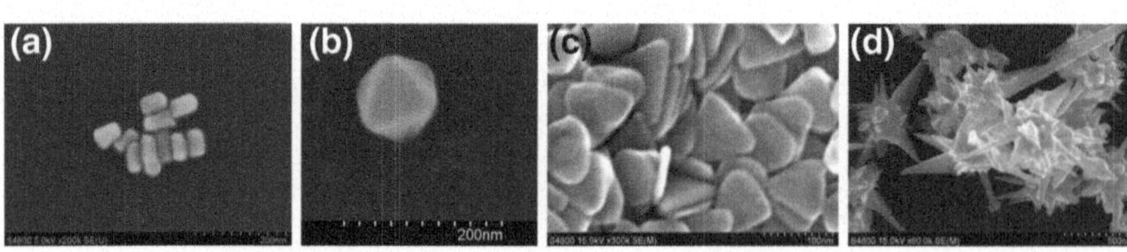

Fig. 7.1 A few examples of different shaped noble metal nanoparticles or nanostructures: **a** Au rods, **b** single crystalline Au octahedron, **c** triangular Ag nanoprisms, **d** branched Au structures

coloring glass since many centuries. The nature of these color effects was explained by Michael Faraday about one and a half century ago. He was able to generate gold nanoparticles by a well-controlled chemical reaction inside the aqueous phase and to obtain a stable colloidal solution. Meanwhile, it is understood that the excitation of localized surface particle plasmons by an outer electromagnetic field is responsible for the characteristic colors of gold or silver nanoparticle solutions [18, 19]. This resonant interaction between the electronic system of the particle and the exciting radiation causes a comparatively sharp absorption band. The resonance wave length is dependent on the composition, the size, and the shape of the metal nanoparticles [1, 20–24].

Plasmonic nanoparticles can be regarded as converter for optical radiation (Fig. 7.2). They convert a certain part of the optical spectrum into heat and release scattered light of certain spectral characteristics. Both aspects are very interesting for nanotechnology. The energy conversion can be used for example for thermal excitation of very small volume elements leading to an apparent focusing of electromagnetic excitation far below optical wavelengths and help to overcome the limitation of focusing by refraction in case of such an activation step [25].

The narrow and intensive plasmonic absorption resonance peak makes nanoparticles interesting for labeling. [1, 26, 27] In principle, this function can be regarded in analogy to labeling by dyes. But, in contrast to dyes, the absorption peaks are higher in case of nanoparticles and more narrow, in general, and the photochemical activity of nanoparticles and, associated therewith, the risk of degradation and bleaching is much lower. In addition, single nanoparticles can react with single biomolecules or other analyst molecules with similar specificity as molecular reaction partners in 1:1 reactions if a corresponding surface functionalizing is used. At the same time, a single nanoparticle can be much more easily detected by optical sensors and can be more convenient operated than a single molecule.

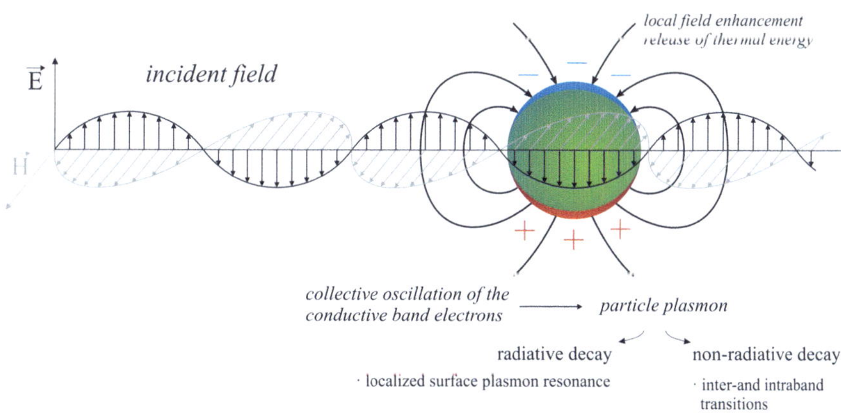

Fig. 7.2 Response of sufficient small noble metal nanoparticles on an incident electromagnetic field

A high demand for larger application of plasmonic nanoparticles in future arises from the desire of identical physical properties of the nanoparticles obtained by chemical synthesis. Another important challenge is given by the desire of tuning the particle properties. Both requirements give a strong motivation for the search for new reaction strategies for the synthesis of plasmonic nanoparticles. The micro reaction technology with the ability of realization of fast mass and heat transfer and new process windows is very promising for this search [9, 10, 14, 28–31]. In the following, it will be shown that the technique of micro segmented flow is of particular attractiveness for the synthesis of metal nanoparticles under conditions of micro reaction technology [32].

7.2 Requirements of the Synthesis of Metal Nanoparticles and the Specific Advantages of Micro Fluid Segment Technique Therefore

The synthesis of particles and in particular the generation of metal particles by precipitation reactions were regarded as dangerous with respect to the implementation of the synthesis in micro reactors. It was expected that the high surface-to-volume ratio in micro fluidic systems would promote the deposition of particles at the reactor walls, the chemical fouling and finally the blockage of the micro channels. In particular, the initiation of nucleation at the walls in case of reductive generation of metal nanoparticles was a real problem for the micro continuous-flow synthesis of nanoparticles. First experiments for the synthesis of gold nanoparticles by reduction of tetrachloroauric(III) acid with ascorbic acid in aqueous solution using split-and-recombine micro mixers based on a glass/Si/glass sandwich construction confirmed that nucleation occurred at the inner walls of the hydrophilic micro channels preferably. This chemisorbed seeds grew rapidly by further supply of both reactants in the continuous-flow synthesis after the deposition of first metal inside the channel. The efficient catalytic activity of metal itself for the further reduction of metal ions is responsible for this undesired effect.

But, it could be shown that the problems with side wall deposition could be overcome by reducing the chemical affinity of the channel surfaces against the metal ions and the charged metal nanoparticles formed inside the reaction mixture by a surface hydrophobization [28]. The undesired deposition of gold at the reactor walls was strongly reduced by treating the micro channels by silanization agents. In this way, the active OH-groups at the surface were substituted by alkoxygroups. In result, very reproducible micro continuous-flow syntheses of gold nanoparticles leading to stable colloidal solutions have become possible. The size homogeneity of nanoparticles was improved in comparison with conventional batch methods. The diameter of the gold nanoparticles could be tuned by a variation of the reactant ratios and additionally by variation of the flow rate with certain limits [28].

Alternatively to alkyl silanized glass and silicon devices, micro reactors made of hydrophobic and reactant-resistive polymers or of perfluorinated polymers are well suited for the reductive generation of metal nanoparticles from metal salt solutions. Beside micro fluidic chip devices, micro tubes, interconnectors, injectors and other components of organic material, or—even better—of perfluorinated polymers are very useful for nanoparticle synthesis, if the reaction temperature is low enough to avoid softening or corrosive reactions of the polymer materials.

The application of micro reactors and micro tubes with hydrophobic or hydrophobized surfaces is particular interesting for the application of the segmented flow technique in nanoparticle synthesis due to the control of the wetting behavior [33]. Alkanes or perfluorinated hydrocarbons are mainly applied as inert carrier solutions for segmented flow synthesis using aqueous solutions. Therefore, these liquids are perfectly wetting hydrophobized channel walls and micro reactors of hydrophobic materials [34]. In contrast, the aqueous reactant solution and the reaction mixture show high contact angles, which means low wetting of these surfaces. The consequence is that aqueous segments, flowing through such channels are well embedded inside the carrier solution. The carrier solution causes a "chemical insulation" of the reactant mixture from the solid surfaces of the tubes, channels and reactors. This insulation is an additional efficient effect for suppressing the adsorption of nanoparticles or nucleation at the channel walls.

7.3 General Aspects of Particle Formation and Partial Processes of Noble Metal Nanoparticle Synthesis

Monodispersity is the most fundamental challenge in the synthesis of nanoparticles for a transduction of signals and for nanotechnical applications [35, 36]. This challenge is not only driven by the reduction of dispersive effects in particle distribution, particle transport, and chemical activity, but is mainly motivated by the fact that the physical and namely the electronic and optical properties of the particles are affected by the shape and the size of the nanoparticles [37, 38]. Beside the chemical composition, geometrical parameters have a strong influence on charge mobility and on the resonance properties of the metal nanoparticles [23, 39–42].

Homogeneity can either be achieved by a thermodynamic or kinetic control. The deciding advantage of a process with a thermodynamic control consists in the self-optimizing principle. All process trajectories are moving towards the equilibrium. The stable state will work as a dominating attractor for all individual process paths and lead to a leveling of different starting conditions and fluctuations. But, thermodynamic stable structures demand for process at higher temperatures and with longer reaction times, which is not typical for microreaction technology, on the one hand. On the other hand, the types of nanoparticle structures, which are formed under thermodynamic control, are limited [12, 43, 44].

In contrast, the variability in shapes, compositions and structures of nanoparticles is much higher under a kinetic control. Process steps can be fast and strongly move away from the thermodynamic equilibrium. The process steps as well as the character of intermediates and final products can be dominated by transport effects as well as by the rate ratios of competing chemical reactions. In result, a huge spectrum of compositions, shapes and sizes can be realized [8]. But, the significant draw-back of this kinetic control is the absence of any self-controlling mechanisms in the particle formation. The combination of near-equilibrium process steps with far-equilibrium steps can be a strategy for reducing the problems of process control. But, even in the case of only one involved kinetically controlled step in the whole process chain, the realization of homogeneous and reproducible conditions is the key challenge for a kinetically controlled synthesis.

The use of kinetic control in nanoparticle synthesis demands for a strict control of reaction conditions, transport mechanisms and for an enormous high reproducibility in all process parameters. Such a control can only be achieved if the mechanism is sufficient understood.

The formation of metal nanoparticles by reduction of metal ions from a homogeneous liquid can be subdivided into the following steps (Fig. 7.3):

1. Formation of primarily formed reduced atomic species (atoms or clusters) by a redox reaction in the homogeneous phase. Coordination compounds and ligand-stabilized metal clusters can assist the first process step.

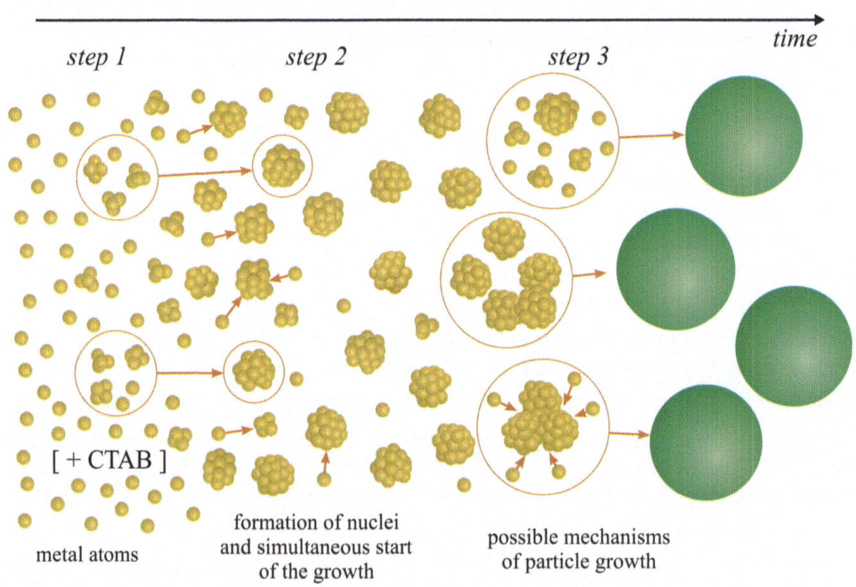

Fig. 7.3 Possible mechanisms for the wet chemical formation of metal nanoparticles by reduction of an ionic species and subsequent nuclei formation and particle growth

2. Nucleation step: formation of stable nuclei by regular assembling of metal atoms or clusters or by growth of clusters.
3. Particle growth by one of the three following mechanisms:
   (a) reaction of atoms, ions, molecular species or well-defined clusters with the growing particle.
   (b) aggregation between the particles.
   (c) simultaneous particle-particle aggregation and deposition of species from the molecular/atomic disperse state.

The first step can be regarded as familiar to a typical chemical reaction in homogeneous phase. It is to assume that in most cases this step is not too far from the thermodynamic equilibrium and therefore it can be regarded as a more or less reversible part of the process chain.

The second step, the nucleation, is more complex and related to higher activation thresholds for the backward reaction. In consequence, this step is to be assumed as the critical and mostly irreversible part of the process chain.

The reversibility of the third step is strongly dependent on the mechanisms and conditions of material deposition and aggregation. In case of metal deposition by reduction of metal ions, it is assumed that the particle growth by reaction with ions and/or molecular species is a typical electrochemical process with the formation of a mixed potential (Fig. 7.4): the growing particle has the function of a mixed electrode. It acts as an anode for the oxidation of the reducing agent, as well as a cathode for the reduction of the ionic or complex species to elemental metal.

In many homogeneous redox reactions, there is a high activation threshold for the oxidation of the reducing agent corresponding to the anodic partial process on the surface of the metal nanoparticle. This threshold is lowered by the anodic activity of the particle. In result, the electrochemical activity of the particle has

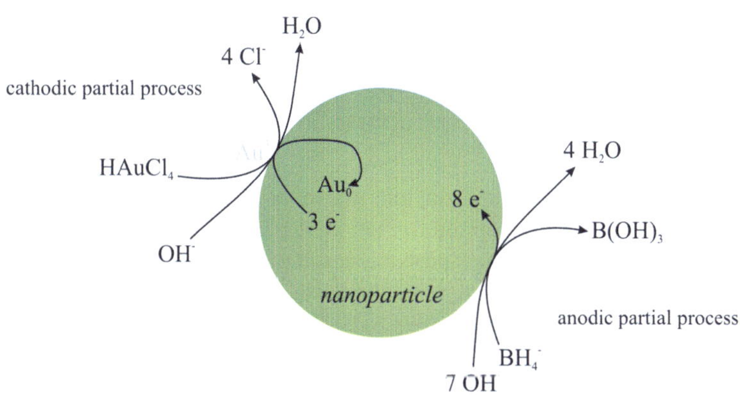

Fig. 7.4 Formation of a mixed potential at the nanoparticles surface as for example during the synthesis of gold nanoparticles using sodium borohydride as reducing agent. The nanoparticle itself is in the role of a mixed electrode

the effect of a catalyst: the further growth of the particle by metal deposition is a nanoparticle-catalyzed reduction of the metal ions to elemental metal. The electrochemical potential of the nanoparticle is determined by the superposition of the anodic and the cathodic partial process. The strength of the effect of the single partial processes is determined by their individual standard potentials and by the concentrations of the involved chemical species according to Nernst law. So, the electrochemical process determines the electrochemical potential of each nanoparticle and their intensities are—vice versa—dependent on the common potential. The spontaneous adjusting of the electrochemical potential of the nanoparticle is determined by the condition of balanced currents: The amount of anodic current has to be identical with the amount of cathodic current in a given short interval with negligible potential drift.

The particle potential is not only important for the electrochemical behavior and therefore the cathodic growth, but is also determining the particle charge and thus an important parameter for possible particle aggregation [45, 46]. Apart from solvation mechanisms in some special cases, the electric charges and the resulting repulsive electrostatic forces between homonymous charged particles are mainly responsible for the avoidance of precipitation and thus for maintaining a stable colloidal solution (Fig. 7.5a). Aggregation and precipitation depend on the probability that the velocity of the moving particles relative to each other is high enough for an overcompensation of the Coulomb repulsion. The particle velocity is mainly controlled by particle mass and temperature, the repulsive forces by particle charge and ion strength. This is the reason for an uncontrolled aggregation and precipitation if the dispersed nanoparticles are discharged (Fig. 7.5b), and it is also the reason for the higher tendency to aggregation in case of smaller particles, which have a higher mobility at the same temperature compared to larger particles, if they carry a certain charge (Fig. 7.5c). From this effects can be concluded that particle aggregation as well as reproducible assembling can be controlled by controlling the electrochemical potential during the particle synthesis.

7.4 Addressing of Size and Shape in a Micro Segmented Flow-Through Metal Nanoparticle Synthesis

In case of a fast reduction of the metal ions due to the use of a strong reducing agent, a fast nucleation, which leads to the formation of a large number of small particles with a spherical shape, is the most probable result of the particle growth. In order to obtain shape anisotropic nanoparticles with high aspect ratios, seed mediated syntheses are usually the method of choice [21, 41, 47]. The basic approach is a two-step process. At first, small crystallization seed particles with the utmost homogeneity are synthesized in the presence of a suitable ligand (e.g. CTAB in case of gold nanorods or PSSS, PVP for silver nanoplates) by the help of a strong reducing agent. In a second step, the shape anisotropic growth of the seed nanoparticles is promoted by an additional supply of the corresponding metal salt under application of a soft reduc-

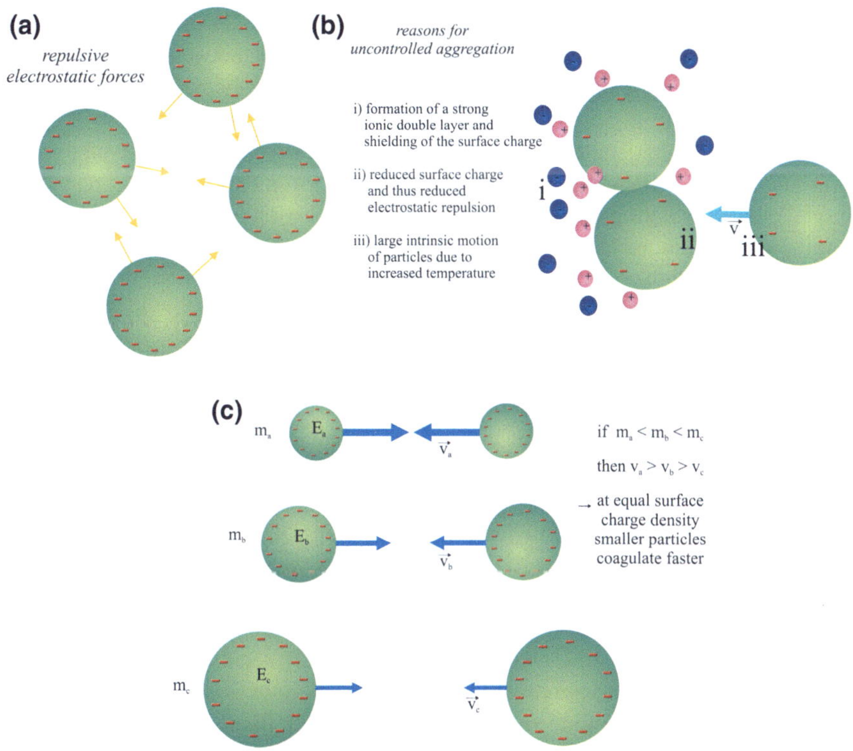

Fig. 7.5 **a** Homonymous charged particles repel each other at sufficiently high surface charge due to repulsive electrostatic forces. **b** Schematic illustration of reasons of uncontrolled nanoparticle aggregation. **c** Smaller particles with lower masses have an increased tendency to aggregation due to a higher relative speed at the same temperature than larger particles with the same surface charge density

ing agent [48]. The reasons of the shape anisotropic growth are not yet satisfactorily explained. The protocols that can be found in literature for the synthesis of these kinds of nanomaterials were found empirically and the proposed explanations for the directed growth differ significantly in terms of the suggested mechanisms. Common explanatory approaches refer to the coverage of specific crystallographic planes by e.g. halide ions or ligand molecules, which inhibit or prevent the deposition of material in these lattice orientations. Another scientifically established approach is the method of underpotential deposition, which is based on the use of a minimum quantity of a species with a less negative standard electrode potential than the equilibrium potential of the reduction of the desired (substrate) material [49]. This method is primarily used for the growth of a homogeneous crystal structure, which is of great importance for the formation of a spatially non-isotropic shape. Another very recent explanatory approach is based on the consideration of the electronic conditions at the nanoparticles surfaces. Here it is assumed that cathodic and anodic areas form on the particle surface as a result of a non-isotropic charge density distribution. At least it could be shown that the particle-molecule interaction between Ag nanoplates and the poly anionic ligand poly(styrene sulfonic acid) is decisively responsible for the shape stability of these flat, areally extended nanoprisms [50].

In the following, a detailed description of the synthesis strategy for homogeneous silver nanoprisms with a triangular base area will be given exemplary for the preparation of shape anisotropic nanomaterials under application of the micro segmented flow-through technique [51].

The implementation of the silver nanoprism synthesis into a micro flow-through process was based on the literature-known batch protocol published by D. Aherne and co-workers [48]. First, the known protocols were calculated down to milliliter scale and both necessary steps of seed particle synthesis and nanoprism growth were carried out in a conventional batch procedure. These reference experiments were used later for comparison of the results of both methods. The batch processed product particles as well as the microfluidically generated nanoparticles of both synthesis steps—preparation of crystallization seeds and subsequent nanoprism growth—were characterized by differential centrifugal sedimentation (DCS, DC 20000, CPS Instruments Inc., Newtown, USA/PA), UV-vis spectrophotometry (Specord 200, Analytik Jena AG, Jena, Thuringia, Germany), and scanning electron microscopy (Hitachi S-4800 FE-SEM, Hitachi High Technologies America, Inc., Schaumburg, USA/IL).

However, the reproducible working batch protocols needed to be recalculated regarding reactant concentrations and required volume flows for a successful transfer of the reaction into micro fluid segments. As it was initially described for the batch process, silver ions are reduced by sodium borohydride in the presence of the poly anionic effector poly(sodium styrenesulphonate) for the generation of small silver seed particles. In the batch protocol it is explicitly specified that the addition of silver nitrate into a reaction solution containing an excess of reducing agent should happen drop-wise and not exceed a limit of 2 mL $AgNO_3$ per minute. By this method, the concentration of the Ag^+ species is kept low for all time intervals of the reaction process and thus, an otherwise inevitable increase of the particle surface potential due to the adsorption of silver ions is avoided (Fig. 7.6). In case of a fast addition of the silver salt, a compensation of the negative, repulsively acting surface charge of the nanoparticles could occur, which can result in an uncontrollable aggregation and a breakdown of the colloidal state (Fig. 7.7).

By transferring the synthesis of the seed nanoparticles into a micro flow-through process, this problem of colloid stability cannot be neglected. Time dependent or drop-wise additions of reactant solutions can hardly be implemented into a continuous flow process. That is why all locally occurring concentration gradients, which will influence the quality of the obtained particles in a negative way, have to be reduced as fast as possible. This demand cannot be met satisfactory in a homogeneous, laminar flow process (Fig. 7.8). A possible solution is the running of the synthesis in a heterogeneous two-phase process. Under application of the micro segmented flow and under conditions of sufficiently high volume flow rates and a regular segmentation and dosage of the educts, the reactant mixing and the initial phase of nucleation runs fast enough that particle aggregation is prevented (Fig. 7.9).

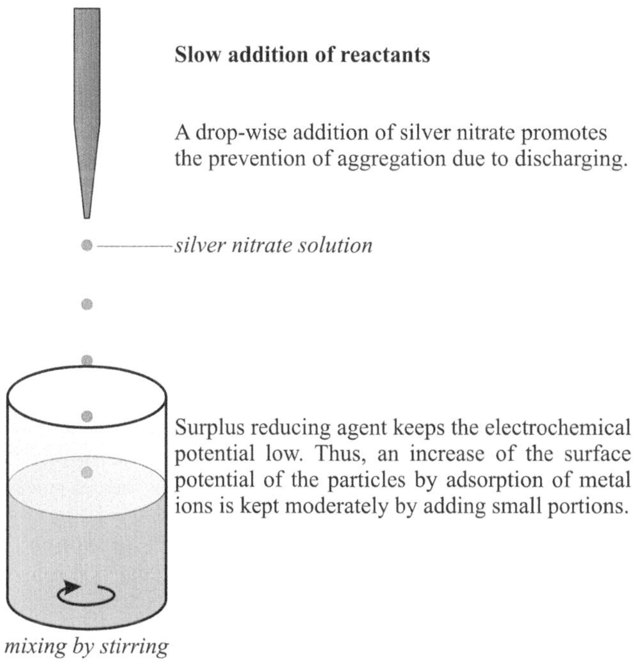

Fig. 7.6 Schematic illustration of careful reactant addition in batch syntheses

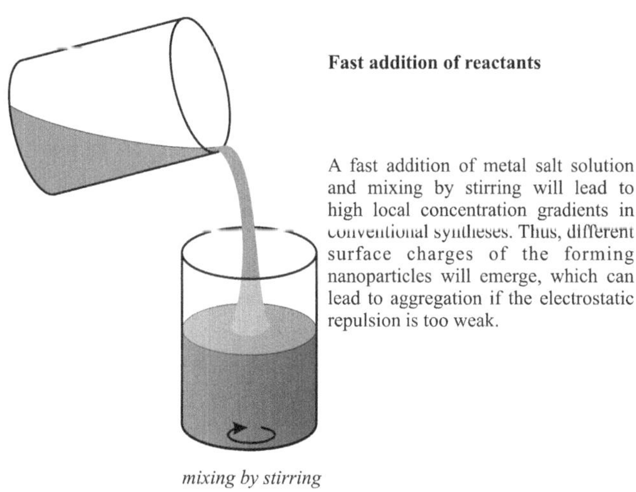

Fig. 7.7 Schematic illustration of a fast reactant addition in batch syntheses

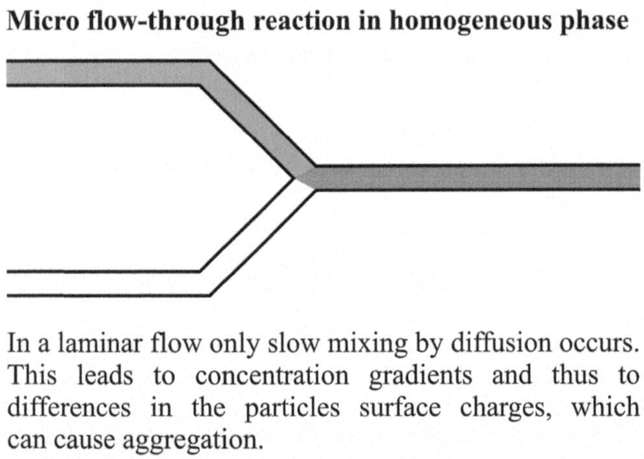

Fig. 7.8 Schematic illustration of the reactant fusion in laminar flow-through processes

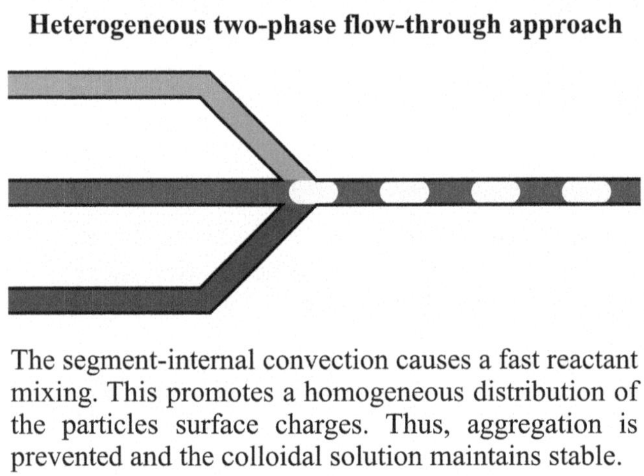

Fig. 7.9 Schematic illustration of the micro segmented flow-through synthesis strategy

The preparation of the silver seed nanoparticles in a micro continuous flow-through process could be realized in an experimental arrangement as it is shown in Fig. 7.10. The five syringes, which were required to run the synthesis, were mounted on an electronically controlled syringe pump system (Cetoni NeMESYS, Cetoni GmbH, Korbußen, Germany) with servo drive and the option to individually adjust the flow rates of the axes. The syringes were connected via PTFE fittings and tubular material with an inner diameter of 0.5 mm (Bohlender GmbH, Gruensfeld, Germany) to standard PEEK fluid components (IDEX Health & Science LCC, Oak Harbor, USA/WA). All of the above mentioned polymer reactor components are available as standard HPLC equipment.

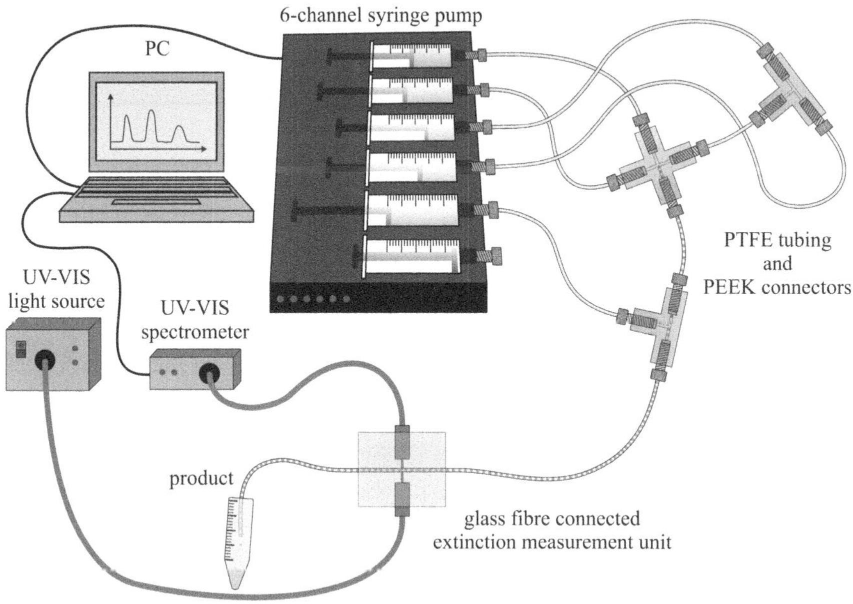

Fig. 7.10 Experimental arrangement for the synthesis of silver seed nanoparticles in micro fluid segments

In order to generate the segmented flow, the immiscible and reaction inert organic carrier medium Perfluoromethyldecalin (F2 Chemicals Ltd, Lea, Lancashire , England) was dosed firstly into the reactor channel. The micro fluid segments were then generated at the first injection unit by the addition of sodium citrate (Merck KGaA, Darmstadt, Germany, purity: 99 %) and a mixture of poly(sodium styrene-sulphonate) (PSSS) (ACROS Organics, Morris Plains, USA/NJ, MW 70.000) and sodium borohydride (Merck KGaA, Darmstadt, Germany, purity: 99 %) at the opposite reactor inlet.

Now, if the generation of these segments happens regularly and the intermediate segment distances remain constant during the whole continuous process, silver nitrate (Merck KGaA, Darmstadt, Germany, purity: 99 %) can be dosed with respectively identical volumes into the preformed segments. Thus, under stable fluidic conditions and regular addition of the aqueous solutions, uniform segments with equal sizes, distances, and reactant ratios are formed. All reactant concentrations as well as the individual flow rates are given in Table 7.1. The listed parameters correspond to an approximate time for the dosage of $AgNO_3$ of about 10 ms in case of 0.3 μL segment volume.

As it was mentioned before, optimum mixing conditions are required in case of such a rapid reaction as the Ag^+ reduction by sodium borohydride to produce homogeneous nanoparticles with a minimum half-width value of the size distribution and a minimum average diameter. Therefore the high total flow rate of 1.6 mL/min was

Table 7.1 Overview of the distribution of chemicals among the syringes, the applied concentrations, and flow rates for the flow-through synthesis of Ag seed particles

Syringe	Chemicals	Concentration	Flow rate
1	Perfluoromethyldecalin	–	1000 μL/min
2	$C_6H_5Na_3O_7$	2.5 mM	474 μL/min
3	$NaBH_4$	10 mM	14.2 μL/min
4	PSSS	500 mg/L	23.6 μL/min
5	$AgNO_3$	2.5 mM	94 μL/min

chosen for the seed nanoparticle synthesis, because it is known, that the segment internal convection is accelerated with increasing flow rate [52]. In the synthesis of Ag nanoparticles by using borohydride as reducing agent obviously a fast nucleation occurs. The segment internal mixing is fast enough for the initiation of a rapid nucleation. By virtue of the fast nucleation process, a rapid consumption of free Ag^+ ions occurs. Thus, the micro flow-through process is robust with respect to a potential increase caused by high Ag^+ concentrations. While the conventional synthetic strategy requires a slow dosing of silver nitrate to prevent particle aggregation, no time-critical synthesis steps occur during the micro fluidic synthesis of silver seed nanoparticles or during the subsequent growth at all. The reason for this simplification can be found in the effective segment internal convection, which leads to high mixing rates, an avoidance of concentration gradients in the reactant solution, and thus to a homogeneous growth of well dispersed nanoparticles. High quality seed nanoparticles are the basis for a regular growth of the desired triangular silver nanoprisms. Silver seed nanoparticles, which were synthesized in the above shown micro fluidic system display an average particle diameter of 3.8 nm with a full width at half maximum (FWHM) of 2.3 nm. A direct comparison of the seed nanoparticles obtained from the conventional batch method and from the micro flow-through synthesis can be found in Fig. 7.11. The differential size distribution spectra were obtained from weight based DCS measurements. The FWHM of the silver seed nanoparticles from the batch method is with 7.5 nm significantly larger than the half-width of the microfluidically obtained nanoparticles.

Another positive effect of homogeneous mixing conditions is reflected by the average particle diameter. The mean particle diameter of the batch synthesized Ag seeds is with a main peak at 4.2 nm larger than the average diameter of the flow-through processed particles. With the batch-generated particles, a second maximum occurs in the weight-based analysis of the centrifugal sedimentation data at about 9 nm. This is either a second population or a partial aggregation of a constant small number of small nanoparticles or it is a mixture of both. In any case, this observation can be drawn back to inhomogeneous mixing conditions during the synthesis in a conventional flask. This means that the main differences in the homogeneity can be explained by the assumption of a slower interdiffusion of the reactant solutions in the batch process (Fig. 7.12). Due to the slower shift of the concentrations, a broader nucleation interval results and early forming particles are already growing while

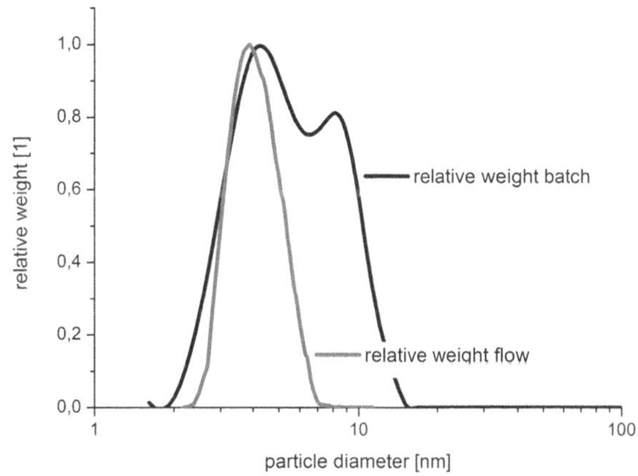

Fig. 7.11 Comparison of the weight based DCS data for batch (*black line*) and flow-through (*red line*) synthesized silver seed nanoparticles

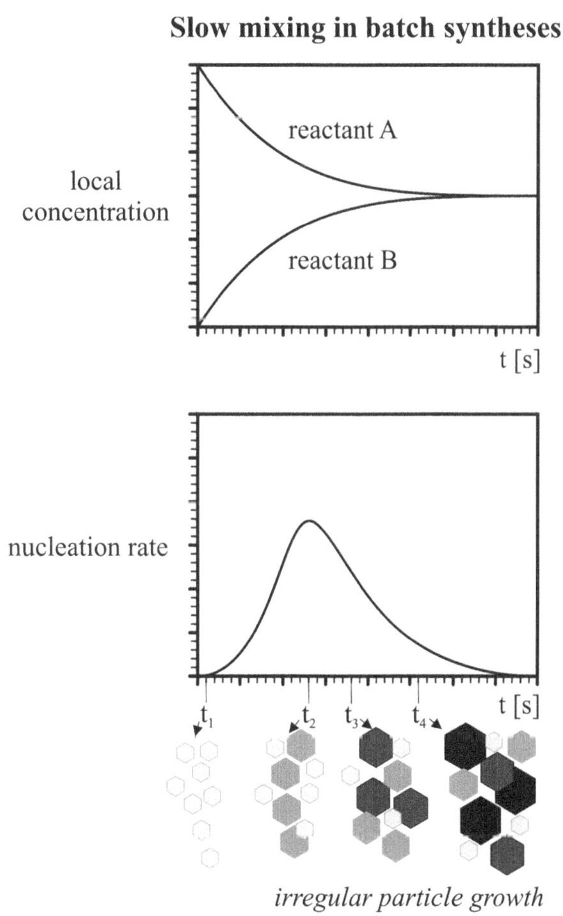

Fig. 7.12 A slow reduction of concentration gradients leads to broad nucleation intervals and thus to an inhomogeneous particle growth

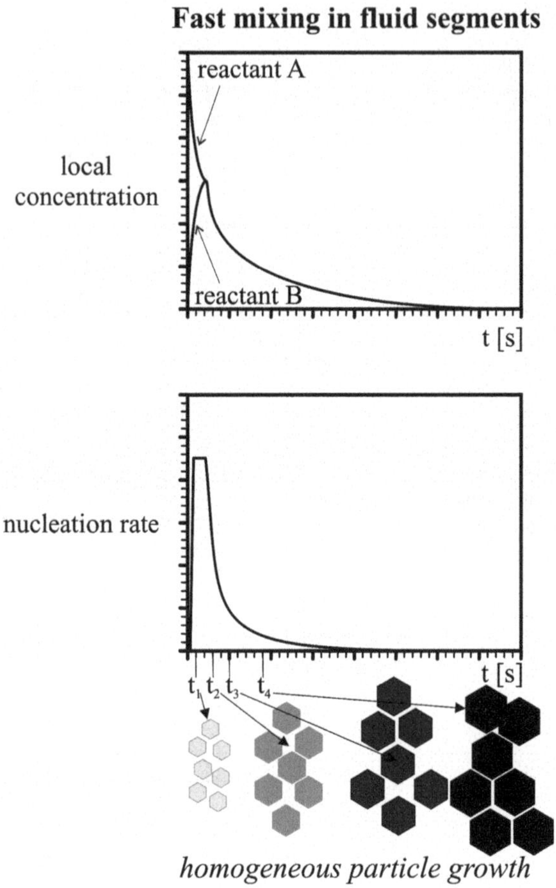

Fig. 7.13 The high mixing rates in fluid segments lead to short nucleation intervals and to a homogeneous particle growth

the nucleation of further nanoparticles is still in process. Due to the high mixing efficiency in micro fluid segments, an increase of the nucleation rate takes place in all volume elements of the fluid compartment (Fig. 7.13). The initiated nucleation leads to a fast and homogeneous decrease of the silver ion concentration and thus to a fast fall below the nucleation threshold. Then no further nuclei are formed and the only ongoing process is the particle growth. This effect leads to the formation of a higher number of smaller particles compared to conventional synthesis methods.

Once homogeneous Ag seed nanoparticles were prepared, the growth step of triangular Ag nanoprisms can be carried out. The growth step requires an experimental setup, which is designed according to the requirements of the synthesis procedure. The utilized experimental arrangement is shown in Fig. 7.14. For the synthesis of Ag nanoprisms, three syringe pumps and two PEEK T-junctions are required. Since

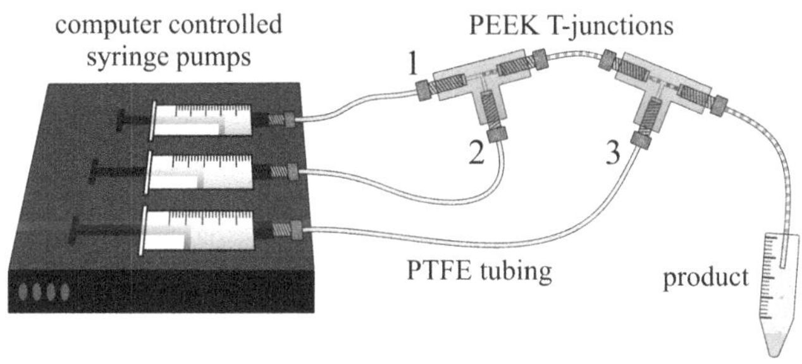

Fig. 7.14 Microfluidic setup for the generation of different sized triangular silver nanoprisms

Table 7.2 Distribution of the used chemicals among the syringes, applied concentrations and microfluidic reaction conditions for the growth of silver nanoprisms

Syringe	Chemicals	Concentration	Flow rate
1	Carrier medium	–	100 μL/min
	5 mL H_2O		
	0.075 mL $C_6H_8O_6$	–	
2	[0.5 mL/0.4 mL / 0.3 mL	20 mM	60 μL/min
	/ 0.2 mL/0.1 mL] Ag seed NP	–	
3	$AgNO_3$	1 mM or 0.4 mM	40 μL/min

the shape anisotropic nanoparticle growth is a slow process, a soft reducing agent, as for example ascorbic acid (Merck KGaA, Darmstadt, Germany, purity: 99.7 %) is required. A complete list of the used chemicals and their distribution among the syringes can be found in Table 7.2. Fluid segments are generated by dosing a pre-mixed solution of water, ascorbic acid and silver seed nanoparticles into the presented carrier stream. The growth reaction starts by dosing silver nitrate into the preformed segments at a second injection unit. A total flow rate of 200 μL/min provides both: a segment internal convection for homogenous reactant mixing and a sufficient residence time for the segments inside the reaction channel.

As it was published by Aherne and co-workers, the planar expanse of the Ag nanoprisms can be tuned by the number of Ag seed nanoparticles in the growth solution [48]. If the metal salt concentration is kept constant and the seed particle density is varied, particles with a large edge length will be generated in case of a low amount of seed nanoparticles and smaller Ag nanoprisms result in case of a high seed nanoparticle density. In any case, the thickness of the obtained product particles is mainly determined by the diameter of the utilized seed particles, since the nanoprism growth takes place almost exclusively in lateral direction and the growth of the particle thickness is effectively suppressed due to the interaction between the nanoparticle and the poly anionic ligand PSSS. To obtain Ag nanoprisms of different sizes during the microfluidic synthesis, the amount of added Ag seed particles was varied likewise.

It is well known in literature that the optical properties and especially the spectral position of the plasmon resonance wavelength depend strongly on the size of the noble metal nanoparticles. Thus, with an alteration of the edge length in a well-defined way, the spectral position of the main plasmon absorption band can be tuned exactly.

The high homogeneity of the microfluidically synthesized Ag nanoprisms of different sizes is well reflected by the size distribution spectra obtained from DCS analysis and by the UV-Vis extinction spectra of the colloidal solutions.

The stokes equivalent sedimentation diameter, which is the result of the DCS analysis, was continuously enhanced from 21 to 24, 27, 30, up to 39 nm in case of a silver nitrate concentration of 0.4 mM and under conditions of a decreasing initial seed nanoparticle density (Fig. 7.15). These volume equivalent sphere diameters correspond according to the SEM analysis to edge lengths of 35, 60, 80, 130 and 180 nm. Due to the variation of the particle expansion, the long-wavelength in-plane dipole mode of the optical spectrum was shifted from 528 to 832 nm (Fig. 7.16). Comparable findings were noted for a silver nitrate concentration of 1 mM. Here, the Stokes equivalent sedimentation diameter, which was derived from DCS analysis, was increased from 24, over 26, 29, 34, until 45 nm. In this case the corresponding SEM-determined edge length of the analyzed nanoprisms was respectively 40, 70, 90, 145, and 200 nm (Fig. 7.17). Due to the increasing size, the main optical absorption band was shifted from 564 to 858 nm (Fig. 7.18). These findings demonstrate the possibility for an exact tuning of the optical properties of the generated nanoparticles by a simple adjustment of the chemical conditions in the micro flow-through system. It is noticeable upon closer inspection of the DCS spectra, which represent the absolute number of nanoparticles per unit diameter that the height of the size distribution peak decreases in case of both examined $AgNO_3$ concentra-

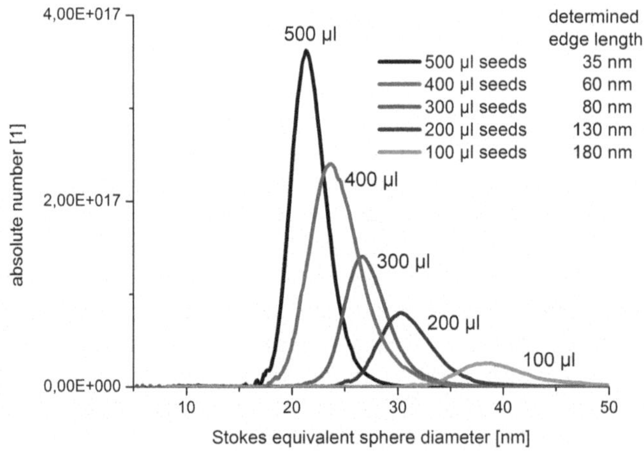

Fig. 7.15 DCS determined Stokes equivalent sphere radius for Ag nanoprisms, which grew under application of 0.4 mM silver nitrate

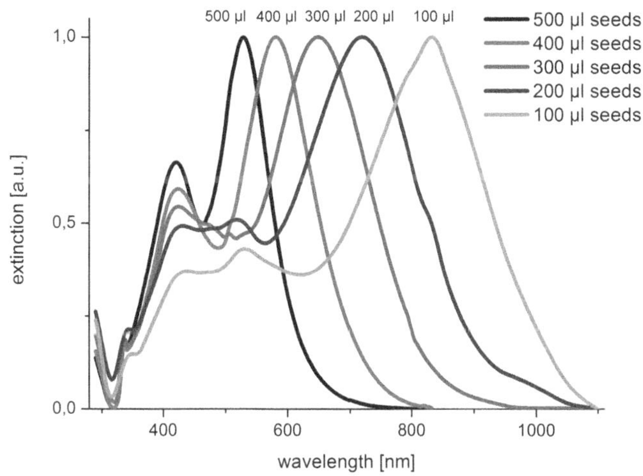

Fig. 7.16 Corresponding UV-Vis extinction spectra to those nanoprisms whose sedimentation sphere radii are shown in Fig. 7.15

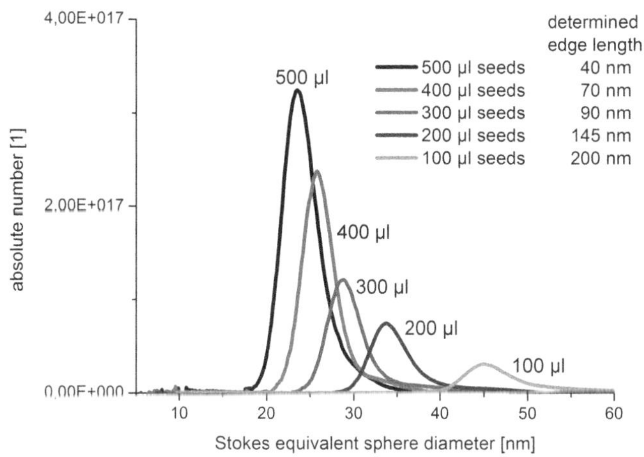

Fig. 7.17 Number based DCS data for silver nanoprisms which were synthesized using 1 mM silver nitrate solution

tions with decreasing seed nanoparticle concentration. Here, the decreasing amount of added seed nanoparticles and the associated increase of the lateral expansion of the nanoprisms are reflected. Furthermore it can be seen that the exponent of the dependence of the number of particles on the seed nanoparticle density is higher than one. This can be interpreted as an increasing probability of prism formation at an increased number of seed nanoparticles.

The same chemical conditions, which were set in the micro flow-through synthesis, were also applied in a conventional batch synthesis in order to compare the

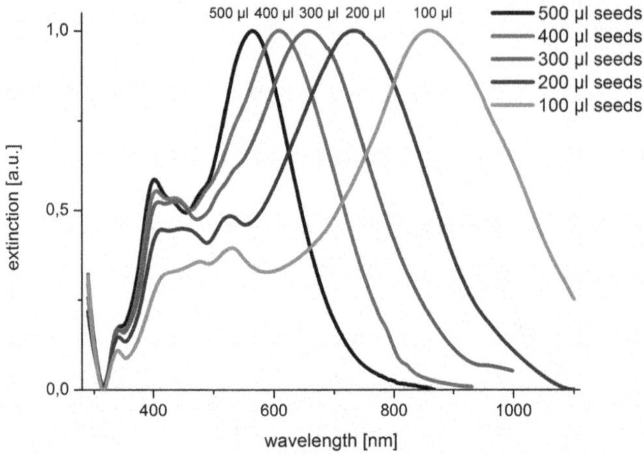

Fig. 7.18 Corresponding UV-Vis spectra to the different sized Ag nanoprisms, whose DCS data is given in Fig. 7.17

resulting product particles according yield of the desired shape and homogeneity of the size distribution. For selected samples the direct comparison of the SEM data was carried out. In Fig. 7.19a the silver nanoprisms obtained from the microfluidic synthesis and in Fig. 7.19b the respective counterparts from the conventional synthesis strategy are shown. On the basis of the SEM analysis it is clearly shown that the micro flow-through synthesis is enormously advantageous with respect to the homogeneity of the desired shape. Especially in case of large Ag nanoprisms with high aspect ratios a significant improvement of shape uniformity was observed. The presented silver nanoprisms obtained by a microfluidic synthesis method under application of the micro segmented flow technique are also withstanding any comparison against other methods presented in literature [21, 41, 53–55].

In summary, it can be stated that the preparation of triangular silver nanoprisms could be successfully implemented into a micro flow-through synthesis route. The underlying, time-critical batch method was transferred into the microfluidic system without any consideration of the initial time dependent addition of the metal salt. A clear improvement of the particle quality was found in case of silver seed nanoparticles as well as for silver nanoprisms of generated sizes. The size and thus the optical properties of the product particles can be easily tuned in a very precise way by altering the chemical conditions of the reactant solutions. It was found that for both, the formation of seed nanoparticles as well as for the subsequent growth step, best results for the particle growth are obtained under conditions of an effective segment-internal mixing. Thus, a five step synthesis route consisting of two separate flow-through processes was applied to generate homogeneous triangular silver nanoprisms:

Fig. 7.19 Comparison of SEM images of (**a**) flow-through generated Ag nanoprisms and (**b**) their counterparts from conventional batch synthesis

First flow-through process at a high total flow rate—generation of Ag seed nanoparticles:

Step 1: Micro fluid segments are formed in a PTFE tube by addition of sodium citrate solution and a mixture of sodium borohydride and poly(sodium styrene-sulphonate) solution into the organic carrier stream.

Step 2: The reaction is started by dosing silver nitrate solution into the preformed segments that contain the remaining reactants.

Second flow-through process at a moderate total flow rate—Ag nanoprism growth:

Step 3: The generation of micro fluid segments is carried out by dosing an aqueous reactant mixture containing a defined amount of silver seed nanoparticles and ascorbic acid into the carrier stream.

Step 4: The growth of the Ag nanoprisms is initiated by dosing silver nitrate into the segments, which are containing the soft reducing agent and silver seed nanoparticles.

Subsequent batch process:

Step 5: The particle growth is completed in presence of a sodium citrate solution after collection of both phases from the flow-through process in an appropriate vessel.

7.5 Micro Segmented Flow Synthesis of Composed Metal Nanoparticles

Apart from shape anisotropic silver nanoparticles, the micro fluidic synthesis strategy has also proven to be advantageous for the synthesis of composed binary metal nanoparticles [56]. In the following section, the strategy and the results of the transfer of the conventional synthesis route of core/multishell noble metal nanoparticles, which was previously released by the group of Luis Liz-Marzán into a micro flow-through system, which is based on the microfluid segment technique is described [57].

Starting from homogeneous, spherical gold nanoparticles with an average diameter of about 10 nm, which were utilized as crystallization seed nanoparticles, an alternating, onion-like metal shell structure was achieved by reduction of the respective metal salt on the surface of the presented seed nanoparticles by the help of an appropriate reducing agent. In this way, gold/silver core/shell and gold/silver/gold core/double shell nanoparticles were obtained. The preparation of the mentioned kind of composed nanoparticles was carried out in both, a conventional batch synthesis as well as in a micro continuous flow-through synthesis under respectively identical chemical conditions. All of the obtained product particles colloidal solutions were characterized by DCS sedimentation spectroscopy, UV-Vis spectrophotometry, as well as by SEM-, and TEM-imaging for the purpose of comparability of both methods.

Once it was confirmed that the batch protocols are working reproducibly and successfully, the reactant concentrations and volumes of these protocols were then recalculated proportional to the volume of a single segment in order to realize the segmented flow reaction. For the implementation of the reaction in micro fluid

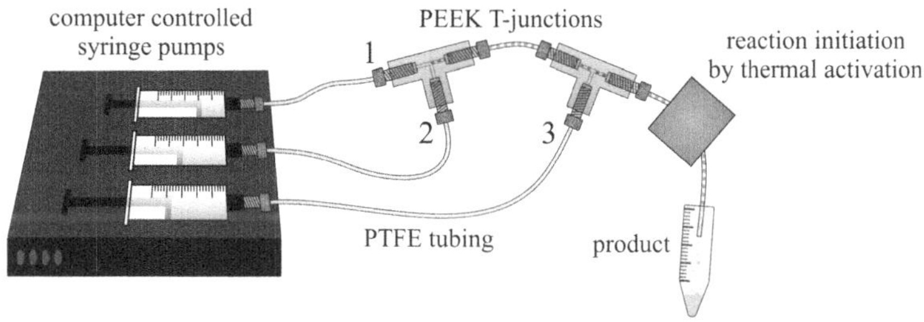

Fig. 7.20 Experimental setup for the microfluidic generation of Au/Ag core/shell and Au/Ag/Au core/double shell nanoparticles

segments, a strategy was developed, which allowed a pre-mixing of the reactants from the batch synthesis such that the reactant mixtures could be distributed to a minimum of syringes in order to keep the system complexity as low as possible. The experimental arrangement, which was utilized for the microfluidic synthesis of gold/silver core/shell nanoparticles, is shown in Fig. 7.20. Three syringes were fixed on individual axes of the syringe pump system and connected with PEEK fluid connectors to standard PTFE tubing. The reactor components and materials are exactly the same as in the above described process for the generation of triangular silver nanoprisms. The only difference is an additional resistive heating element, wherein 1.5 m of the PTFE reactor channel was placed meandering inside a solid tempering block that was set to 80 °C, since the synthesis of gold/silver core/shell nanoparticles requires thermal activation due to the chosen chemical conditions. However, the flow-through synthesis of gold/silver/gold core/double shell nanoparticles could be carried out at room temperature.

For both species of desired product nanoparticles, the pre-mixtures of the respective reactants, their distribution in the syringes as well as the micro fluidic process parameters are given in Table 7.3. Tetradecane (Acros Organics, Geel, Belgium, purity: 99.0 %) served in both cases as immiscible organic carrier medium.

For the microfluidic synthesis of gold/silver core/shell nanoparticles, the fluid segments were generated by dosing the reaction mixture consisting of gold seed nanoparticles, dissolved CTAB (Merck KGaA, Darmstadt, Germany, purity: 99.9 %), silver nitrate (Merck KGaA, Darmstadt, Germany, purity: 99.0 %), and ascorbic acid (Merck KGaA, Darmstadt, Germany, purity: 99.7 %) into the presented carrier stream. At a second injection unit, sodium hydroxide (Merck KGaA, Darmstadt, Germany, purity: 99.0 %) was dosed into the preformed segments, since its presence promotes the reduction reaction of silver ions on the gold nanoparticles surface.

The used gold seed nanoparticles were prepared by heating a mixture of 1 mL 0.5 mM tetrachloroauric(III) acid trihydrate (Carl Roth GmbH, Karlsruhe, Germany, purity: 99.5 %) and 1 mL of a 1.7 mM sodium citrate (Merck KGaA, Darmstadt, Germany, purity: 99 %) to boiling. The thus obtained gold nanoparticles were the basis of all experiments for both applied methods in order to guarantee identical starting conditions.

To cover the obtained gold/silver core/shell nanoparticles with a further shell of gold, the same setup as shown above could be utilized. For this reaction no thermal activation was necessary, very satisfactory results could be obtained at room temperature. The carrier medium was again presented first and the generation of segments took place by dosing a mixture of gold/silver core/shell nanoparticles and ascorbic acid into the stream of tetradecane. The reaction was initiated by an addition of tetrachloroauric acid into the already existing segments. After a simple phase separation, both micro flow-through generated types of product particles were directly analyzed or further used without any additional cleaning step.

In Fig. 7.21 an optical image of the colloidal solutions of the microfluidically generated nanoparticle species are shown. The dispersion of pure Au nanoparticles in water, which was prepared by the sodium citrate reduction method, was reddish-pink in color. This corresponds to a spectral position of the plasmon resonance wavelength at about 520 nm in the optical absorption spectra. The optical properties of the colloidal solution of gold change towards the characteristics of colloidal silver, once the gold nanoparticles are coated with a silver shell after the second process step. Here, the spectral characteristics of nanosized silver dominate the absorption spec-

Table 7.3 Distribution of the used chemicals among the syringes, applied concentrations and flow rates for the synthesis of Au/Ag core/shell and Au/Ag core/double shell nanoparticles

Desired particle structure	Syringe	Chemicals	Volume/Concentration	Flow rate
Au/Ag	1	Tetradecane	Full /-	300 μL/min
core/ shell	2	Au NP	100 μL / -	150 μL/min
		CTAB	1000 μL / 50 mM	
		$AgNO_3$	25 μ L / 10 mM	
		$C_6H_8O_6$	50 μ L / 100 mM	
	3	NaOH	full / 11 mM	150μL/min
Au/Ag/Au	1	Tetradecane	Full /-	300 μL/min
core/double shell	2	Au/Ag NP	500 μL / -	150 μL/min
		$C_6H_8O_6$	25 μ L / 10 mM	
	3	$HAuCl_4$	Full / 1 mM	150 μ L/min

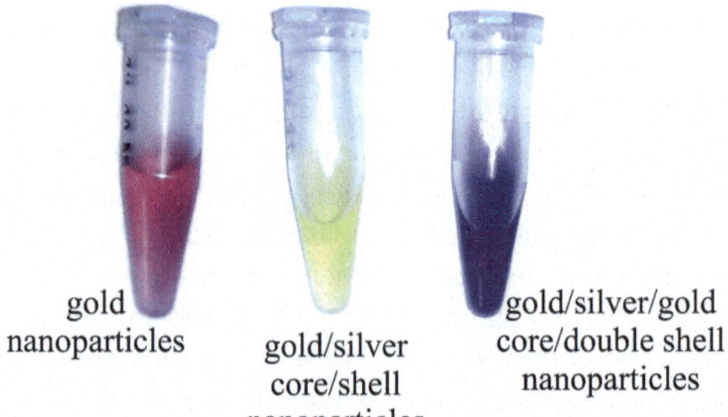

Fig. 7.21 Optical image of the microfluidically generated colloidal solutions of (*red*) gold nanoparticles, (*yellow*) gold/silver core/shell nanoparticles, and (*blue*) gold/silver/gold core/double shell nanoparticles

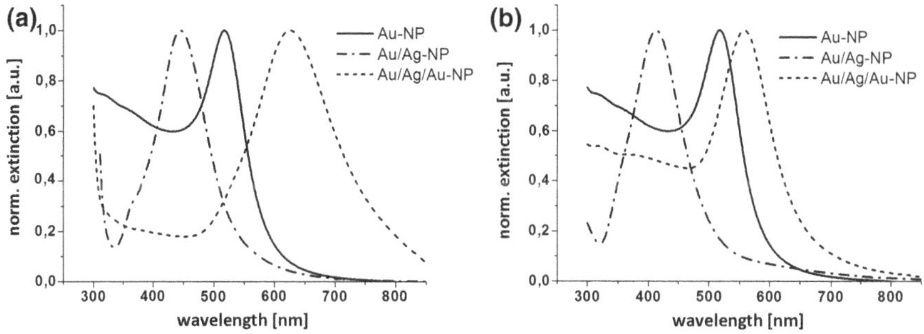

Fig. 7.22 **a** UV-Vis extinction spectra of conventionally synthesized Au, Au/Ag core/shell and Au/Ag/Au core/double shell nanoparticles. **b** Extinction spectra of microfluidically obtained Au/Ag and Au/Ag/Au core/double shell particles. The resonance wavelengths are clearly blue shifted and the FWHM of the core/double shell particles is visibly smaller than in case of the batch generated nanoparticles

tra, which are reflected in a distinct plasmon resonance peak at about 400 nm. The following deposition of a further shell of gold changes the color of the colloidal solution into a deep blue, which indicates a complete change in the resonance conditions. It was reported by Liz-Marzán and colleagues how the spectral position of the plasmon resonance peak can be used for a qualitative characterization of the shell thickness in case of multishell nanoparticles [57]. In Fig. 7.22a direct comparison of the optical extinction spectra of each particle species obtained in batch processes (a) and by the micro flow-through syntheses (b) is shown. By direct comparison it is noticeable that the wavelength of the absorption peaks of the gold/silver core/shell and gold/silver/gold core/double shell nanoparticles obtained from the micro segmented flow syntheses are blue shifted in contrast to the spectral characteristics of the nanoparticles prepared in a conventional batch. This clearly points to the formation of thinner metal shells around the core particles during the synthesis. The narrower bandwidths of the microfluidically processed nanoparticles suggest more homogeneous particle shapes and sizes. For both synthesis methods, the different plasmonic absorption peaks and their subsequent spectral blue and red shifts confirm the material of the outer shell, which is then working like an optical fingerprint for colloidal solutions of (multiply) coated nanoparticles.

To gain more information on the physical properties of the individual nanoparticle samples from the different process steps, the respective nanoparticle solutions of both methods were subjected to the differential centrifugal sedimentation system. Thus, knowledge of the particle size distributions and connected thereto an indication of the quality of the particle ensembles could be achieved. In Fig. 7.23 the weight based DCS data is shown as a direct comparison of the results of the batch syntheses (a) and the microfluidic syntheses (b). In both cases the particle diameter is plotted against the total weight per diameter unit of the detected metal in an absolute form.

From these size distribution spectra can be seen that the initial gold seed nanoparticles have an average particle diameter of 8.5 nm with a FWHM of 2.6 nm. The

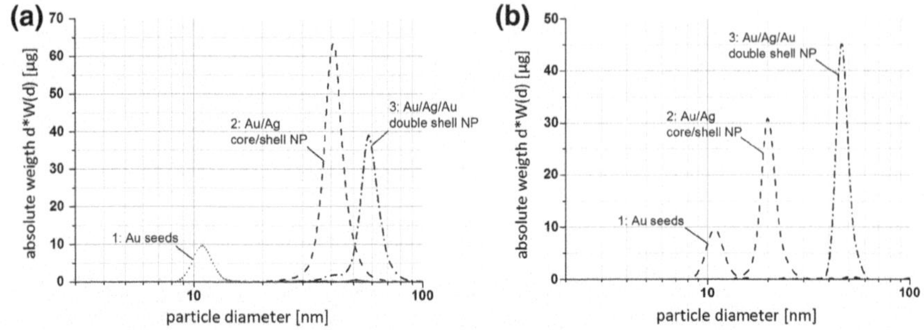

Fig. 7.23 **a** DCS size distribution spectra of the three product particle species synthesized in a batch process. **b** The regarding nanoparticle species obtained in a micro flow-through process

gold seed particles were then coated with the first shell of silver. The average diameter of the batch obtained particles was 41 nm with a size distribution half width of 8.7 nm, which corresponds to a silver shell thickness of about 15 nm. The respective microfluidically obtained counterparts had an average particle size of 20 nm with a FWHM of 3.8 nm, which is about 5 nm smaller than in case of the batch synthesized gold/silver core/shell particles. A silver shell thickness of approximately 4 nm can be derived from the DCS data for the flow-through obtained core/shell nanoparticles. The Au/Ag core/shell particles were subsequently coated with a second shell of gold. The core/double shell particles obtained from the conventional synthesis hat an average diameter of 58 nm with a FWHM of 11.9 nm corresponding to an average shell thickness of 8.5 nm. The size distribution peak of the core/double shell particles generated in the micro flow-through system had its maximum at 46.4 nm with a half width of 7.4 nm, which is 4.5 nm smaller compared to the double shell particles of the batch synthesis. From this data can be seen clearly that the micro segmented flow-through syntheses lead to smaller size distributions. With the first silver shell a significant decrease in the shell thickness of a value of 10 nm could

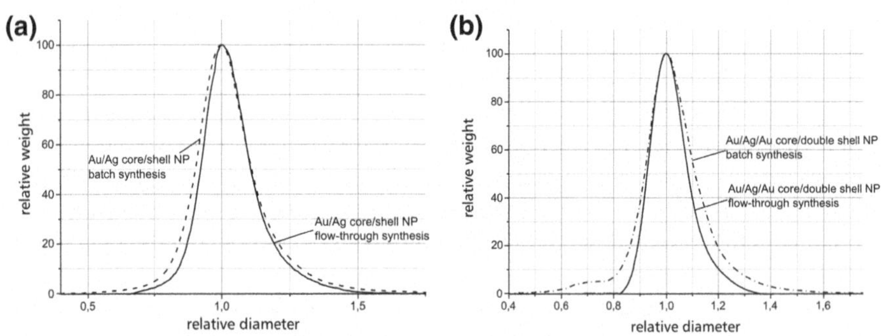

Fig. 7.24 **a** Comparison of the size distribution width of both types of Au/Ag core/shell nanoparticles. **b** Same comparison for both kinds of core/double shell nanoparticles

be achieved for the core/shell particles produced in the microfluidic system. At an equal amount of silver nitrate in both synthesis methods it means that more gold seed nanoparticles are covered and a more regular growth of the core/shell nanoparticles occurred. The half width of the size distribution also dropped markedly from 8.7 to 3.8 nm, which hints for extremely homogeneous product particles. Also in case of gold/silver/gold core/double shell nanoparticles smaller and more homogeneous particles were obtained by the help of the micro flow-through method. In Fig. 7.24a, b a direct comparison of the FWHM of the batch and the flow-through obtained results are shown. On the left side, the normalized weight based size distribution data is plotted against a relative diameter to compare the data obtained for both populations of gold/silver core/shell nanoparticles. On the right side, the comparison of the gold/silver/gold particles is shown respectively. These graphs clearly demonstrate at one glance that the flow-through syntheses lead in all investigated cases to a smaller FWHM and thus to more homogeneous nanoparticles with a more regular and thinner deposited shell. The SEM pictures in Fig. 7.25 of the particles of the three different process steps confirm these findings and show a high uniformity and regular growth in case of the flow- through obtained nanoparticles. A direct comparison is given on the left side: the batch processed core/shell and core/double shell nanoparticles are shown, their flow-through obtained counterparts are to be found on the right side. That there is actually a core/shell or a core/double shell structure was proven by high resolution TEM imaging (see Fig. 7.26). A EDX line scan analysis, which is exemplary shown in Fig. 7.27 for Au/Ag core/shell nanoparticles, confirms the presence of two different metals within the particles.

Table 7.4 gives an overview of the added (theoretical) and measured masses of metal in 0.1 mL nanoparticle solution, of the weight based and number based size distribution data (all data derived from DCS measurements), as well as of the corresponding polydispersity index (PDI) of the three different product particles obtained from the micro flow-through synthesis steps. The PDI specifies the ratio between the weight-based and the number-based DCS data. While the weight-based average particle diameter is more sensitive to particles of higher masses, the number-based particle diameter reflects the absolute particle number per unit diameter. For all flow-through obtained nanoparticle dispersions, the PDI is very close to one, which is another indication for very homogeneous particle populations and narrow size distributions.

The investigations have shown that the micro continuous flow-through synthesis is also in case of complexly structured noble metal nanoparticles very advantageous regarding an increase of homogeneity and particle quality. This can again be attributed to the homogeneous mixing conditions which are provided by the micro segmented flow-through synthesis strategy due to the effective segment internal convection, which is described in detail in Sect. 7.4. The avoidance of concentration gradients due to the effective intermixing of the reactants supports a regular and thinner shell growth, because the deposition of the shell metal is initiated at more seed nanoparticles in a smaller time interval for the heterogeneous nucleation process, which happens at the surface of the inserted seed nanoparticles. Whereas in the batch synthesis some seeds grow very strongly and other seed particles remain uncovered or

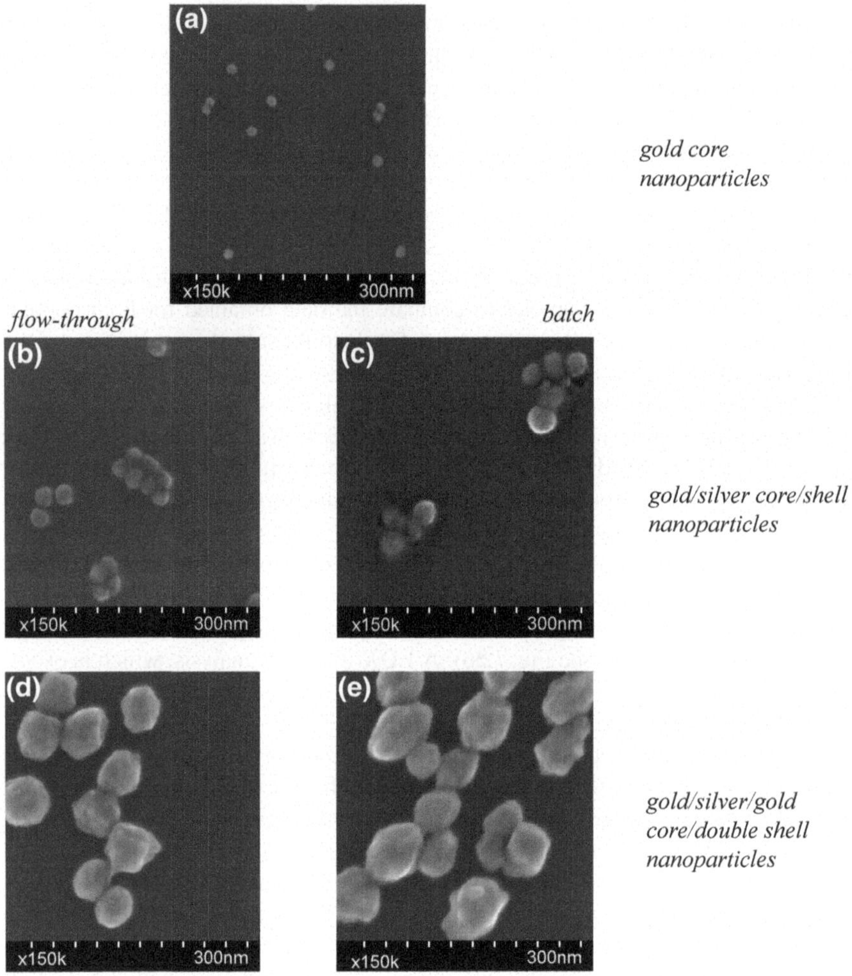

Fig. 7.25 SEM images of (**a**) gold core nanoparticles, (**b**) flow-through processed Au/Ag core/shell nanoparticles, (**c**) batch generated Au/Ag core/shell nanoparticles, (**d**) Au/Ag/Au core/double shell nanoparticles from micro flow-through synthesis, and (**e**) their batch processed counterparts

nearly uncovered in solution, because of the broad interval of the nucleation phase in combination with a rather long time superposition of the (heterogeneous) nucleation and growth intervals, there is in case of the flow-through synthesis a simultaneous start of the deposition process at a significant higher number of seed particles. It is known, that the deposition of metal happens preferably there, where atoms of the same species already have been deposited. This can be considered as the reason for the more inhomogeneous shell growth in case of the conventional syntheses and a regular formation of thinner shells for the microfluidic strategy.

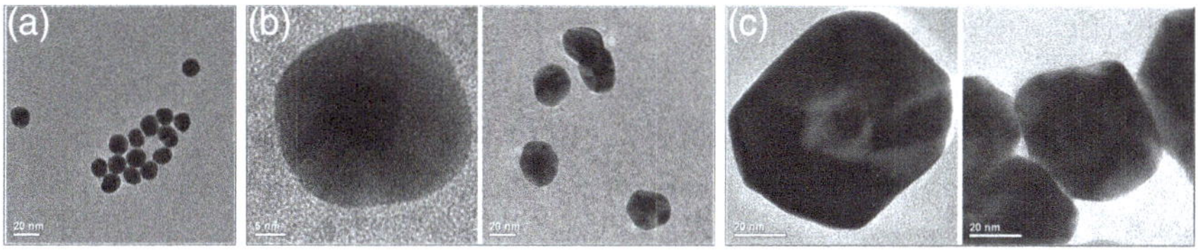

Fig. 7.26 HR-TEM analysis of (**a**) gold core particles and microfluidically obtained gold/silver core/shell particles (**b**) and Au/Ag/Au multishell nanoparticles (**c**)

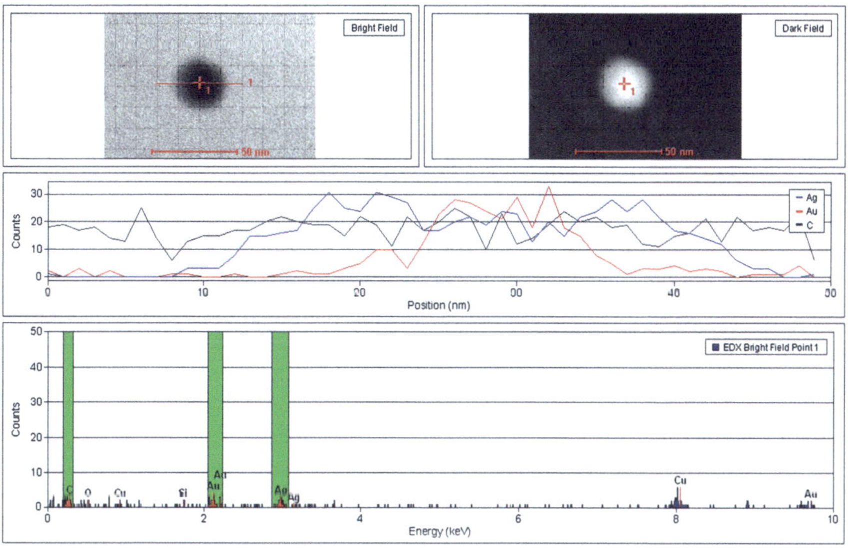

Fig. 7.27 EDX line-scan on Au/Ag core shell particles

Table 7.4 Overview about measured* and theoretical data on the flow-through processed particle species (* determined by DCS analysis)

Particle species	d_w[nm]*	d_{num}[nm]*	PDI (d_w/d_{num})	corresponding mass of metal [μg]* (V = 0.1 ml)	Calculated mass of metal [μg] (V = 0.1 ml)
Au core	11.1 ± 1.3	11.0 ± 1.2	1.01	9.69	9.8
Au/Ag core/shell	20.0 ± 2.0	19.7 ± 1.9	1.02	31.0	36.0
Au/Ag/Ag core/double shell	46.4 ± 3.7	45.6 ± 3.6	1.02	45.2	46.2

7.6 Automated Synthesis Experiments in Large Parameter Spaces for a Variation of the Plasmonic Properties of Nanoparticles by Varied Reactant Composition in Fluid Segment Sequences

This section focuses first the fundamental basics of the combinatorial synthesis of different types of plasmonic nanoparticles. After an overview of the versatile screening possibilities of different process parameters and a detailed explanation of the experimental method, two examples for combinatorial syntheses in microfluidic segment sequences with selective modification of the reactants compositions for each individual segment sequence will be presented. One example will show the precise tunability of the spectral position of the main dipole resonance wavelength of triangular silver nanoprisms in a spectral range which covers about 400 nm and the other example will inform on the adjustability of the silver shell thickness during the combinatorial synthesis of gold/silver core/shell nanoparticles in which discrete silver shell thicknesses between 1.4 and 20 nm were deposited around a gold core particle in one single microfluidic run.

To take full advantage of the possibilities and the capability of the method, the options for parameter variation shall be clarified first. It is known, that different process parameters during the synthesis have great impact on the physical and chemical properties of the obtained products. Also in case of noble metal nanoparticle syntheses, the process parameters are of crucial importance [51].

A promising method of implementing a combinatorial synthesis in a micro segmented flow-through system is for example the variation of the flow rate ratios of the reactant solutions during the segment formation. This can be achieved by a syringe pump system with computational control and independently adjustable axes. By the help of a flow rate program, which is usually specially adapted to the needs of the investigation, the composition of the segments of each combinatorial variation step is set. For purposes of statistical analysis and the proof of reproducibility, it is recommended to generate a number of segments per combinatorial step that is in the order of some 10. These n segments of the individual combinatorial step will all have the same reactant composition. If the conditions switch towards the next combinatorial step, again n segments are generated with constant chemical compositions but which differ, however, from the compositions of the previous segment sequence. If A is the number of gradations of the initial reactant concentration, D is the number of the reactants, whose concentrations are varied, and N is the total amount of the generated segments, then N is given by:

$$N = n \cdot A^D. \tag{7.1}$$

The parameter D also equals the dimension of the concentration space. From Eq. (7.1) can be seen, that either highly-resolved combinatorial screenings in a low-dimensional concentration space or lower-resolved synthesis screenings in a high-dimensional concentration space can be realized. In Table 7.5 the dependence between resolution (number of gradations and thus the number of combinatorial steps) and the total number of generated segments is shown.

Table 7.5 shows the need for a reasonable balance between the resolution of the parameter space screening and the dimension of the parameter space. Experimentally easy to handle are total numbers of generated segments up to about 15000. In micro toxicological studies, two and three dimensional screenings were carried out utilizing micro fluid segment sequences with a total number of about 1000–5500 segments [58, 59].

Beside flow rate changes, a further influencing factor, which is of interest for screening experiments, is the residence time. In fields of nanoparticle synthesis, the residence time between the initiation of nucleation and an addition of any kind of effector or a further metering of a metal salt solution has a crucial influence on the particle quality as well as on the physical properties of the product particles. Since the residence time is reciprocal to the flow rate, the simplest way to screen the residence time is a variation of the total flow rate. If the flow rate ratios remain constant, the chemical conditions inside the segments will also remain constant. Here, conclusions can be drawn on the impact of the segment internal convection on the physical properties of the obtained product particles.

Another interesting parameter, which can be screened, is the reaction temperature. The reaction, which leads to the formation of small metal nuclei as well as the subsequent nanoparticle growth are strongly temperature-dependent processes. A variation of the temperature will have an effect on the shape, the crystallinity, the size, and on the state of aggregation of the nanoparticles [60]. The interaction between the nanoparticle surfaces and ligands is also strongly affected by the temperature. Fast temperature changes could for example initiate or terminate nucleation processes or influence the particle growth in a certain direction. To alter the temperature, the easiest way is the use of a heating or cooling bath or heat exchangers, which are adapted to microfluidic systems. A possible alternative for the heating of polar liquids as for example the aqueous reactant solutions, which form the segments, is the application of microwaves in the electromagnetic near field using a wave conductor. The high field constant of water supports a focusing of microwave power on the segment and leads to a direct thermal activation of the segment liquid. The high permittivity of water causes a focusing of the microwave power on the segment and leads to a direct thermal activation of the aqueous segment. An interesting challenge would be a combination of concentration variations and the screening of different temperatures. From this kind of experiments useful knowledge about the reaction kinetics, which determine the nanoparticle formation could be gained.

In the experiments described below different strategies for combinatorial concentration variations were carried out at constant temperature and constant total flow rate. The next section will describe the characteristics of the utilized programmed two dimensional flow rate courses.

The variation of the educts concentration by changing the respective flow rates can be performed in a linear or in a non-linear way. Non-linear flow rate changes are of interest in case of screening the reaction conditions in connection with a desired variation of the concentration ratios in several orders of magnitude. A linear-

Table 7.5 At a constant number of segments per combinatorial step *n*, an increasing dimension necessitates a decrease in resolution if the total amount of generated segments is supposed to stay in a reasonable order of magnitude and the expenditure experimentally manageable

Concentration variation (%)	Gradation *A*	Number of segments per step *n*	Dimension *D*					
			2	3	5	2	3	5
			Number of combinatorial steps A^D			Total number of segments *N*		
5	21	50	441	9261	4084101	22050	463050	2.042 E8
10	11	50	121	1331	161051	6050	66550	8052550
20	6	50	36	216	7776	1800	10800	388800
50	3	50	9	27	243	450	1350	12150

linear flow rate run is shown in Fig. 7.28. During the combinatorial experiment, the complete two-dimensional parameter space is covered in steps of 20 %. As it can be seen from this figure, the functioning of this method is based on two times two gradually changing flow rates, while the total flow rate is kept constant at any time of the experiment. The first pair of flow rates is slowly shifting and is in turn overlain by two fast changing flow rates. Due to the always constant total flow rate, constant fluidic conditions are ensured, which is enormously important since it is known that a change in the total flow rate has great impact on the mixing conditions and thus on the particle properties. From the chart in Fig. 7.28, the individual flow rates of each stream of reactant solution, which determine the chemical conditions of each combinatorial step, can be derived directly. The same description applies for the flow rate course shown in Fig. 7.29. The difference here is a linear-logarithmic variation of the reactant concentration ratios. For the two reactant streams, which are dosed with fast changing flow rates, the flow rate variation is programmed logarithmic to the base 1.8. In both cases, 36 combinatorial steps each with different chemical conditions within the according segment sequences are generated.

The combinatorial synthesis screening demands a microfluidic setup, which is specially adapted to the requirements of the method. The experimental arrangement utilized for both below presented experiments is schematically shown in Fig. 7.30. The design of the reactor was carried out using the same materials and reactor material as described above. Two PEEK cross-junctions served as injection units for the two pairs of reciprocally in flow rate changing reactant streams. After a residence loop of a length of 1270 mm, the segments passed the flow-through cells for optical analysis. A 4-channel photometer unit served for the monitoring of the segment lengths and the distances between the segments. An UV-Vis (AvaLight-D(H)-S, Avantes Inc., Broomfield, USA/CO) source and an optical spectrophotometer (AvaSpec-1024, Avantes Inc., Broomfield, USA/CO) were coupled by an optical fiber system to the fluid capillary in order to get highly resolved extinction spectra.

By the help of this technology, the optical properties of the colloidal solution of each single segment generated during the whole combinatorial run were recorded and saved.

The first presented experiment for the variation of reactant ratios in a microfluidic system under application of the segmented flow technique will be the combinatorial synthesis of triangular silver nanoprisms. It was already shown above, that the synthesis of silver nanoprisms in micro fluid segments on the basis of homogeneous, optimized seed nanoparticles is advantageous with regard to a homogeneous size distribution and a high yield of the desired shape [51, 61]. The possibility of a tuning of the nanoprisms size by shifting the ratio between the amount of seed nanoparticles and silver nitrate was clearly shown in the work of Aherne et al. [48] in their studies on the batch reaction. This principle was also applied to the combinatorial micro flow-through synthesis. For these experiments, a linear-linear variation of the reactant concentrations was chosen (Fig. 7.28). During the combinatorial experiment, the seed nanoparticle density was varied through the use of two input flows. The used seed particles were obtained in the same way as it was described in Sect. 7.4. The first input flow was a mixture of water, ascorbic acid and 500 μl silver seed nanopar-

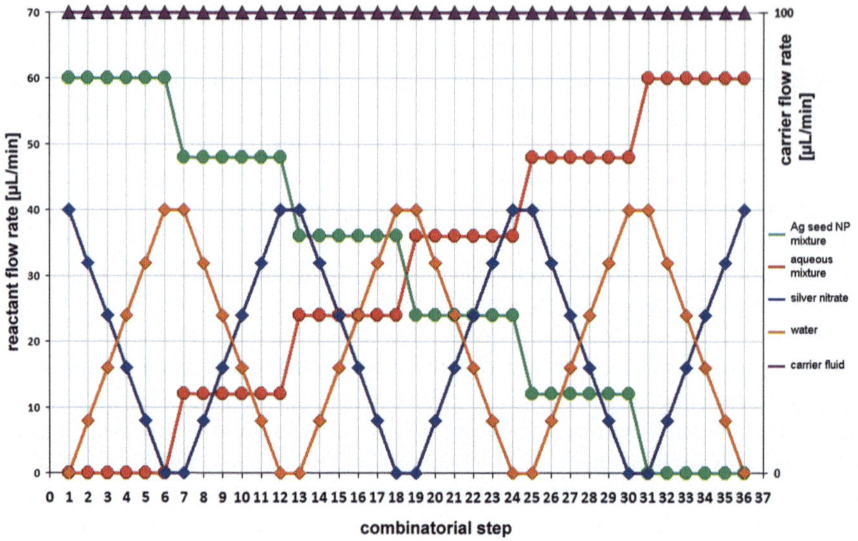

Fig. 7.28 Programmed flow rate run for a combinatorial screening of the tunability of the main plasmon resonance wavelength of triangular silver nanoprisms. The spanned parameter space is covered completely in steps of 20 %

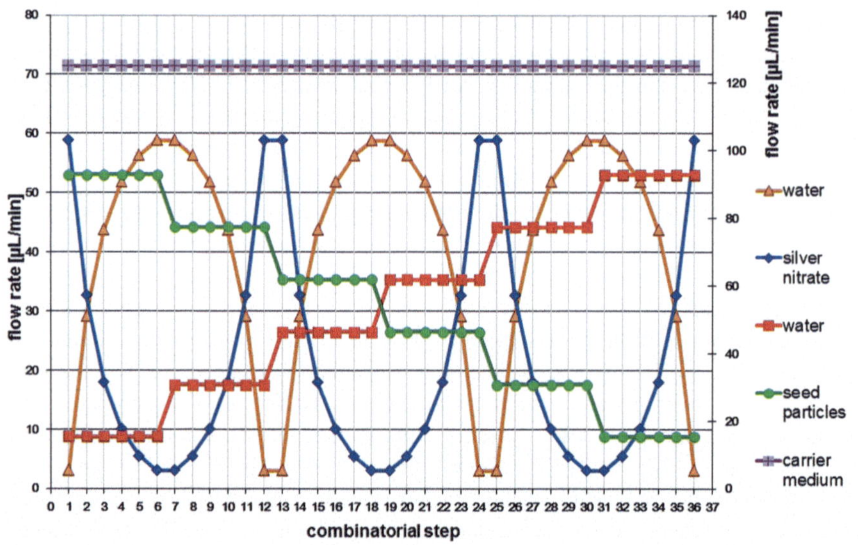

Fig. 7.29 Flow rate course for a combinatorial synthesis screening of silver shell deposition around gold core particles with a non-linear variation of the flow rate of silver nitrate and a linear variation of the additions of the colloidal solution of gold seeds

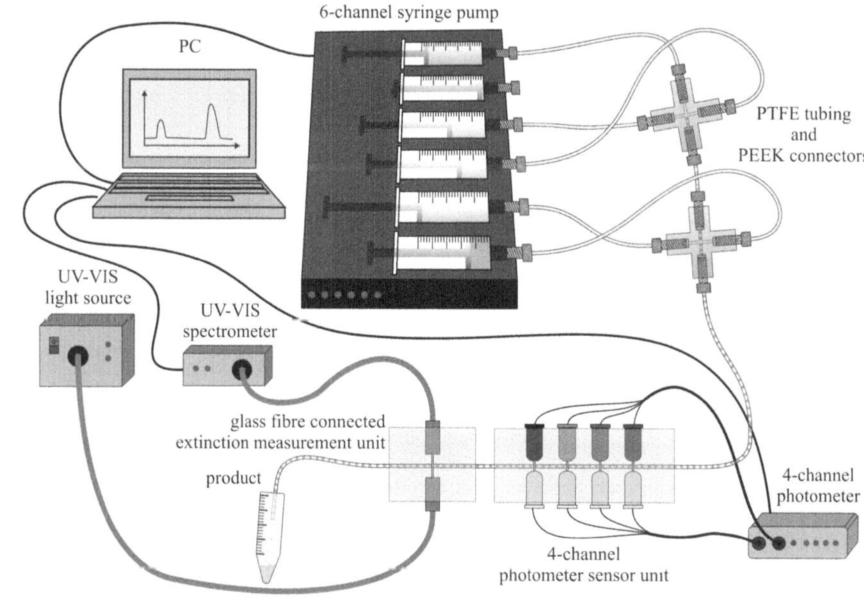

Fig. 7.30 Microfluidical arrangement used for combinatorial concentration variation screenings

ticles. The second stream only contained water and ascorbic acid. The effective seed nanoparticle density can now be altered by simply changing the flow rates of both solutions. The adjustment of the Ag seed nanoparticle density took place at the first cross junction shown in Fig. 7.30. The principle of two opposing inlets is necessary for the realization of the combinatorial variation of the reactant concentration. In the combinatorial flow rate run, which is shown in Fig. 7.28, the flow rate course of the addition of the seed nanoparticle mixture and the respective aqueous ascorbic acid solution, which was used for the purpose of a controlled dilution while maintaining the chemical and microfluidic conditions, is represented by the two slowly shifting flow rates (red and green line). The same strategy is applied at the second cross junction. In the same way as described above, the precise dilution of 1 mM silver nitrate solution to a desired degree is achieved by using a second inlet stream of water. The corresponding flow rate variation can be found in Fig. 7.28 and is represented by the two fast alternating flow rates (blue and orange line). Since shape anisotropic nanoparticle growth processes are determined by a slow deposition of metal atoms, moderate flow rates can be applied. A total flow rate of 200 μL/min supports both: good mixing conditions due to the segment internal convection and a sufficient long residence time of the segments inside the reactor channel for the growth of the silver nanoprisms. Flow rates in the dimension of a few hundred μL/min also facilitate a thorough spectral analysis of the colloidal solution in each micro fluid segment. With a flow rate of 200 μL/min, a volume element of the stream will be transported with a velocity of 17 mm/s. The growth process of the silver nanoprisms is initiated

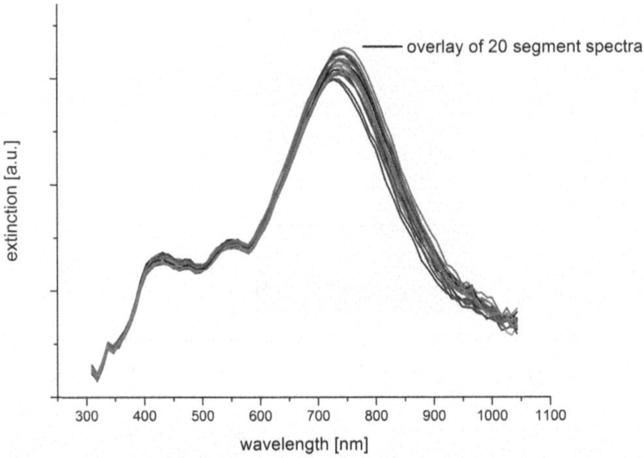

Fig. 7.31 Subsequently recorded segment spectra from combinatorial step 13

by dosing silver nitrate solution in its set dilution into the preformed segments. The segments that contain the complete growth solution will reach the optical detection unit after a time interval of 75 s.

As it can be derived from Fig. 7.28, silver nanoprisms with different physical properties will be obtained from each combinatorial step due to the changing flow rates and thus concentration ratios. Within one combinatorial step a constant dosing of reactants and the formation of homogeneous segments is necessary to generate a segment sequence, which consists of compartments of the same colloidal solution. Thus, the physical properties of the nanoprism solution will not differ within the same step but noticeably if two or more combinatorial steps are compared. In Fig. 7.31 the optical absorption spectra of the colloidal nanoprism solution from 20 sequential segments, which were generated under conditions of combinatorial step 13 are shown.

It is known that the spectral position of the main dipole resonance peak as well as the shape of the optical spectrum is very sensitive towards small changes in the environment. From Fig. 7.31 can be seen that the optical spectra of the single segments are sufficiently congruent. In that way, the extinction spectrum of each generated segment from each combinatorial step was taken. For comparing the data from each step, the average spectrum of a sequence of segments was taken and is used as representative for the regarding combinatorial step. Thus, the described method of the in situ analysis of the optical properties is highly statistical and reproducible.

In the described experiment, the duration of one combinatorial step was set to 45 s. This results in a total time of 28.5 min for the whole combinatorial experiment. The total number of generated segments amounts to 7473, resulting in 206 segments per combinatorial step. A great additional advantage of this method of parameter screening is the very low consumption of chemicals. In total 554 μL of silver nitrate and 831 μL of the silver seed nanoparticle containing mixture was required. The individual combinatorial steps can be considered as individual experiments and the single segments are equivalent with replicate experiments. Thus, with the above mentioned low amount of chemicals exactly 36 different experiments with each 206 repetitions could be carried out in one single combinatorial micro flow-through run. On closer inspection of Fig. 7.28 it is noticeable that during the flow rate shift of the reactant streams conditions emerge, in which the flow rate of at least one crucial reactant stream was set to zero. This applies exactly to 11 of the set flow rate ratios. Thus, from 25 of 36 experiments silver nanoprisms with respective different physical properties emerged. The remaining 11 combinatorial steps can be considered as reference experiments. This concerns the combinatorial steps C6, C7, C18, C19, and C30–36. After completion of the in situ analysis, some chosen fluidic conditions were manually adjusted and operated continuously to receive a sufficient large amount of sample volume for further analysis. The obtained product particles were additionally characterized by SEM, offline UV-Vis, and DCS analysis.

As described above, the silver seed nanoparticle density is adjusted by both slowly shifting flow rates by dilution. The slowly shifting flow rates are overlain by two fast alternating flow rates. Thus, according to the resolution of the screening procedure, each concentration of silver nitrate was dosed to each adjusted seed nanoparticle density. This allows versatile possibilities for comparison of the different conditions. The highly regular dependence of the silver nanoprisms size from the different chemical conditions is reflected by the UV-Vis spectra. In Fig. 7.32a–e the extinction spectra of the colloidal solutions of all relevant combinatorial steps are presented. As already explained above, the spectral position of the main dipole resonance wavelength is sensitively dependent on the edge length of the nanoprisms. The different diagrams of Fig. 7.32 show the optical response of the colloidal solutions in case of a per diagram constant amount of silver seed particles but a changing silver nitrate concentration. Based on the highest possible seed nanoparticle density under the chosen chemical conditions is the percentage of the seed particle density 100 % for Fig. 7.32a, 80 % in case of (b), 60 % for (c), 40 % for (d), and 20 % in case of Fig. 7.32e. From the single diagrams can be seen that the spectral position of the main plasmon resonance mode is blue shifted with decreasing silver nitrate concentration. This corresponds to a decreasing edge length of the silver prisms. By the comparison of the different diagrams (a)–(e), a general red shift of the main dipole mode can be recognized, which occurs due to the increasing particle size due to the decreasing seed particle density.

These diagrams demonstrate the possibility of a precise fine tuning of the spectral position of the main dipole resonance wavelength of microfluidically obtained silver nanoprisms only by a simple flow rate variation of the reactant streams in a spectral range between 550 and 900 nm. The strong dependence of the particle size from the preset reactant ratios was confirmed by DCS analysis. Exemplary, the DCS data for 5 different seed nanoparticle densities at a constant presence of 1 mM silver nitrate is displayed in Fig. 7.33. The DCS size distribution spectra indicate the Stokes equivalent sedimentation diameter of the flat, triangular nanoplates and not the true edge length. Since the particles sediment in the centrifugal force field in a sphere equivalent behavior, the actual edge length of the nanoprisms can only be derived from SEM-imaging. However, the DCS data are of great importance for an evaluation of the particle quality. An absence of secondary peaks in combination with a narrow FWHM hints at highly homogeneous product particles. The high yield of the desired particle shape is confirmed by the characteristic course of the optical extinction spectra and also by SEM analysis. Three examples of different sized silver nanoprisms that were obtained from different combinatorial steps are shown in Fig. 7.34.

In the following, a combinatorial experiment with automated variation of two reactant concentrations during the flow-through synthesis of gold/silver core/shell nanoparticles is described as a second example for a two-dimensional parameter variation [62]. As described in Sect. 7.5, the flow-through synthesis of gold/silver core/shell nanoparticles in micro fluid segments is advantageous compared to the batch method [56]. In these experiments, it was shown that the deposition of a second material is possible. Now, the question rises whether the thickness of the silver shell around the gold core particles can be tuned under the condition of a consistently high homogeneity of the product nanoparticles. This challenge was again met by a combinatorial experiment using the experimental setup as shown in Fig. 7.30. The only extension is a resistive heating element downstream the last injection units since the initiation of the silver shell growth requires thermal activation under the chosen chemical conditions and best results were achieved if the synthesis was carried out at 80 °C. To generate the fluid segments, an aqueous mixture of 50 mM CTAB, 11 mM NaOH, 100 mM ascorbic acid and gold seed nanoparticles at one side of the upper cross-junction and a mixture of 50 mM CTAB, 11 mM NaOH, 100 mM ascorbic acid and water at the opposite side of this cross-junction were dosed into the presented

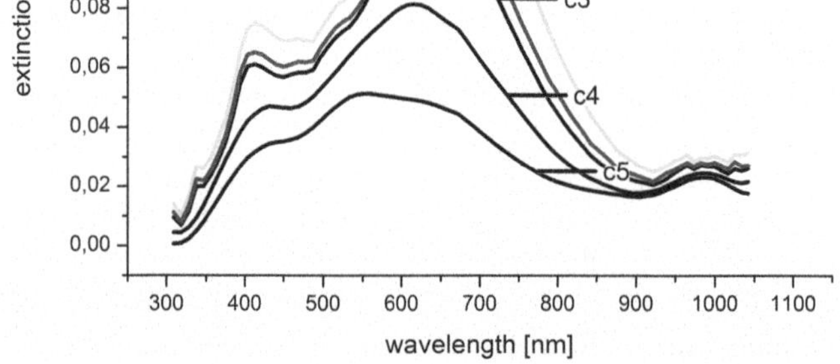

Fig. 7.32 **a** Spectral shift under conditions of maximum seed particle density (100 %) but decreasing silver nitrate concentration. **b** Optical response under conditions of 80 % of initial seed particle density and increasing silver nitrate concentration. **c** Optical response to 60 % of initial seed density and decreasing $c(AgNO_3)$ **d** Optical response to 40 % of initial seed density and increasing $c(AgNO_3)$ **e** The maximum red shift of the main plasmon resonance occurred for all $c(AgNO_3)$ under conditions of 20 % of the initial seed particle concentration

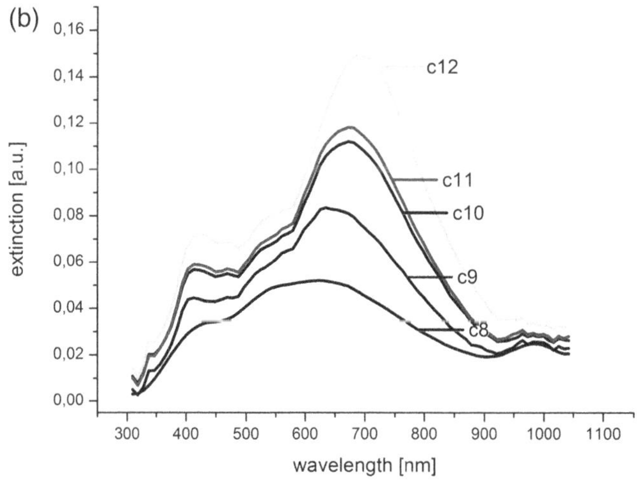

Fig. 7.32 (continued)

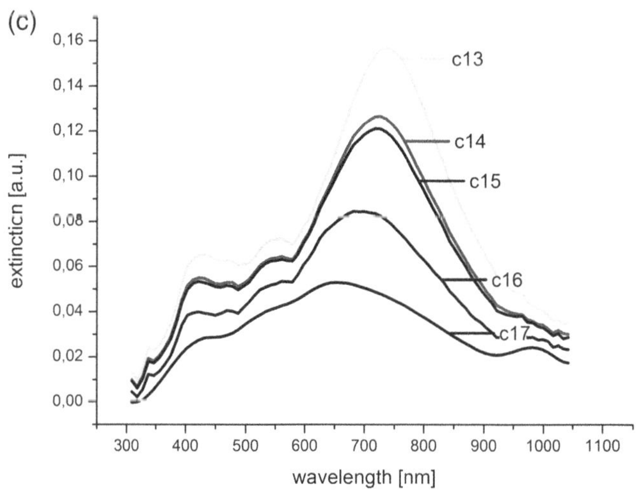

Fig. 7.32 (continued)

carrier stream. A 2.5 mM silver nitrate solution and respectively water were dosed at both opposite inlets of the second cross-junction into the preformed segments. The fast segment internal convection leads to a fast mixing of these reactant streams and just as in the above described example the effective reactant concentration can be precisely adjusted by controlled dilution. The automated flow rate run, which was used for the combinatorial experiments for adjusting the silver shell thickness, is shown in Fig. 7.29. A logarithmic variation of one reactant concentration was chosen in order to investigate the effect of the effect of different ratios of gold core

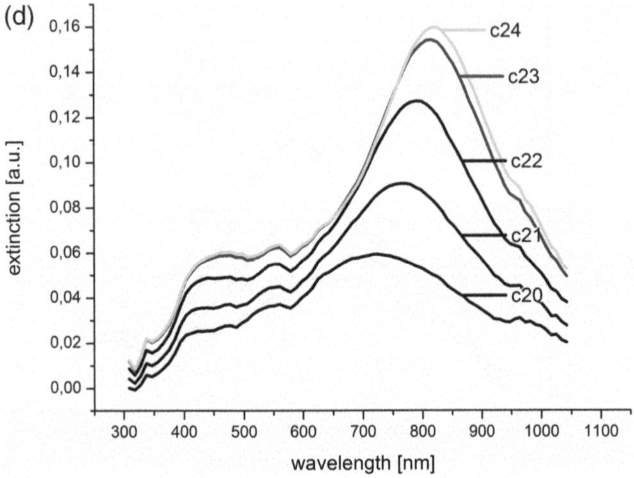

Fig. 7.32 (continued)

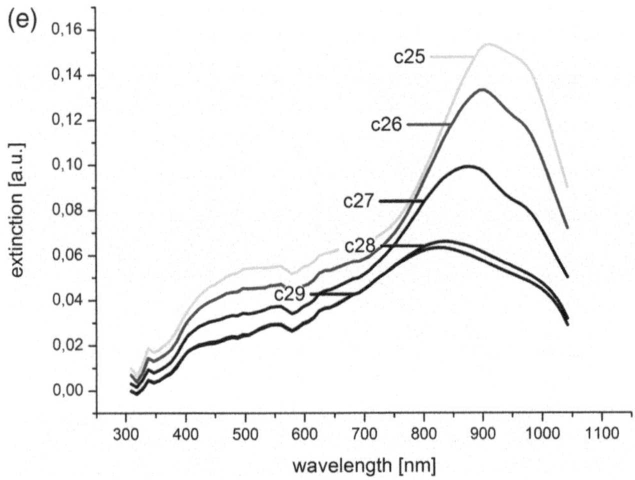

Fig. 7.32 (continued)

particles and silver deposition in the particle shells in an enlarged concentration space and especially the investigation of the possibility of the formation of very thin silver shells. Again, the whole combinatorial experiment was conceived for 36 combinatorial steps, in which six different gold core particle concentrations were combined with six different silver nitrate concentrations. Each sequence of segments contained about 260 segments of the same preset concentration. The experiment started with a flow rate ratio of gold core particle solution to the silver nitrate stream of 1:1.1 and has ended with a ratio of 1:6.6.

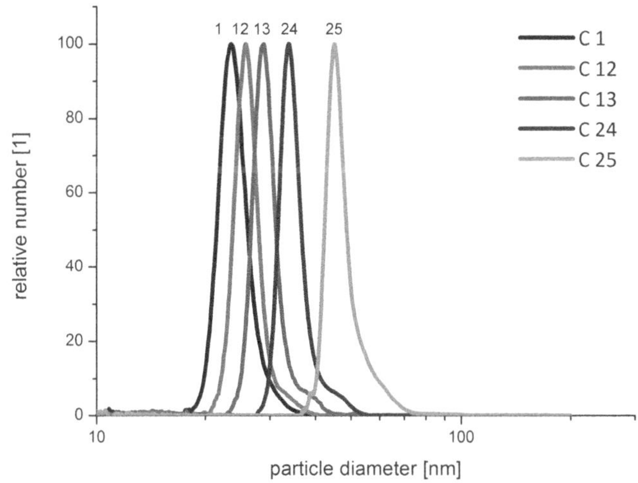

Fig. 7.33 Weight based Stokes equivalent sphere radii of nanoprisms from five different combinatorial steps with each 1 mM silver nitrate addition

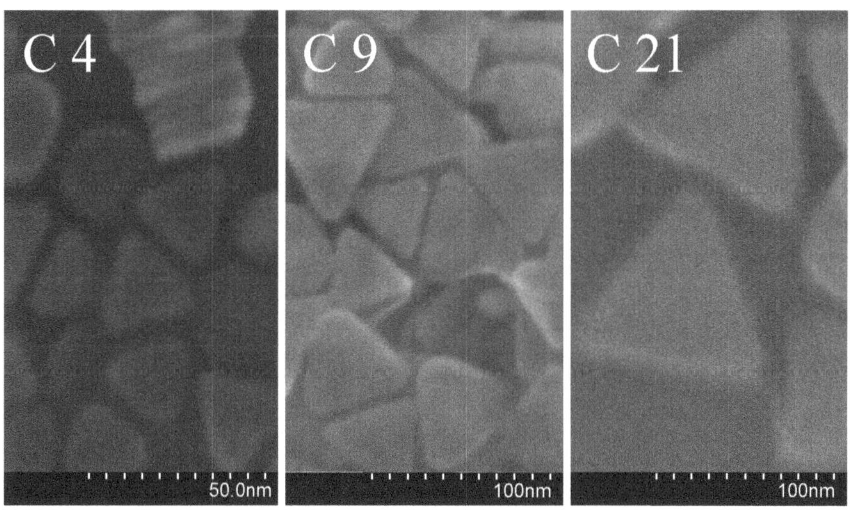

Fig. 7.34 SEM images of three different samples obtained from three different combinatorial steps but under conditions of equal silver nitrate concentration (0.4 mM)

By the help of the flow-through photometers, each segment sequence was controlled regarding segment size and intermediate distances. Again, the optical extinction spectra of each single segment were taken by in situ flow-through spectrophotometry and stored to hard disc. The length of the residence loop between the starting point of the shell growth and the position of spectrophotometry was 1500 mm. This results in a residence time of 70.7 s at a total flow rate of 249 μL/min, which is

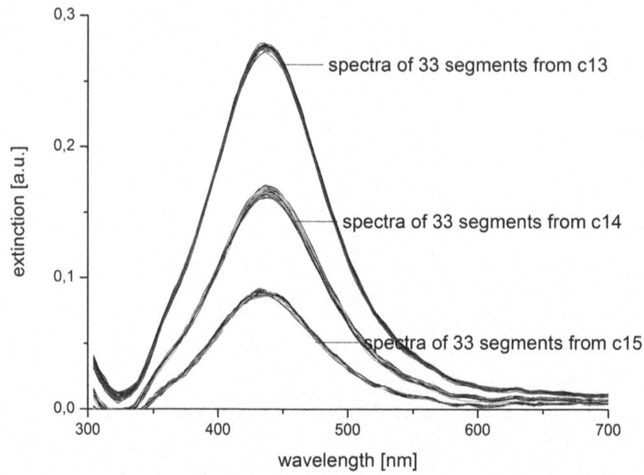

Fig. 7.35 Extinction spectra of each 33 subsequently analyzed segments of three different but successively each other following combinatorial steps

sufficient for a completion of the silver shell growth on the gold core particles under the chosen chemical conditions.

The recorded optical spectra prove a high stability of transport and reaction conditions. From Fig. 7.35 can be seen that the spectra of one combinatorial step are quite reproducible. The deviations between the spectra of one and the same combinatorial step are very small compared to the differences between the steps brought forth by the changed flow rate ratios. To gain a sufficient large sample volume for checking the particle size distribution by means of differential centrifugal sedimentation, 18 different flow rate ratios were chosen for the subsequent micro continuous flow-through synthesis. Additionally, further analyses were carried out such as offline UV-Vis or SEM-imaging.

For a thorough comparison of the effect of the investigated concentration ratios, homogeneous gold core particles were required, which henceforth are used in case of all parameter variations. These gold core particles were obtained according to the well-known sodium citrate reduction method, which was first published by Turkevich et al. [63, 64] By DCS measurement an average diameter of 11.5 nm and a FWHM of 2.6 nm were determined for the utilized gold cores. Their weight based differential size distribution spectrum is shown in Fig. 7.36. The concentration of the gold core particles was varied during the combinatorial experiment in a range between 85 and 14.3 % in relation to the initial concentration of gold nanoparticles in the used reactant mixture. Correspondingly, the silver salt concentration was altered over the whole combinatorial flow rate run in a range between 95 and 9 %. Here, a difference from the above described experiments can be seen. A flow rate offset was applied to prevent the occurrence of segment sequences from which no desired product particles are obtained, because one crucial reactant flow rate was set to zero.

In effect, core/shell nanoparticles with various physical properties result from 36 different reactant concentration combinations.

First, the gold core particles were diluted with ultra-pure water in a 1:1 ratio. This leads to correspondingly larger particles due to the formation of thicker silver shells as is was the case for undiluted gold particles.

By comparing the results of the DCS analysis of those combinatorial steps with a constant amount of added silver ions but a decreasing concentration of gold core particles, the expected shift of particle size of the particle size towards higher average diameters was observed in case of all investigated combinations. In Fig. 7.37 the size distribution spectra for the core/shell particles from the conditions a constant low silver nitrate influx but a decreasing gold nanoparticle density are shown exemplary. A similar consistent behavior was observed in case of an increasing silver nitrate concentration but constant gold core particle density. By the weight based DCS spectra from the Fig. 7.38a, b is demonstrated that an increasing presence of silver ions leads to both, an increase of the average core/shell particle size and an increase of the total sample weight.

The optical spectra of Au/Ag core/shell particles are solely dominated by the optical characteristics of silver in case of such high silver shell thicknesses. Thus, the same combinatorial flow rate run as shown in Fig. 7.29 has also been applied for the synthesis of Au/Ag core/shell particles with very low silver shell thickness. This particle species is of particular interest because the optical properties of silver coated gold particles with Ag shell thicknesses below 3 nm are significantly influenced by a plasmonic contribution of the gold core particles, which results in an absorption band at wavelength around 520 nm. In Fig. 7.39 the optical extinction spectra of the utilized plain gold nanoparticle solution and the core/shell product particles from

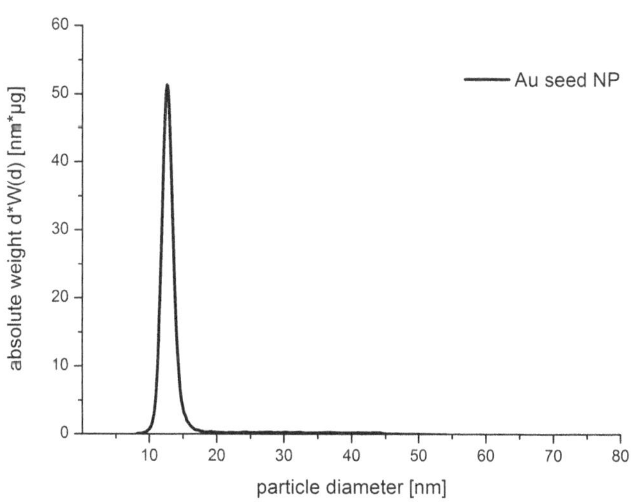

Fig. 7.36 Weight based size distribution spectra of the gold (*core*) nanoparticles used for the generation of gold/silver core/shell nanoparticles

combinatorial Step 1, 4 and 5 are overlain. In Fig. 7.40, the corresponding size distribution spectra are presented. Here, the differences of the average particle size are shown for the comparison of uncovered gold core particles and the products from combinatorial Step 1, in which the $AgNO_3$ supply is at 95 % and for the core/shell particles of Step 5, in which only 9 % of the initial silver nitrate concentration is dosed into the reactor channel. The DCS data confirm that in case of unchanged addition of Au core nanoparticles, but decreasing addition of $AgNO_3$ solution clearly a reduction of the thickness of the Ag shell around the presented Au cores was found. The average particle diameter of the plain Au nanoparticles is 11.5 nm (FWHM: 2.6 nm), the average diameter of the core/shell particles from combinatorial Step 1 is 15.9 nm (FWHM 3.6 nm), which could be reduced to 14.3 nm (FWHM 3.8 nm) in combinatorial Step 5, what lowers the silver shell thickness from 2.2 nm (C1) to 1.4 nm (C5).

From Fig. 7.39 can be seen that in the case of such thin and additionally decreasing silver shell thicknesses, the gold core particle increasingly influences the optical spectrum. In all investigations no residual gold particles have been found by DCS analyses of the colloidal solutions, which indicates a very regular initiation of the shell formation as well a silver shell growth for all inserted gold core particles. In all cases, a very high homogeneity of the product particles was observed, which is reflected by narrow, regular DCS spectra with an absence of additional side populations at other units of diameter. These findings were additionally confirmed by SEM measurements. Some examples for core/shell particles obtained under different reaction conditions from different combinatorial steps are shown in Fig. 7.41.

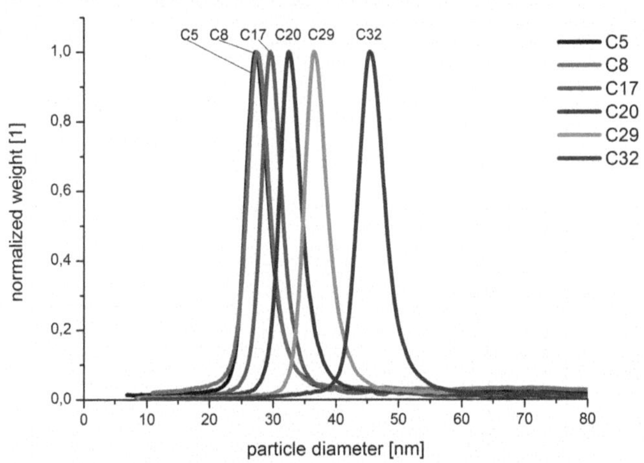

Fig. 7.37 Normalized weight based DCS spectra for conditions of equal low silver nitrate concentration but decreasing amount of gold core particles

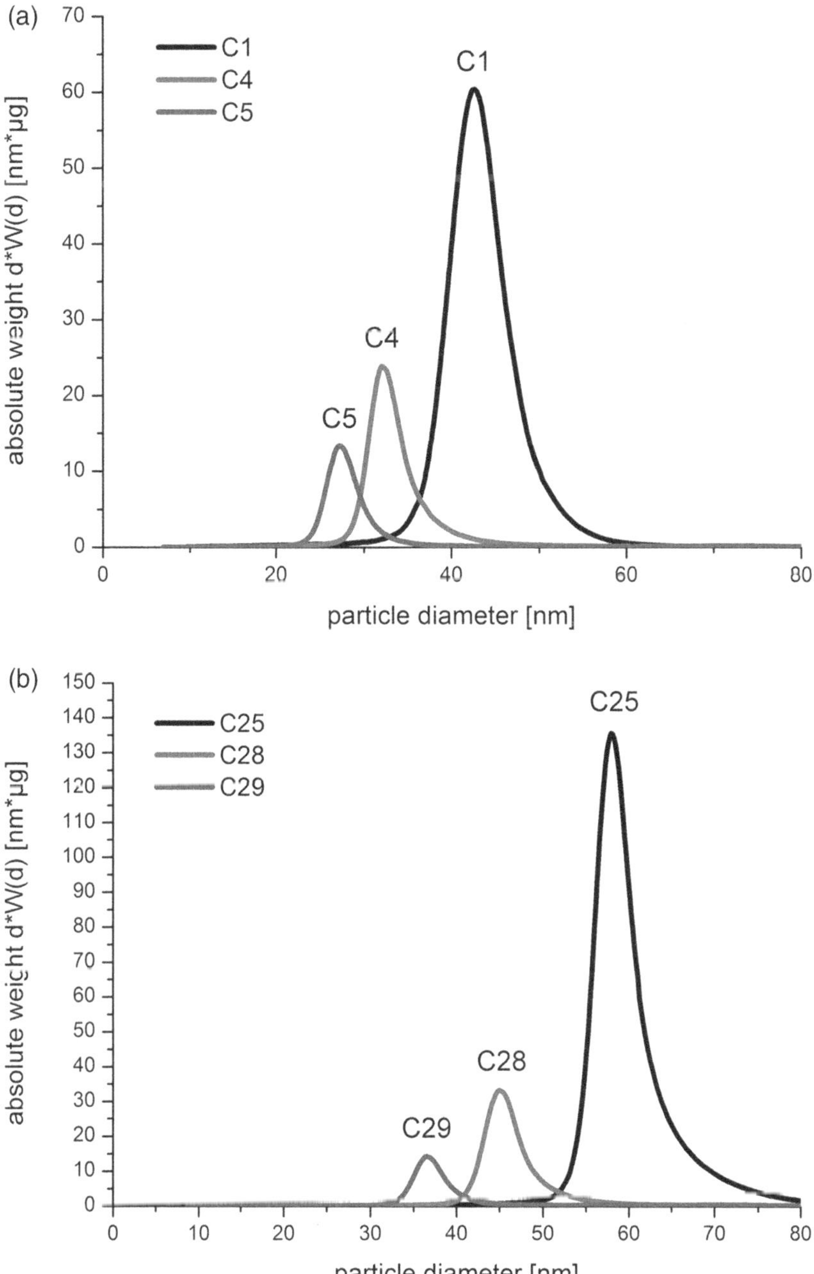

Fig. 7.38 **a** Weight based DCS data for conditions of equal high gold core particle density but decreasing silver nitrate concentration. **b** Weight based DCS data for conditions of equal high gold core particle density but decreasing silver nitrate concentration

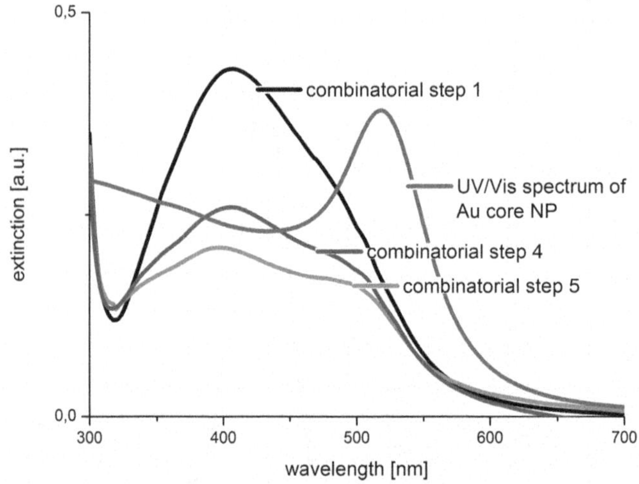

Fig. 7.39 UV-Vis extinction spectra of plain gold nanoparticles and three samples of Au/Ag core shell particles with decreasing shell thicknesses. The relative influence of the gold core increases with decreasing shell thickness

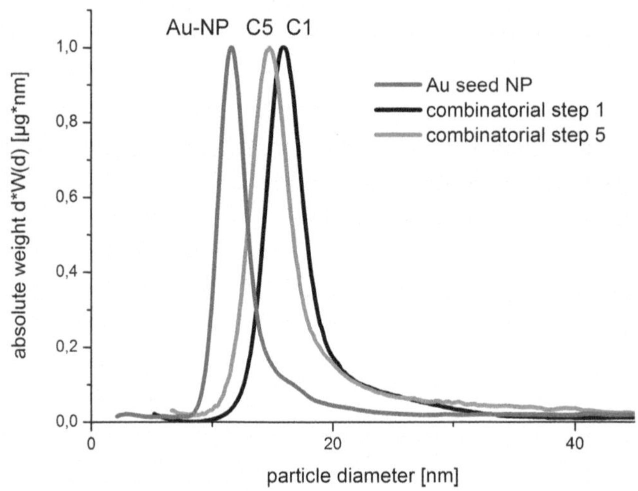

Fig. 7.40 By the weight based DCS spectra of the utilized gold (*core*) particles, the core/shell particles from combinatorial Step 1, and 5 the average diameter of the different particle samples can be compared

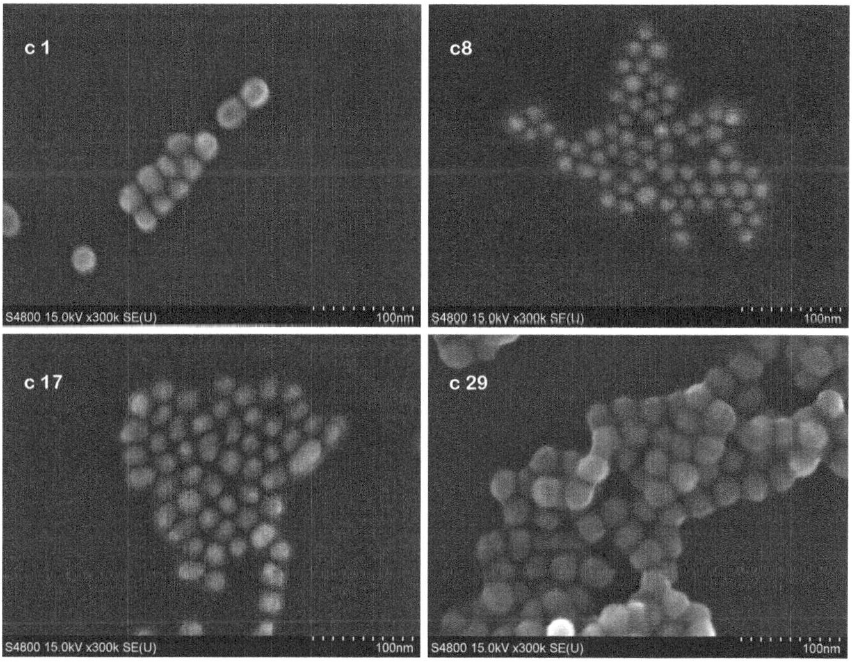

Fig. 7.41 SEM images of particle samples obtained under different chemical conditions from different combinatorial steps

7.7 Conclusion and Outlook on Metal Nanoparticle Formation in Micro Segmented Flow

For all nanoparticle systems presented in this chapter, the synthesis process in a micro continuous flow-through system using the method of the micro segmented flow has proven to be advantageous compared to conventional synthesis methods. Due to the fast segment internal convection and the therefrom resulting high mixing rates, an improvement of the particle quality was achieved in case of binary multishelled spherical and highly shape anisotropical flat, triangular silver nanoparticles. Time critical synthesis protocols could be transferred into the micro flow-through synthesis without any consideration of the initial time dependence of the addition of one reactant. The high mixing rates and the low volumes, which are beneficial properties of the micro segmented flow, lead to a suppression of local concentration gradients and thus to well controllable reaction conditions, which are necessary for a fast nucleation and a regular particle growth. The DCS analyses clearly show a narrower FWHM and a lower average particle diameter for those prepared nanoparticles which were additionally compared to their batch proceeded counterparts, as for example the silver seed nanoparticles and the core/multishell nanoparticles in Sect. 7.5. Actually, in case of the micro flow-through processed nanoparticles no evidence of aggregation could be found in the DCS spectra. The homogeneity of the size distributions is also evidenced by an improvement of the optical spectra. Especially in the analysis of the core/multishell particle systems was found that the microfluidic synthesis route leads towards narrower plasmon absorption bandwidths. The good mixing conditions of the micro flow-through synthesis also promote the formation of the desired shape. The uniformity and the high yield of triangular Ag nanoprisms could be confirmed by SEM measurements.

The experiments on the combinatorial flow-through synthesis have shown that automated combinatorial parameter screenings in large sets of extremely reduced volumes can be realized highly reproducible if the technique of the micro segmented flow is applied. This method can be performed equally for linear or non-linear changes of the reactant flow rates. The change of the physical properties of the product particles due to changing reactant concentration ratios can be monitored by in situ spectrophotometry. The optical responses of the individual segment sequences corresponded very well with the expected shifts in the particle size or composition. Selected concentration combinations were checked by further methods of analysis after a micro continuous-flow synthesis. Subsequent DCS measurements confirmed both, the possibility of fine tuning the particle size and extremely high size homogeneity in case of both combinatorial experiments. By means of combinatorial syntheses, the spectral position of the main dipole resonance wavelength of triangular silver nanoprisms could be precisely tuned in a range between 550 and 900 nm and the thickness of the silver shells which were deposited around regular gold core nanoparticles could be adjusted between 1.4 and 20 nm.

Now that the method of two dimensional combinatorial parameter screenings has proven to be successful for the synthesis of noble metal nanoparticles with either a complex structure or a non-spherical shape, it can be considered to go a few steps further. Much more information about the investigated particle system could be gathered from one single microfluidic run in three or higher dimensional parameter spaces. Also the integration of further process parameters, such as the addition of effectors or ligand molecules, a screening of the influence of different reaction temperatures with conclusion to the reaction kinetics, and the screening of different residence times between critical reaction steps could have a great impact on further developments of nanomaterials. The micro segmented flow technique allows the screening of very large parameter fields with a simultaneous low consumption of chemicals but high reliability. By this technique not only new particle architectures and compositions could be screened, it also enables the option for a controlled particle assembling and post-synthetic modification. From such kind of investigations, new synthesis protocols or parameter optimizations for novel nanoscale materials could be achieved. Also, information on kinetic effects or growth processes in case of non-spherical nanoparticles with high aspect ratios and regular lattice structures could be gained. It would also be very important, to understand the initial phase of particle nucleation, which is a field of knowledge about which to date only little is known.

Acknowledgments The authors gratefully acknowledge the funding from the DFG (FKK. 1403/22-1). The help of Henry Romanus with TEM measurements is highly appreciated. Furthermore, we would like to thank Frances Möller and Steffen Schneider for their technical support. A special gratitude for the useful discussions is also addressed to our colleagues from the IPHT Jena, Wolfgang Fritzsche and Andrea Csáki.

References

1. J. Zhao, L.J. Sherry, G.C. Schatz, R.P. Van Duyne, Molecular plasmonics: Chromophore-Plasmon coupling and single-particle nanosensors. Ieee J. Sel. Top. Quant. **14**, 1418–1429 (2008)
2. M. Muniz-Miranda, SERS-active Ag/SiO2 colloids: photoreduction mechanism of the silver ions and catalytic activity of the colloidal nanoparticles. J. Raman Spectrosc. **35**, 839–842 (2004)
3. K. Liszewski, Applying nanoparticles as molecular tags—fundamental shift in biological labeling. Genet. Eng. News **21**, 9–10 (2001)
4. J.M. Köhler, L. Abahmane, J. Wagner, J. Albert, G. Mayer, Preparation of metal nanoparticles with varied composition for catalytical applications in microreactors. Chem. Eng. Sci. **63**, 5048–5055 (2008)
5. K.B. Crozier, A. Sundaramurthy, G.S. Kino, C.F. Quate, Optical antennas: resonators for local field enhancement. J. Appl. Phys. **94**, 4632–4642 (2003)
6. L. Abahmane, J.M. Köhler, G.A. Gross, Gold-Nanoparticle-Catalyzed Synthesis of Propargylamines: the traditional A(3)-Multicomponent reaction performed as a two-step flow process. Chem-Eur. J. **17**, 3005–3010 (2011)
7. M. Homberger, U. Simon, On the application potential of gold nanoparticles in nanoelectronics and biomedicine. Philos. T R Soc. A **368**, 1405–1453 (2010)
8. Y.N. Xia, Y.J. Xiong, B. Lim, S.E. Skrabalak, Shape-controlled synthesis of metal nanocrystals: simple chemistry meets complex physics? Angew. Chem. Int. Ed. **48**, 60–103 (2009)
9. C.X. Zhao, L.Z. He, S.Z. Qiao, A.P.J. Middelberg, Nanoparticle synthesis in microreactors. Chem. Eng. Sci. **66**, 1463–1479 (2011)
10. C.A. Serra, Z.Q. Chang, Microfluidic-assisted synthesis of polymer particles. Chem. Eng. Technol. **31**, 1099–1115 (2008)
11. K. Mallick, M.J. Witcomb, A. Dinsmore, M.S. Scurrell, Fabrication of a metal nanoparticles and polymer nanofibers composite material by an in situ chemical synthetic route. Langmuir **21**, 7964–7967 (2005)
12. S. Li, G.A. Gross, P.M. Gunther, J.M. Köhler, Hydrothermal micro continuous-flow synthesis of spherical, cylinder-, star- and flower-like ZnO microparticles. Chem. Eng. J. **167**, 681–687 (2011)
13. H. Zhang, E. Tumarkin, R. Peerani, Z. Nie, R.M.A. Sullan, G.C. Walker, E. Kumacheva, Microfluidic production of biopolymer microcapsules with controlled morphology. J. Am. Chem. Soc. **128**, 12205–12210 (2006)
14. J.I. Park, A. Saffari, S. Kumar, A. Gunther, E. Kumacheva, Microfluidic synthesis of polymer and inorganic particulate materials. Ann. Rev. Mater. Res. **40**, 415–443 (2010)
15. X.M. Sun, X. Chen, Z.X. Deng, Y.D. Li, A CTAB-assisted hydrothermal orientation growth of ZnO nanorods. Mater. Chem. Phys. **78**, 99–104 (2003)
16. T.H. Ha, H.J. Koo, B.H. Chung, Shape-controlled syntheses of gold nanoprisms and nanorods influenced by specific adsorption of halide ions. J. Phys. Chem. C **111**, 1123–1130 (2007)
17. U. Kreibig, L. Genzel, Optical-absorption of small metallic particles. Surf. Sci. **156**, 678–700 (1985)
18. U. Kreibig, Systems of small metal particles—optical-properties and their structure dependences. Z. Phys. D Atom. Mol. Cl. **3**, 239–249 (1986)

19. A. Pinchuk, U. Kreibig, A. Hilger, Optical properties of metallic nanoparticles: influence of interface effects and interband transitions. Surf. Sci. **557**, 269–280 (2004)
20. S.K. Dondapati, M. Ludemann, R. Muller, S. Schwieger, A. Schwemer, B. Handel, D. Kwiatkowski, M. Djiango, E. Runge, T.A. Klar, Voltage-induced adsorbate damping of single gold nanorod plasmons in Aqueous solution. Nano. Lett. **12**, 1247–1252 (2012)
21. A.J. Frank, N. Cathcart, K.E. Maly, V. Kitaev, Synthesis of silver nanoprisms with variable size and investigation of their optical properties: a first-year undergraduate experiment exploring plasmonic nanoparticles. J. Chem. Educ. **87**, 1098–1101 (2010)
22. A.I. Henry, J.M. Bingham, E. Ringe, L.D. Marks, G.C. Schatz, R.P. Van Duyne, Correlated structure and optical property studies of plasmonic nanoparticles. J. Phys. Chem. C **115**, 9291–9305 (2011)
23. N. Okada, Y. Hamanaka, A. Nakamura, I. Pastoriza-Santos, L.M. Liz-Marzan, Linear and nonlinear optical response of silver nanoprisms: local electric fields of dipole and quadrupole plasmon resonances. J. Phys. Chem. B **108**, 8751–8755 (2004)
24. L.J. Sherry, R.C. Jin, C.A. Mirkin, G.C. Schatz, R.P. Van Duyne, Localized surface plasmon resonance spectroscopy of single silver triangular nanoprisms. Nano. Lett. **6**, 2060–2065 (2006)
25. K. Konig, I. Riemann, W. Fritzsche, Nanodissection of human chromosomes with near-infrared femtosecond laser pulses. Optic. Lett. **26**, 819–821 (2001)
26. X. Le Guevel, F.Y. Wang, O. Stranik, R. Nooney, V. Gubala, C. McDonagh, B.D. MacCraith, Synthesis. Stabilization, and functionalization of silver nanoplates for biosensor applications. J. Phys. Chem. C **113**, 16380–16386 (2009)
27. J. Reichert, A. Csaki, J.M. Köhler, W. Fritzsche, Chip-based optical detection of DNA hybridization by means of nanobead labeling. Anal. Chem. **72**, 6025–6029 (2000)
28. J. Wagner, J.M. Köhler, Continuous synthesis of gold nanoparticles in a microreactor. Nano. Lett. **5**, 685–691 (2005)
29. Y.J. Song, J. Hormes, C.S.S.R. Kumar, Microfluidic synthesis of nanomaterials. Small **4**, 698–711 (2008)
30. L. Frenz, A. El Harrak, M. Pauly, S. Begin-Colin, A.D. Griffiths, J.-C. Baret, Droplet-based microreactors for the synthesis of magnetic iron oxide nanoparticles. Angew. Chem. Int. Ed. **47**, 6817–6820 (2008)
31. S. Duraiswamy, S.A. Khan, Droplet-based microfluidic synthesis of anisotropic metal nanocrystals. Small **5**, 2828–2834 (2009)
32. A. Aimable, N. Jongen, A. Testino, M. Donnet, J. Lemaitre, H. Hofmann, P. Bowen, Precipitation of nanosized and nanostructured powders: process intensification and scale-out using a Segmented Flow Tubular Reactor (SFTR). Chem. Eng. Technol. **34**, 344–352 (2011)
33. R.L. Hartman, J.R. Naber, N. Zaborenko, S.L. Buchwald, K.F. Jensen, Overcoming the challenges of solid bridging and constriction during Pd-Catalyzed C-N bond formation in microreactors. Org. Process Res. Dev. **14**, 1347–1357 (2010)
34. C. de Bellefon, N. Tanchoux, S. Caravieilhes, P. Grenouillet, V. Hessel, Microreactors for dynamic, high throughput screening of fluid/liquid molecular catalysis. Angew. Chem. Int. Ed. **39**, 3442-+ (2000)
35. J.W. Kim, A.S. Utada, A. Fernandez-Nieves, Z.B. Hu, D.A. Weitz, Fabrication of monodisperse gel shells and functional microgels in microfluidic devices. Angew. Chem. Int. Ed. **46**, 1819–1822 (2007)
36. S. Xu, Z. Nie, M. Seo, P. Lewis, E. Kumacheva, H.A. Stone, P. Garstecki, D.B. Weibel, I. Gitlin, G.M. Whitesides, Generation of monodisperse particles by using microfluidics: Control over size, shape, and composition (vol. 44, p. 724 2005). Angew. Chem. Int. Ed. **44**(2005), 3799–3799 (2005)
37. V. Bansal, V. Li, A.P. O'Mullane, S.K. Bhargava, Shape dependent electrocatalytic behaviour of silver nanoparticles. Crystengcomm **12**, 4280–4286 (2010)
38. K.P. Charle, F. Frank, W. Schulze, The optical-properties of silver microcrystallites in dependence on size and the influence of the matrix environment. Ber. Bunsen. Phys. Chem. **88**, 350–354 (1984)

39. A.K. Singh, D. Senapati, A. Neely, G. Kolawole, C.J. Hawker, P.C. Ray, Nonlinear optical properties of triangular silver nanomaterials. Chem. Phys. Lett. **481**, 94–98 (2009)
40. G.P. Lee, A.I. Minett, P.C. Innis, G.G. Wallace, A new twist: controlled shape-shifting of silver nanoparticles from prisms to discs. J. Mater. Chem. **19**, 8294–8298 (2009)
41. W.Z. Lai, W. Zhao, R. Yang, X.G. Li, Preparation and optical properties of triangular silver nanoplates by a dual-reduction method. Acta. Phys. Chim. Sin. **26**, 1177–1183 (2010)
42. Y. Chen, C.G. Wang, Z.F. Ma, Z.M. Su, Controllable colours and shapes of silver nanostructures based on pH: application to surface-enhanced Raman scattering. Nanotechnology **18**, 325–602 (2007)
43. C. Xue, G.S. Metraux, J.E. Millstone, C.A. Mirkin, Mechanistic study of photomediated triangular silver nanoprism growth. J. Am. Chem. Soc. **130**, 8337–8344 (2008)
44. J. Polte, T.T. Ahner, F. Delissen, S. Sokolov, F. Emmerling, A.F. Thunemann, R. Kraehnert, Mechanism of gold nanoparticle formation in the classical citrate synthesis method derived from coupled In Situ XANES and SAXS evaluation. J. Am. Chem. Soc. **132**, 1296–1301 (2010)
45. A. Fernandez-Nieves, F.J.D. Nieves, The role of zeta potential in the colloidal stability of different TiO2/electrolyte solution interfaces. Colloid Surf. A **148**, 231–243 (1999)
46. M. Mirnik, Electric potential in theories of colloid stability. Kolloid Z Z Polym. **185**, 66 (1962)
47. J. Boleininger, A. Kurz, V. Reuss, C. Sonnichsen, Microfluidic continuous flow synthesis of rod-shaped gold and silver nanocrystals. Phys. Chem. Chem. Phys. **8**, 3824–3827 (2006)
48. D. Aherne, D.M. Ledwith, M. Gara, J.M. Kelly, Optical properties and growth aspects of silver nanoprisms produced by a highly reproducible and rapid synthesis at room temperature. Adv. Funct. Mater. **18**, 2005–2016 (2008)
49. M.L. Personick, M.R. Langille, J. Zhang, C.A. Mirkin, Shape control of gold nanoparticles by silver underpotential deposition. Nano. Lett. **11**, 3394–3398 (2011)
50. A. Knauer, N. Visaveliya, J.M. Koehler, Spontaneous transformation of polyelectrolyte-stabilized silver nanoprisms by interaction with thiocyanate. J. Colloid Interface Sci. **394**, 78–84 (2013)
51. A. Knauer, A. Csaki, F. Moller, C. Huhn, W. Fritzsche, J.M. Köhler, Microsegmented flow-through synthesis of silver nanoprisms with exact tunable optical properties. J. Phys. Chem. C **116**, 9251–9258 (2012)
52. D. Malsch, M. Kielpinski, R. Merthan, J. Albert, G. Mayer, J.M. Köhler, H. Susse, M. Stahl, T. Henkel, mu PIV—analysis of taylor flow in micro channels. Chem. Eng. J. **135**, S166–S172 (2008)
53. X.Y. Dong, X.H. Ji, J. Jing, M.Y. Li, J. Li, W.S. Yang, Synthesis of triangular silver nanoprisms by stepwise reduction of sodium borohydride and trisodium citrate. J. Phys. Chem. C **114**, 2070–2074 (2010)
54. R.C. Jin, Y.W. Cao, C.A. Mirkin, K.L. Kelly, G.C. Schatz, J.G. Zheng, Photoinduced conversion of silver nanospheres to nanoprisms. Science **294**, 1901–1903 (2001)
55. X.M. Wu, P.L. Redmond, H.T. Liu, Y.H. Chen, M. Steigerwald, L. Brus, Photovoltage mechanism for room light conversion of citrate stabilized silver nanocrystal seeds to large nanoprisms. J. Am. Chem. Soc. **130**, 9500–9506 (2008)
56. A. Knauer, A. Thete, S. Li, H. Romanus, A. Csaki, W. Fritzsche, J.M. Koehler, Au/Ag/Au double shell nanoparticles with narrow size distribution obtained by continuous micro segmented flow synthesis. Chem. Eng. J. **166**, 1164–1169 (2011)
57. B. Rodriguez-Gonzalez, A. Burrows, M. Watanabe, C.J. Kiely, L.M. Liz-Marzan, Multishell bimetallic AuAg nanoparticles: synthesis, structure and optical properties. J. Mater. Chem. **15**, 1755–1759 (2005)
58. J.L. Cao, D. Kursten, S. Schneider, A. Knauer, P.M. Gunther, J.M. Köhler, Uncovering toxicological complexity by multi-dimensional screenings in microsegmented flow: modulation of antibiotic interference by nanoparticles. Lab. Chip. **12**, 474–484 (2012)
59. A. Funfak, J.L. Cao, A. Knauer, K. Martin, J.M. Köhler, Synergistic effects of metal nanoparticles and a phenolic uncoupler using microdroplet-based two-dimensional approach. J. Environ. Monitor. **13**, 410–415 (2011)

60. M. Grzelczak, J. Perez-Juste, P. Mulvaney, L.M. Liz-Marzan, Shape control in gold nanoparticle synthesis. Chem. Soc. Rev. **37**, 1783–1791 (2008)
61. A. Knauer, J.M. Koehler, Screening of Multiparameter Spaces for Silver Nanoprism Synthesis by Micro segmented Flow Technique. Chem. Ing. Technik **85**(4), 467–475 (2013)
62. A. Knauer, S. Schneider, F. Möller, A. Csáki, W. Fritzsche, J.M. Koehler, Screening of plasmonic properties of composed metal nanoparticles by combinatorial synthesis in micro-fluid segment sequences. Chem. Eng. J. **227**, 80–89 (2013)
63. J. Kimling, M. Maier, B. Okenve, V. Kotaidis, H. Ballot, A. Plech, Turkevich method for gold nanoparticle synthesis revisited. J. Phys. Chem. B **110**, 15700–15707 (2006)
64. J. Turkevich, P.C. Stevenson, J. Hillier, A study of the nucleation and growth processes in the synthesis of colloidal gold. Discuss Faraday Soc. **11**, 55–75 (1951)

Part III
Biological Application: Cell-Free Biotechnology, Cell Cultivation and Screening Systems

Chapter 8
Characterization of Combinatorial Effects of Toxic Substances by Cell Cultivation in Micro Segmented Flow

J. Cao, D. Kürsten, A. Funfak, S. Schneider and J. M. Köhler

Abstract This chapter reviews the application of micro segmented flow for the screening of toxic effects on bacteria, eukaryotic microorganisms, human cells and multicellular systems. Besides, the determination of complete dose/response functions of toxic substances with a minimum of cells and chemicals, it is reviewed how two- and multi-dimensional concentration spaces can be screened in order to evaluate combinatorial effects of chemicals on cells. The challenge for the development of new and miniaturized methods is derived from the increase of the number of different used substances in technique, agriculture and medicine, from the increasing release of new substances and nanomaterials into our environment and from the improvement of the insight of toxicity of natural substances and the interferences between different substances resulting in toxic effects on different organisms, cells and tissues. The application of two-dimensional toxicological screenings on selected examples of effector combinations is described. Examples for the detection of an independent, an additive and a synergistic interference between two substances are given. In addition, it is shown that the screening for toxicological effects in complete two-dimensional concentration spaces allows the detection of complex response behaviour—for example, the formation of tolerances and stimulation peaks—which thereby can be characterized. The characterization of interference of toxic organic substances with silver nanoparticles is reported as an example for the potential of micro segmented-flow technique for evaluating the toxicological impact of new materials. Finally, it is demonstrated that the technique can be applied for different organisms like simple bacteria, single cell alga such as *Chlorella vulgaris* and multicellular systems up to the development of complete organisms beginning from eggs.

J. Cao (✉) · D. Kürsten · A. Funfak · S. Schneider · J. M. Köhler
Ilmenau University of Technology, Institute of Chemistry and Biotechnology, PF 10 05 65, 98684 Ilmenau, Germany
e-mail: jialan.cao@tu-ilmenau.de

J. M. Köhler and B. P. Cahill (eds.), *Micro-Segmented Flow*, Biological and Medical Physics, Biomedical Engineering,
DOI: 10.1007/978-3-642-38780-7_8,

8.1 Introduction: Miniaturized Techniques for Biomedical, Pharmaceutical, Food and Environmental Toxicology

The industrialized world and a high level of daily life are unimaginable without a large spectrum of materials and substances for different purposes. The spectrum of produced and used substances reaches from chemicals for the production of food (agriculture and related fields) over a huge number of different materials used in industry to the large variety of special chemicals for households, for medicine and in scientific applications.

A reasonable and risk-free use of substances is only possible if the effect of these substances on organisms is known. This problem is addressed by the toxicology. Meanwhile, thousands of substances have been evaluated for their effect on living beings and, in particular, the danger of a lot of poisons for humans was investigated. But, toxicology is confronted with very serious problems of individual response and combinatorial effects [1–4]. At first, different species respond differently to toxic substances. In addition, the effect of poisons is dependent on the path of uptake; different organs and different tissues show different critical doses and different tolerances for a certain substance. Furthermore, the sensitivity for a toxic substance depends on the stage of individual development, on health status and genotypes. Finally, cells and tissues, organs and complete organisms are not only exposed to one substance, but are always in changing interactions with different substance mixtures (see for example: [5, 6]). As result, a huge number of toxicological problems has to be solved for a complete understanding of risks in an environment filled with a very broad spectrum of different chemicals from natural and technical sources. There is currently no way that this complex task can be solved by a conventional approach. Therefore, new strategies and concepts for evaluation of the effect of chemical mixtures have to be developed [7] and new methods and systems for experimental investigations and systematic screenings are required. An important step in this direction is the substitution of animal tests by in vitro testing using human cell lines or other target organisms and to implement the toxicological assays in robotic high-throughput screening systems [8, 9].

Beside the evaluation of toxicological risks, the concentration-dependent and combinatorial effects of drugs for therapeutic purposes are a challenge for the development of new efficient methods in toxicological testing. Therefore, it is possible to use synergistic effects of drugs from plant constituents [10].

During the last years, several investigations on the use of microfluidics for toxicological studies have been investigated in order to get larger sets of information on toxic effects with test methods consuming only small amounts of test chemicals and the target material. It was shown that a cultivation of cells and microorganisms under microfluidic conditions and their use for assays is possible [11–15]. Toxicological investigations can be executed on three-dimensional tissue fragments and cell cultures [16–19]. Microfluidic systems can be applied for different types of screenings and are suitable for investigations of the dose-dependent response of cells on drugs. A special procedure is applied in order to create concentration gradients for the investigation of dose-dependent responses from molecular or cellular systems [20]. In addition, these miniaturized techniques promise possibilities of enhancements in developing parallelized systems, combinations of microchannel flow systems and arrays and the application of automated procedures [21–25], whereby they contribute significantly to a strict reduction of costs per single assays.

Unfortunately, the complexity of this problem is even more enlarged by the fact that toxic effects of one substance can be affected by other substances. Thus, the problems of toxicology are multiplied by a very high number of possible binary and ternary or even higher combinations, which then generate special dose/response functions of different organisms. Despite a broad and detailed knowledge on the toxicological effect of a lot of substances on cells and organisms, the problem of interaction of cells and pollutants in the presence of other substances or other organisms is unsolved in most cases. In recent work on droplet-based microfluidics, it was shown, that this technique is particular suitable for biological screenings [26–29] and for the realization of larger parameter fields, for kinetic investigations and for the evaluation of highly resolved dose/response relationships [30–37]. Biotechnical microsystems and droplet-based microfluidic systems and micro fluid segments are not only suited for biochemical operations and PCR [38, 39] and for handling of cell suspension but can also be used for assays using multicellular objects like fishes [40, 41] and nematodes [42–45]. In the following, the suitability of this technique for one upto three-dimensional microtoxicological screenings with single test volumes in the nanolitre range is described.

8.2 Advantages of Micro Segmented Flow for Miniaturized Cellular Screenings

Some of the general advantages of micro fluid segment technique in micro reaction technology are also relevant for cellular screenings [46, 47]. First, the inclusion of cells in fluidic compartments with small volumes embedded in a hydrophobic separation liquid leads to an efficient separation and decoupling of cultivation volumes (Fig. 8.1a). Highly parallelized procedures become possible with very small total volumes [48–52]: A sequence of about 2,000 microfluidic segments with a volume of 0.2 μL each means that only a total of 0.4 mL of cultivation liquid is required.

The second important aspect is the partial or complete decoupling of the cultivation liquid from the channel walls. The problem of interaction of biomolecules and cells with the walls is particular serious in microfluidic systems because the ratio of surface to volume is dramatically enhanced in comparison with conventional cultivation systems. In micro segmented flow, the biofouling can be significantly reduced under conditions with a high wetting of the walls by the carrier solution and a low wetting by the cultivation liquid. In general, hydrophobized wall surfaces or tubes from lipophilic materials are used for micro segmented flow. Perfluorinated tubes and device materials are combined frequently with liquid perfluorcarbons used for the

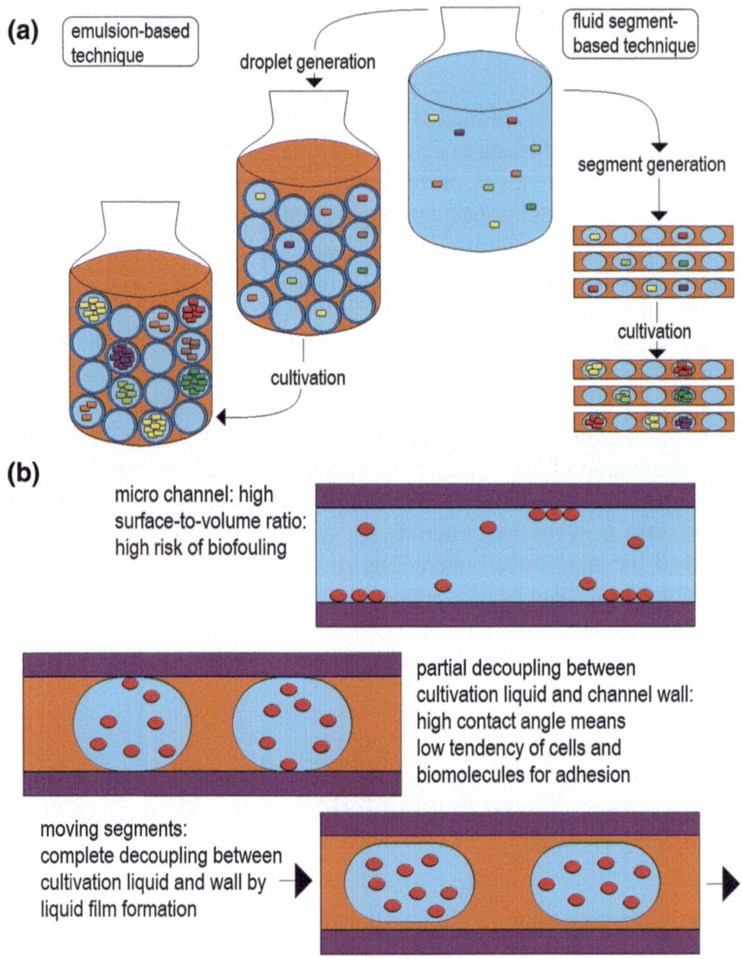

Fig. 8.1 Functional decoupling by microfluidic compartimentation: **a** decoupling of microorganism populations by separation in droplets and micro fluid segments, **b** decoupling between cultivation volume and channel walls by controlling the wetting behaviour

carrier phase. In these cases, the carrier solution always has a low contact angle with the wall surface, but the cultivation liquid forms a high contact angle corresponding to low wetting. During transport a liquid film of carrier solution is formed in a capillary slit between the wall and the segment leading to a complete separation of the segment content from the wall (Fig. 8.1b). As result, the adsorption of biomolecules and the adhesion of cells at the inner walls of microtubes and microchannels are considerably suppressed.

A third important aspect is the transport of biomolecules, particles, dissolved gases and other molecular components inside the liquid segments (Fig. 8.2). The

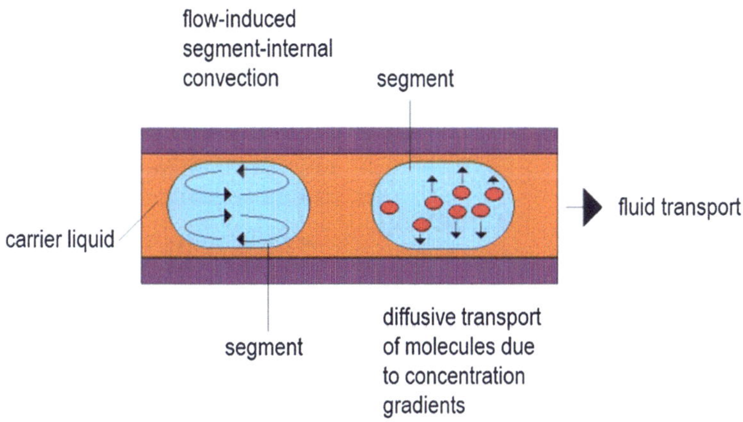

Fig. 8.2 Segment-internal transport processes: coupling between convective transport by flow-induced segment-internal circulation and transport by diffusion

reduction of volume in comparison with conventional cultivation conditions and even in comparison with the conditions in microtiterplates leads to a reduction of diffusion paths which results into a faster transport by diffusion. The application of tubes with lower wall thickness consisting on highly gas permeable perfluorinated materials support the gas exchange between the fluid segments and the environment. Furthermore, the axial transport of segments induces a segment-internal circular convection. This is a considerable difference to homogeneous microfluidic systems in which a laminar flow is observed in general due to the low Reynolds numbers. In segmented flow, the interface tension between the aqueous cultivation phase and the lipophilic carrier phase keeps the segments stable during the transport and induces a radial component in the liquid streaming. This circular convection is very efficient for the fast transport of particles, molecules and is of particular importance for mixing processes with vesicles, sensor particles and cells, which are too large for showing a significant diffusive contribution to their transport. The segment-internal convection supports the mixing after addition of components into segments (mixing, dosing) as well as the homogenization during further procedures [53].

Further, the incorporation of cultivation volumes inside micro channels and micro tubes provide a safe processing of segments and a low loss of solvent and other volatile components on the one hand, but also ensure an ordered mobility for serial processing of all segments on the other hand. Although the typical volumes of micro fluid segments are about two orders of magnitude lower than the typical liquid volumes in microtiterplates, the evaporation rates are much lower than in open microtiterplates. Whereby the guidance of segments by the tubes or channel walls allow a safe and reproducible serial processing of complete segment sequences without any pipetting steps. Several key segment operations such as dosing, mixing, splitting, switching [54–56] and optical or electrical measurements [57–61] can be performed without open surfaces.

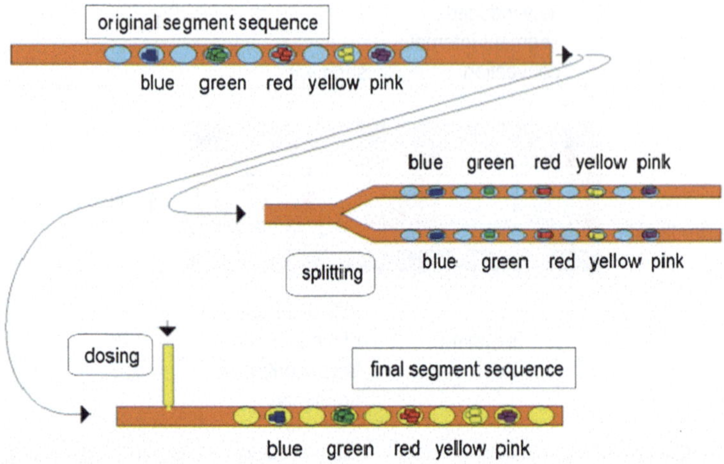

Fig. 8.3 Individuality of segments in biological screenings: the order of segments is kept constant during all segment operations from generation over transport, dosing, measurements up to the splitting and recombination of segments

Finally, the fixed order of segments inside the channels and microtubes allows an addressing of individual segments [62]. Therefore, each segment position is defined by a predetermined starting composition and an individual process history (Fig. 8.3). Every segment can be related to a fixed point in the space of investigated parameters. Thus, it is easily possible to assign data from measurements on single segments to a composition and process data. This is an important precondition for using micro segmented flow for investigation of highly resolved dose/response functions and on toxicological screenings in two and higher-dimensional concentration spaces.

8.3 Miniaturized Determination of Highly Resolved Dose/Response Functions

The applicability of micro fluid segments for the monitoring of bacterial growth and for dose-dependent effects was investigated based on the example of testing the growth of *E. coli* in micro fluid segments at different concentrations of the phenolic uncoupler 2,4-dinitrophenol, antibiotics ampicillin and chloramphenicol (Fig. 8.4). Therefore, a continuous variation of all substances had been applied. The effect on the bacteria inside segments was detected by measuring the increase of the intensity of cellular autofluorescence by use of a microflow-through fluorimeter and measuring the reduction of the intensity of transmitted light by use of a microflow-through photometer over time. On the one hand microflow-through photometer gives information on the cumulative number of cell divisions. On the other hand the autofluorescence intensity indicates a number of physiological active cells.

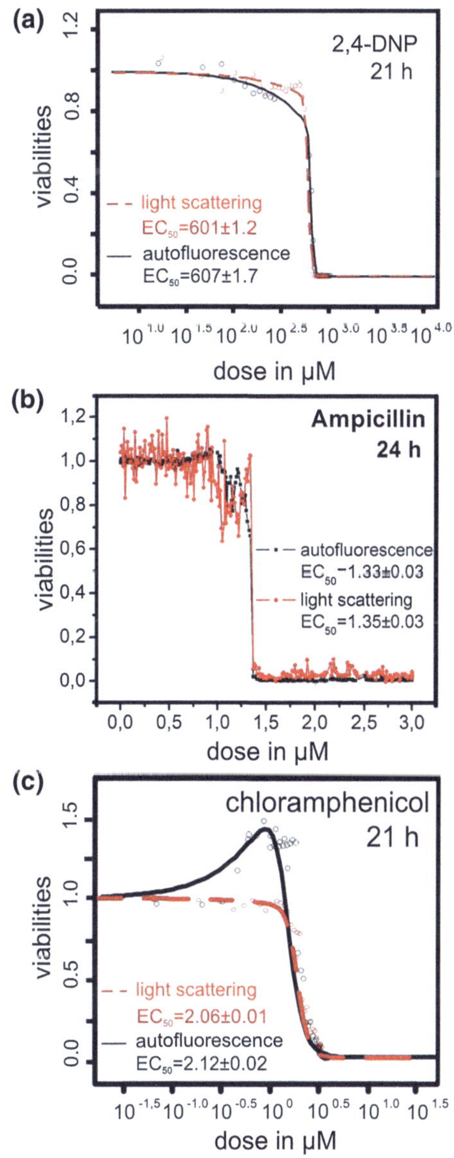

Fig. 8.4 Dose/response functions for the single effectors determined with continuous varied concentrations of toxins regarding the endpoints on the light scattering (*red line*) and autofluorescence intensity (*black line*) of *E. coli* after one day cultivation in micro fluid segments: **a** 2,4-dinitrophenol, **b** ampicillin, **c** chloramphenicol

Large sequences each comprising about 300 individual segments with gradually varying concentrations of single effectors were generated. The EC(50) values, defined as the concentration which cause 50 % reduction in autofluorescence and light scattering signal compared with the controls, were determined. The graphs show that the autofluorescence signal as well as the light scattering signal supply the same EC(50) values, but the shapes of the both dose/response curves are different. In the experiments with phenolic uncoupler 2,4-DNP after an incubation time of 21 h an EC(50) value for endpoint cell density of 601 μM and for autofluorescence at 607 μM have been found (Fig. 8.4a).

In the same way, the dose/response function for the effect of antibiotic ampicillin was investigated. The observed EC(50) value of 1.33 μg/mL by autofluorescence and 1.35 μg/mL by light scattering is in good agreement with the expected value as known from conventional cultivation experiments (EC(50) = 1.25 μg/mL) (Fig. 8.4b) [63]. In Fig. 8.4c the dose/response relationships of chloramphenicol on the growth (red line: EC(50) $= 2.06 \pm 0.01$ μg/mL) and autofluorescence intensity (black line: EC(50) $= 2.12 \pm 0.022$ μg/mL) after 21 h *E. coli* cultivated inside microfluid segments are shown. A stimulation of bacterial growth in the sublethal range is known in the case of the antibiotics chloramphenicol [64]. The investigation of *E. coli* in micro fluid segments confirmed this stimulation (Fig. 8.4c). In our experiments, this stimulation is well reflected by a significant increase (more than 30 % compared with the controls) in the endogenous autofluorescence signal observed in all micro fluid segments at sublethal concentrations in the range of 0.6–1.2 μg/mL (Fig. 8.4c black line). This hormesis effect could be caused by the stimulation of the physiological processes in the sublethal region.

The micro fluid segment technique was also used for the evaluation of the effect of silver nanoparticles on *E. coli*. An EC(50) value of 0.09 pmol/L nanoparticles was found for silver nanopatricles with a particle size of 20 nm (Fig. 8.5a). Besides the determination of EC(50) values, the micro segmented flow technique allows a comfortable and rapid measuring of highly resolved dose/response functions. An example of such a dose/response functions with very small steps of concentration variation is shown in Fig. 8.5b for the incubation of *E. coli* with Au/Ag core/shell nanoparticles over a time of 24 and 48 h, respectively. The curves supply several information: At first, both curves reflect the same sharp transition from bacterial growth to non-growth indicating the critical concentration of nanoparticles. This indicates that no adaptation on higher nanoparticle concentration takes place in any micro fluid segment. Secondly, the curves reflect a nearly constant bacterial growth at lower nanoparticle concentrations, but larger differences in growth within the sublethal range. In this range, not only segments with reduced growth but also some segments with enhanced growth had been observed, what indicates the stochastic effects of inhibition and stimulation. This effect became more visible after an incubation time of 48 h.

The micro fluid segment technique was also used for the evaluation of the effect of an antibacterial peptide on *E. coli* [65]. The dose-dependent growth of bacteria inside fluid segments could be detected both by measuring the optical density (increasing light scattering due to increasing cell density) and by measuring the fluorescence intensity of pH-sensitive micro beads (reflecting the metabolic activity). Both methods returned similar results for the biological activity in dependence on the concentration of the antibiotic substance (Fig. 8.6). In this experiment, a considerable shift of EC(50) value with increasing incubation time was observed. This effect is possibly caused by either the adaptation of microorganisms to the active agent or by

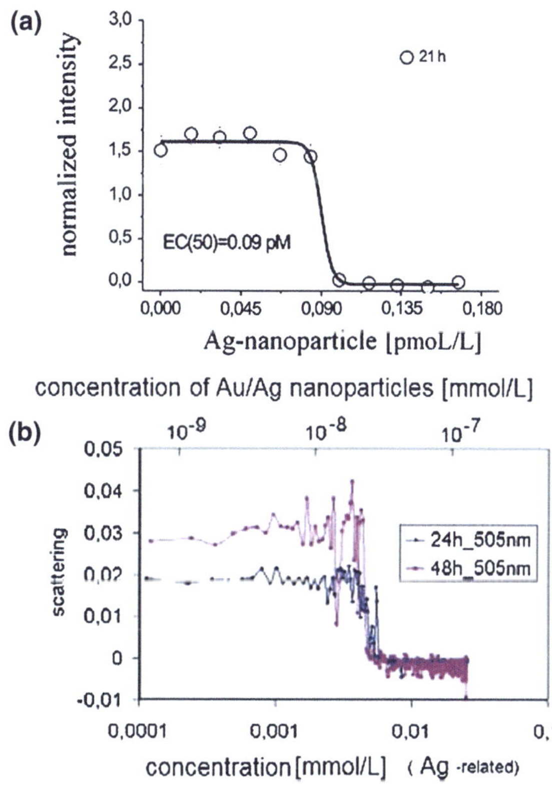

Fig. 8.5 Response of *E. coli* on colloidal solutions of toxic metal nanoparticles: **a** dose/response function against silver nanoparticles (size: 20 nm diameter), **b** highly resolved dose/response function against Au/Ag core/shell nanoparticles

the metabolic decomposition through the substance with the increasing number of bacteria inside the fluid segments.

Despite drugs, the effect of food components on bacterial growth is of interest as well. Therefore, the effects of ethanol and caffeine have been analysed by the determination of dose/response functions inside the micro fluid segments. The experiments prove a steep transition between growth and lethality for *E. coli* at 2.65 % of ethanol (Fig. 8.7a). This value is close to typical observed values in the digestive tract after consumption of alcoholic beverages and indicates the effect of ethanol on the bacterial flora inside human bodies. A very sharp transition between the sublethal and the lethal range was also found for the application of caffeine (Fig. 8.7b). A small stimulation effect in the sublethal range was detected by autofluorescence intensity by measuring highly resolved dose/response functions.

In the field of ecotoxicology, the technique of micro segmented flow was also successfully applied for the determination of the dose/response function with *Chlorella vulgaris* against copper chloride [66]. Therefore, a linear variation of concentration of copper ions was realized by a linear sweep of flow rate ratios in a sequence of about 350 segments. The CO_2 supply was supported by the addition of sodium hydrogen carbonate to the cultivation solution. The EC(50) values of about 36 μg/mL (Fig. 8.8)

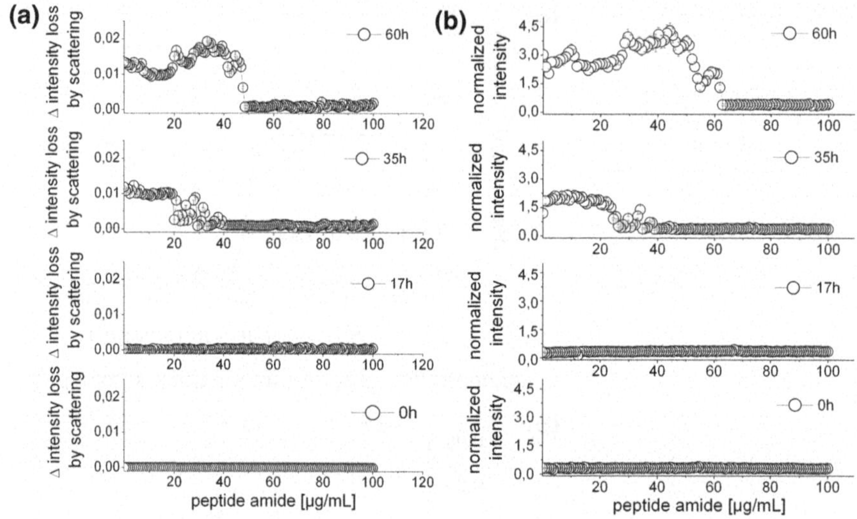

Fig. 8.6 Effect of the antibacterial peptide amide on the growth and the metabolic activity of the *E. coli* cultivated inside micro fluid segments. **a** Monitoring of the parameter light scattering as an indicator of the cell culture growth at different cultivation time intervals by photometric measurement. **b** Monitoring of the parameter fluorescence intensity of pH dependent microbeads as an indicator for the metabolic activity of the cell culture by fluorimetric measurement. Reprinted with permission from ref [65]. Copyright 2009 Elsevier

have been found with a high reproducibility. The measurements in micro fluid segments resulted in a much higher quality of dose/response functions as compared to the reference measurements in microtiterplates.

8.4 Strategy and Set-Up for Generation of 2D- and 3D-Concentration Programs

A small group of typical classes of combination effects can be observed in binary mixtures of effectors (Fig. 8.9). The first case is represented by two effectors, which have an independent effect on the target. In this case, a toxic effect can be observed when the critical concentration of one of the both components is passed (Fig. 8.9a). In the second case, the effectors can substitute each other. The effects of both components are added. A pure additive behaviour is observed when a linear function in the two-dimensional does/response maps is present (Fig. 8.9b). A concave function means that mixtures have a higher toxic effect then the single substances. In this case, the effect of one substance is enforced by the activity of the second one. As result, a synergistic behaviour can be observed (Fig. 8.9c). When convex response functions are observed then the effect of the combination of both substances is smaller than in

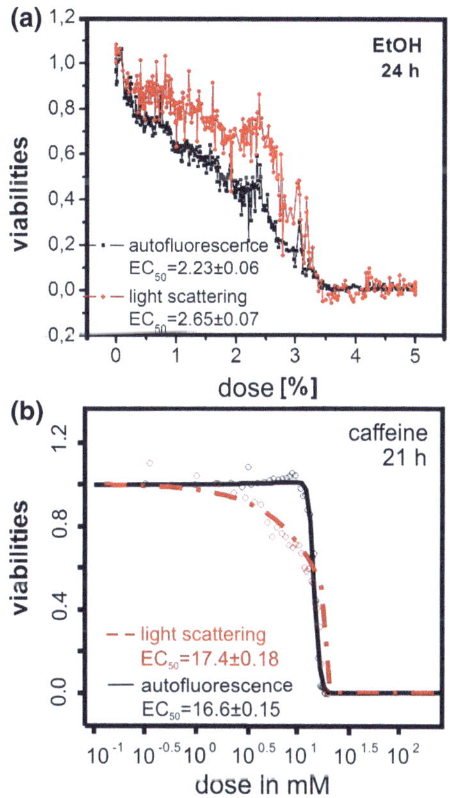

Fig. 8.7 Effect of components of human food on bacteria *E. coli* after one day incubation inside micro segments measured with micro flow-through photometry and fluorimetry: **a** dose/response relation of ethanol, **b** dose/response relation of caffeine

the additive case but stronger than in the independent case. An antagonistic behaviour leads to convex dose/response functions, which exceed the values for the critical concentration for one or for both substances (Fig. 8.9d).

The strategy for the investigation of 2- and 3-dimensional concentration spaces was based on the easy possibility of variation of composition of micro fluid segments by a continuous shifting of flow rate ratios. Such a shift of flow rates can be realized down to flow rates of less than one μL/min using computer-controlled, nearly pulsation free syringe pumps. The only important boundary condition for the realization of reproducible concentration programs is the strict control of size and distance of the generated fluid segments. This requirement has been met by the principle of constant total flow rates. A constant value for the flow rate of the inert carrier liquid and for the flow rate of the aqueous mixture of test solution can easily be realized if any change in the flow rate of one component of the test mixture is compensated by the complementary change in one or a group of the other flow rates of aqueous solutions.

The complete addressing of two-dimensional concentration spaces was realized by low pulsating PC-controlled syringe pumps (Cetoni GmbH, Germany).

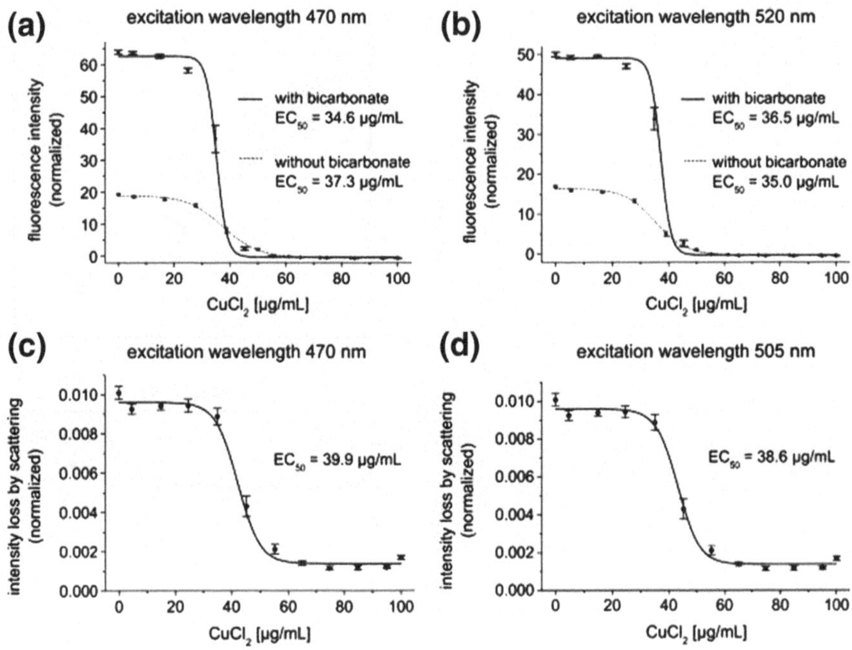

Fig. 8.8 Ecotoxicological screening in micro fluid segments: high coincidence of dose/response functions for different endpoint detection and in presence and absence of bicarbonate addition; determination of the effect Cu(II) ions on *Chlorella vulgaris* detected by light scattering and fluorescence at two different wave lengths: **a** dose/response function obtained by autofluorescence under excitation at 470 nm, **b** autofluorescence under excitation at 520 nm, **c** transmission measurement at 470 nm, **d** transmission measurement at 505 nm. Reprinted with permission from ref [66]. Copyright 2011 WILEY-VCH Verlag GmbH & Co. KGaA

Continuous shifting of the flow rate of one effector solution and a cycling change of the solution of the second effector lead to the systematic realization of a complete binary concentration field in fluid segment sequences between some hundred and about thousand well-separated liquid compartments. The realization of a three-dimensional concentration space by flow-rate programs is shown in Fig. 8.10. Three solutions with different effectors were applied through the channels 1–3. Different concentration triples have been realized by different variations of flow rates: The flow rate of the solution with effector one was varied slowly, whereby the flow rates of the effector solution two and three was varied in intermediate and fast cycles each. In a fourth channel, a solution without any effector was actuated with a flow rate complementary to the total of the first three flow rates. Therefore, the total flow rate was kept constant and equal hydrodynamic conditions were realized for the formation and transport of all generated segments within one experimental run. The realization of a three-dimensional concentration space with a resolution of 20 % was well illustrated by the variation of three dyes with different absorption maxima. In this experiment, a segment sequence with about 5000 segments was generated.

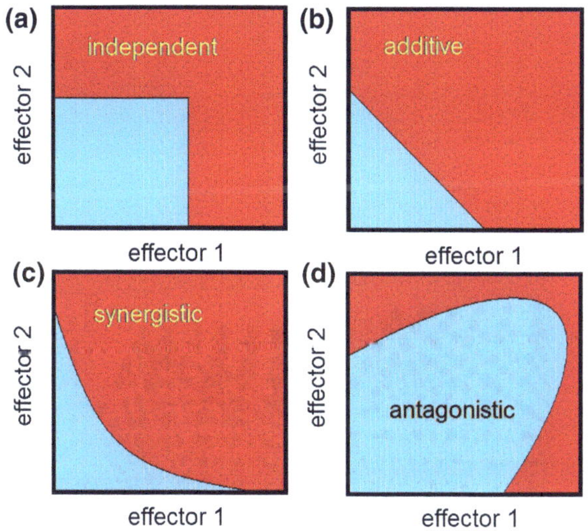

Fig. 8.9 Types of combinatorial effects presented by bolographic maps (schematically): **a** independent behaviour, **b** additive behaviour, **c** synergistic behaviour and **d** antagonistic behaviour

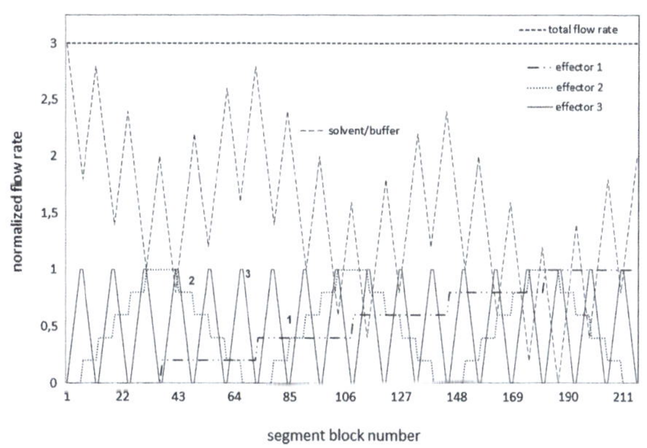

Fig. 8.10 Flow rate program for the realization of a three-dimensional concentration spaces

Three dyes in six different concentrations were combined in segments resulting in 216 different combinations of the dye-concentrations. Therefore, each concentration point was presented by a set of about 22–23 segments showing the high reliability of the method [67]. The theoretical placement of the 216 concentration points in the three-dimensional space is illustrated by the upper row of Fig. 8.11a–c. The lower row (Fig. 8.11d–f) reflects the experimentally realized concentrations in all segments and the success of the applied concept. Each concentration point is marked by a cloud of concentration triples. However, the deviations of the practically achieved concentrations from the theoretical values are significantly smaller than the concentration steps what proves the reliable addressing of the whole concentration space.

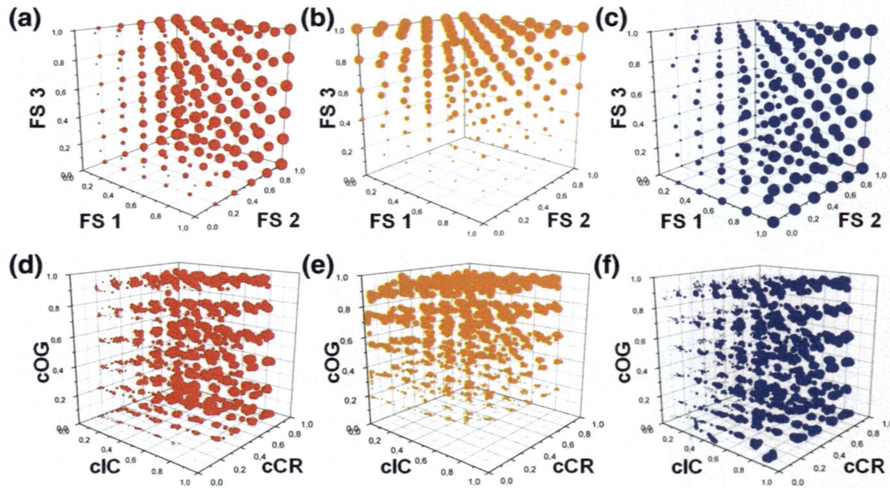

Fig. 8.11 Three dimensional addressing in 6 concentration steps resulting in 216 concentration points with stepwise varied compositions of cochineal red (CR), orange G (OG) and indigo carmine (IC) in a segment sequence with over 5400 segments by in-situ flow-through photometry. **a–c** Expected values, **d–f** measured values. Reprinted with permission from ref [67]. Copyright 2012 Royal Society of Chemistry

The responses of cultures of *E. coli* and *Chlorella vulgaris* on different phenols, heavy metal ions and metal nanoparticles have been characterized by optical measurements of cell density and physiological activity. The measurements allow the precise determination of critical concentrations of single effectors as well as for all binary, ternary combinations. It was shown, that different types of responses against combinatorial effects can be distinguished such as additive or synergistic behaviour [65–67].

The results underline the advantages of the use of the micro fluid segment technique in the future for also investigating related combinatorial problems such as the evaluation of combination effects of pharmaceutical drugs, pesticides and the decomposition products made of synthetic materials. In addition, the technique is very promising for studying cellular interactions down to the single cell level due to the small involved liquid volumes. Thus, this technique is not only of interest for microtoxicological, but also for microecological investigations under strictly controlled well-defined conditions in larger parameter fields.

8.5 Determination of Combinatorial Effects by Characterization of Dose/Response Functions in Two-Dimensional Concentration Spaces

The response behaviour of organisms against combinations of two effectors can be well characterized by map-like representations. In these so-called bolographic maps the growth-related optical intensity is shown as the third dimension using a color code ranging from blue to red versus the both effector concentrations displayed on x and y axis. A nearly independent activity of two toxic effectors was observed for the combination of silver nanoparticles and atrazine on *Chlorella vulgaris* (Fig. 8.12a). The critical concentrations of both effectors are nearly non-effected by each other. The corresponding presentation in the bolographic map is a nearly rectangular plateau in the lower left part. The observed behaviour hints to different ("orthogonal") mechanisms for the effect of the toxic substances.

In contrast, a continuous reduction of the critical concentration of one effector by increasing concentration of the other effector was found for the example of combination of the two antibiotics chloramphenicol and tetracycline. In this case, one effector can be directly substituted by the other one. In the bolographic map, a triangular shaped area for the growth of bacteria is the typical representation of this additive behaviour (Fig. 8.12b).

A synergistic behaviour was observed for the combination of silver nanoparticles and 2,4-dinitrophenol applied in micro fluid segments containing *E. coli* (Fig. 8.13a). The addition of small concentrations of silver nanoparticles—far below the toxic threshold—led to a drastic reduction of the EC(50) value of the phenolic uncoupler. Vice versa, a low dose of the phenolic uncoupler lowered the EC(50) value for silver

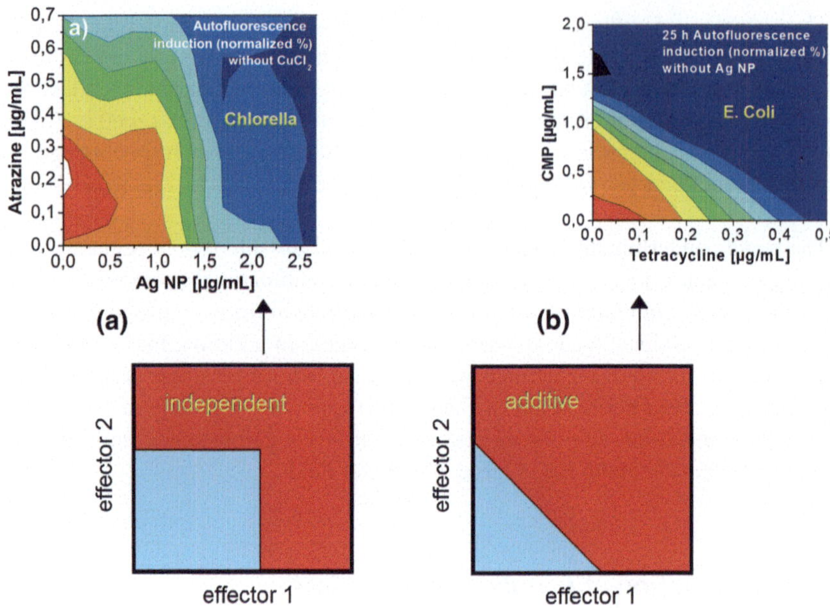

Fig. 8.12 Linear combinatorial effects by combination of two effectors: **a** nearly independent effect of silver nanoparticles and atrazine on the growth of *Chlorella vulgaris*—it is to assume that the effect mechanisms are different resulting in an orthogonal behaviour, **b** additive behaviour in the activity of two translation inhibitors (chloramphenicol and tetracycline) on the growth of *E. coli*—parallelism in effect mechanism

nanoparticles drastically. This behaviour is marked by a concave shape of growth region in the isobolographic map.

An antagonistic behaviour takes place when one effector stimulates the growth of the target organisms above the lethal dose of the other effectors. Such a stimulation effect was observed in case of the combination of caffeine with the frequently used blood pressure drug captopril in the presence of ampicillin. In the single effector screening experiment, no growth of *E. coli* was observed at a concentration of 1.2 µg/mL ampicillin. But the addition of caffeine led to a strong stimulation of bacterial growth. This stimulation was enforced by the addition of captopril (Fig. 8.13b). As a result, a hill-like response pattern was found on the bolographic map.

8.6 Multi-Endpoint Detection under Microfluidic Conditions

The response of target organisms on toxic substances results in different dose-dependent signals when using different detection channels. Thus, the reduction of light transmission by the increasing number of scattering objects is related to cellular growth, whereby dead cells and cell fragments also contribute to the signal

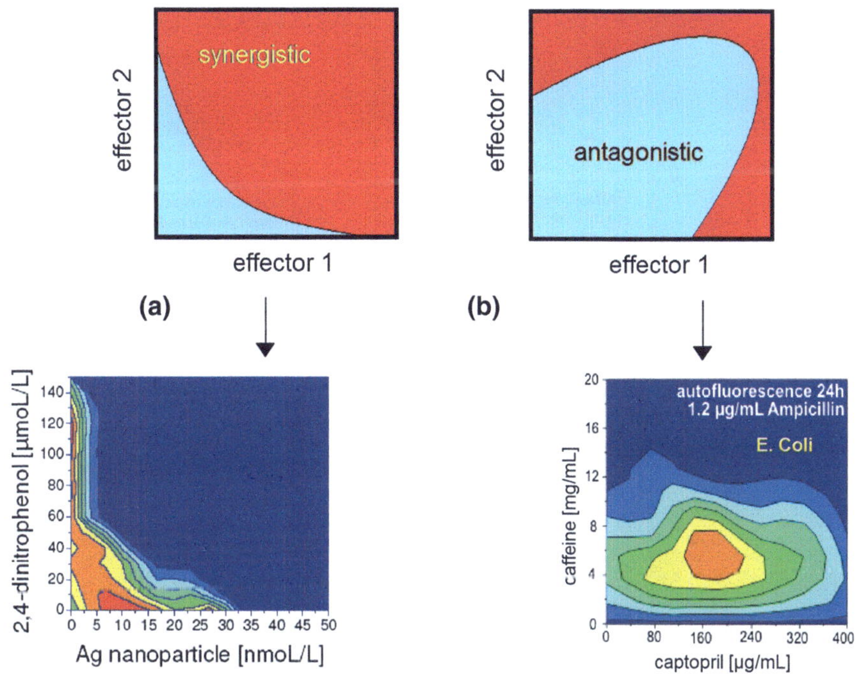

Fig. 8.13 Non-linear combinatorial effects: **a** strong synergism in the combined effect of a phenolic uncoupler (2,4-dinitrophenol) and silver nanoparticles on the growth of *E. coli*, **b** complex behaviour: partially antagonistic effect of caffeine and captopril on the antibiotic activity of ampicillin during *E. coli* cultivated in micro segments

strength beside the living cells. In contrast, the autofluorescence of cells indicates some viability and can additionally be related to a certain physiological activity. The light transmission provides no information about the physiological activity of cells, but a change in the pH in the surrounding medium reflects the consumption and metabolization of solvent components and the release of metabolic products and hereby indicates the physiological activity of the cell population. In conclusion, light transmission, autofluorescence and pH-measurements supply informations which are complementary to some extent and should be used accordingly.

An increase in pH was observed on the cultivation of *E. coli* in micro fluid segments in reference experiments of larger cultivation volumes (15 mL falcon tubes) and in micro fluid segments (cultivation volume: 0.5 µL). The pH-measurement inside micro fluid segments have been realized by the use of pH-sensitive fluorescent micro sensor beads [68]. In both cases, an increase of pH from about 6–8.5 was found during a four-day cultivation experiment (Fig. 8.14). This type of measurement allows the detection of the effects of different toxin concentrations on the physiological activity. They are particular suited for the evaluation of changes in the response behaviour in sublethal region.

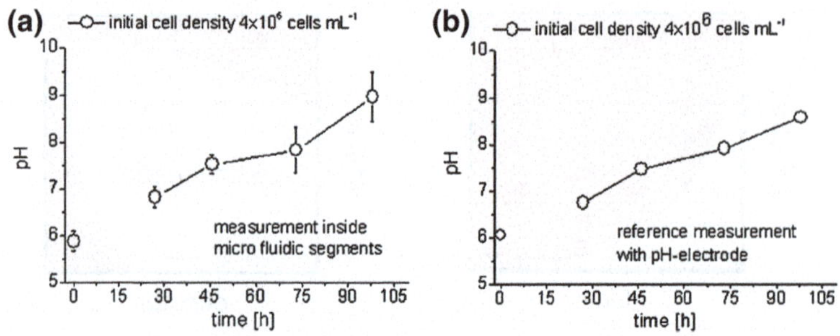

Fig. 8.14 Investigation of the metabolic activity development during a 4-day cultivation of *E. coli*: **a** pH-determination on *E. coli* cultivated inside micro fluid segments by use of pH-sensitive fluorescent micro beads and non-contact optical read-out with microflow-through fluorimeter, **b** reference experiments: pH-determination using pH glass electrode during *E. coli* cultivated in 15 mL falcon tubes. Reprinted with permission from ref [68]. Copyright 2008 Springer-Verlag

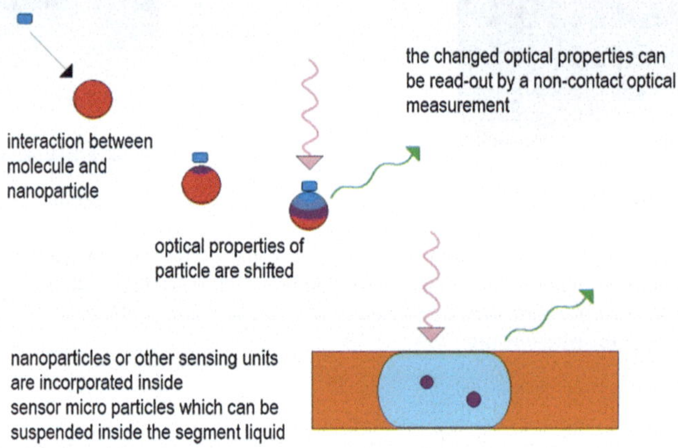

Fig. 8.15 Non-contact monitoring of microbiotechnical processes inside micro fluid segments by use of micro sensor beads

The application of micro sensor beads for primary signal transduction is a promising strategy for the characterization of micro fluid segments, in general [69–71]. This allows the realization of a transduction chain for chemical sensing with a virtually non-contact read-out. The primary and direct interaction between the analyte and the sensor takes place inside the fluid segments. The primary sensor reacts on the change of its optical properties like for example the colour change or the concentration-dependent fluorescence intensity (Fig. 8.15). Then, this change can be read-out without any direct contact by optical devices. The only precondition is the applicability of sensor beads inside the micro fluid segments. The danger of an undesired interference between the target organisms and the sensor particles can be kept low by the use of inert polymer materials for carrying the sensor dyes and through dye molecules being covalently bound to the polymer matrix. The concept of sensor beads provides essential advantages in comparison to dissolved indicator dyes:

1. The indicator system (dye molecules) is partially decoupled from the biological system. Thus, the dye molecules can not be metabolized by the organisms. On the other hand, the molecules are immobilized and can not act as toxins on the target organisms. The tendency of interference with other effectors is strongly reduced.
2. The sensor particles can be applied directly by suspending them inside the cultivation mixture. No additional transducers are required in relation with the chemical and the biological system.
3. The small sensor particles act like disposable devices. Therefore, aging effects can be neglected. The sensor signals are not affected by cross-talk or regeneration procedures. The reproducibility is intrinsic high and referencing is easily possible by the application of the same sensor particles in micro fluid segments with different concentrations of effectors.
4. The sensor bead concept can also be used for multi-channel sensing. Hence, sensor beads working in different spectral ranges can be applied for a parallel optical read-out using different spectral channels.
5. Sensor beads are also suited for an image-based multi-channel read-out.
6. The polymer matrix of sensing particles can be used for enrichment of analyte components (extraction effect) and for the enhancement of sensitivity and selectivity of sensing, like for example in the case of selective adsorption.

The sensor bead concept shows great promise due to the functional decoupling of essential components of the sensing system. The functional decoupling is realized by the high selectivity of the detection and an universal non-contact secondary transduction at the same time: Hereby, the optical read-out represents the universal non-contact secondary transduction principle and the direct contact between the analyte liquid and the sensing material can be used for realizing highly selective molecular interactions, as for example with molecular recognition. This strategy will become very important in the near future for the development of highly specific assays for toxicological investigations which are suited for application in micro fluid segment technology.

The consumption of oxygen or photosynthetic activity can be monitored after the application of oxygen-sensitive sensor particles [72, 73]. In this case a triplet-dye is immobilized in the swellable polymer matrix of the sensor particles: An increase of fluorescence intensity indicates a decreasing luminescence quenching due to the reduction of molecular oxygen.

8.7 Interferences Between Food Components, Nanoparticles and Antibiotics

The effect of food components on the activity of drugs is particular interesting in case of antibiotics. On the one hand, it is of interest how food components affect the antibiotic activity, which is important for the dosing of antibiotics. On the other hand, the effect of antibiotics on the bacterial flora inside the destination tract is important for the treatment of infections, diseases and rehabilitation.

The effect of antibiotics in the presence of a certain food component was investigated in a three-dimensional parameter study. Therefore the interference between chloramphenicol, caffeine and a phenolic uncoupler for *E. coli* as target organism was chosen [74]. The interaction was investigated up to the lethal range of all three components. High resolved dose/response functions of light scattering as well as the autofluorescence for the single components were already shown in Fig. 8.4a for DNP, Fig. 8.4c for chloramphenicol and Fig. 8.7b for caffeine, respectively. The flow rate program for the three-dimensional concentration space is shown in Fig. 8.16a. For the measurement, the concentration of 2, 4-dinitrophenol (DNP) was varied quickly, the concentration of caffeine slowly, and the concentration of chloramphenicol was varied by a mediate flow rate sweep. Figure 8.16b, c show the response of the *E. coli* populations in the segments measured by light scattering and fluorescence after 21 h incubation. No growth takes place at higher concentrations of chloramphenicol (large windows) and on the highest concentrations of the phenolic uncoupler (small windows). A stimulating effect of caffeine close to the critical concentration of DNP

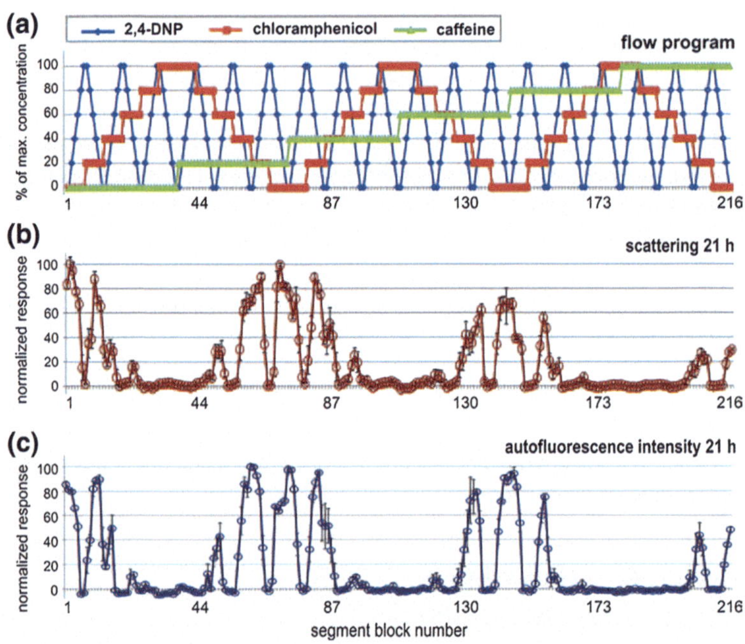

Fig. 8.16 **a** Flow rate program for the generation of 3D addressing with step-wise varied compositions of chloramphenicol, caffeine and 2, 4-DNP in a segment sequence with over 1100 segments. **b** Effects of the ternary mixture on the reduction of light intensity by using microflow-through photometry after 21 h *E. coli* cultivated inside segments. **c** Response on autofluorescence intensity by using microflow-through fluorimetry. Reprinted with permission from ref [74]. Copyright 2012 American Scientific Publishers

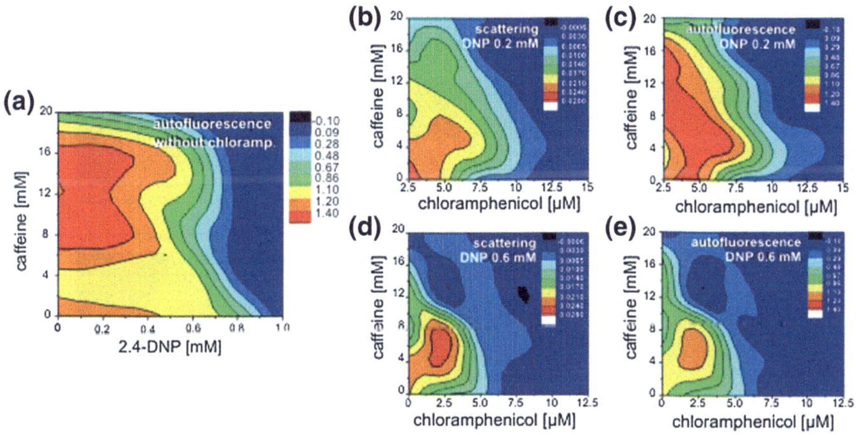

Fig. 8.17 Bolographic maps for the combinatorial activity of caffeine and chloramphenicol at different concentrations of DNP after 21 h *E. coli* cultivated inside micro segments: **a** binary mixture of caffeine and DNP, **b** variation of caffeine and CMP concentration for 0.2 mM DNP obtained from the light scattering, **c** variation of caffeine and CMP concentration for 0.2 mM DNP obtained from the autofluorescence intensity, **d** variation of caffeine and CMP concentration for 0.6 mM DNP obtained from the light scattering, **e** variation of caffeine and CMP concentration for 0.6 mM DNP obtained from the autofluorescence intensity. Reprinted with permission from ref [74]. Copyright 2012 American Scientific Publishers

is well illustrated by the bolometric maps (Fig. 8.17a). The existence of a considerable synergism in the stimulating effect by chloramphenicol and caffeine at higher DNP concentration was found (Fig. 8.17b), which compensates the antibacterial effect of DNP and leads to a significant enlargement of the lethal DNP dose.

It is well known that silver nanoparticles are toxic for bacteria. That is why they are used in bacteriostatic materials and coatings. Hence, there is nearly no knowledge on the interference of these bacteriostatic effects with other toxins. The combinatorial effect of a phenolic uncoupler (2,4-dinitrophenol DNP) in the presence of gold and silver nanoparticles was studied by use of a two-dimensional approach realized by covering a complete concentration space with micro fluid segments of different concentration pairings [75]. A nearly additive effect of gold nanoparticles and the DNP was found and well reflected on the isobolographic maps. In contrast, the presence of silver enforces the effect of DNP strongly. The isobolographic reflect a strong synergistic interaction between the both components (Fig. 8.13b).

Complex relations have been discovered by the investigation of interaction of nanoparticles with two antibiotics. Therefore, a three-dimensional screening of the response of *E. coli* against ternary combinations of silver nanoparticles, chloramphenicol and ampicillin using micro segmented flow technique was realized [67].

The transition between stimulation, compensation and synergism between chloramphenicol, ampicillin and silver nanoparticles against the growth of *E. coli* is well illustrated by the isobolographic maps in Fig. 8.18. The 3-D plot shows the

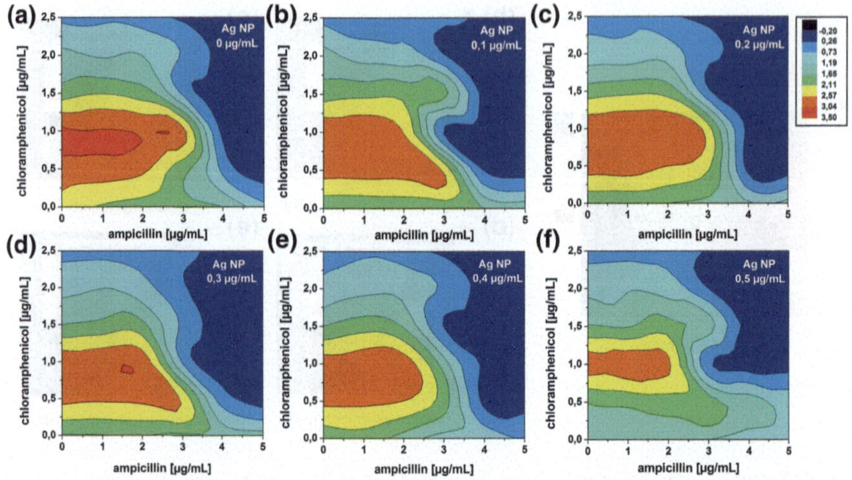

Fig. 8.18 Dose/response curves for a 3D-space with the two antibiotic drugs (ampicillin and chloramphenicol) and silver nanoparticles based on the autofluorescence intensity of *E. coli* after 24 h cultivation inside micro droplets. Reprinted with permission from ref [67]. Copyright 2012 Royal Society of Chemistry

growth-stimulating effect of chloramphenicol which is compensated to some extent by ampicillin. Elevated levels of silver nanoparticles further compensate the stimulatory effect of chloramphenicol and act synergistically with both antibiotics at sublethal concentration to inhibit bacterial growth. This potentiation of the antibiotic action by silver nanoparticles can be explained by the different mode of action of the substances and has already been described by other researchers [76].

8.8 Application of Micro Fluid Segments for Studying Toxic Effects on Multicellular Organisms

Microfluidics can be applied for investigations of multicellular systems when the size of organisms matches with the dimensions of fluidic chambers or channels [44, 45]. Also, the application of micro fluid segment technique is not restricted on cell suspensions, but can also be used for screenings using more complex living objects [42]. Thereby, it was shown that cell ensembles and tissue fragments can be handled and investigated in micro fluid segments. Even multicellular organisms such as the nematode *C. elegans* and fly embryos had been introduced and handled in microfluidic systems [43, 77].

The application of eggs and embryos for minimalized toxicological screenings was already shown 2007 by the introduction of zebra fish eggs into micro fluid segments and the cultivation of fish embryos (Fig. 8.19). It was shown that in most

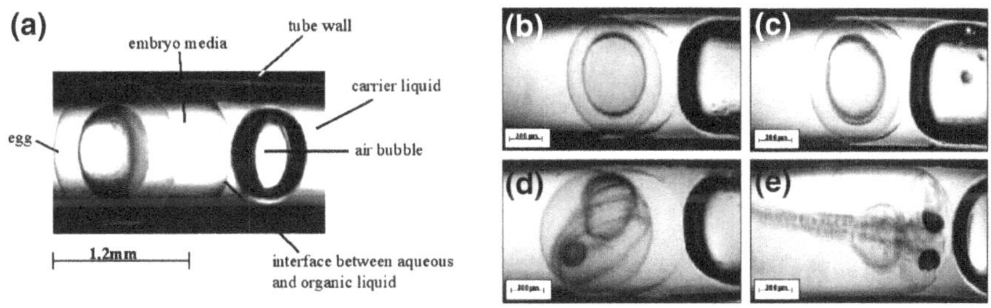

Fig. 8.19 Cultivation of zebrafish embryos inside micro segments: **a** Multicell stage of a *D. rerio* embryo in a micro droplet, **b** 4–5 h old embryo between the multicell stage and beginning of the epiboly stage, **c** 9–10 h old embryo at the end of the epiboly stage, **d** 48–50 h old embryo and **e** Hatched embryo with an age of 74–75 h. Reprinted with permission from ref [40]. Copyright 2007 Royal Society of Chemistry

cases healthy young fishes could be released from the segments after time spans of up to five days [40]. A moderate enlargement of segment volumes was applied in case of these larger multicellular objects in order to meet their space requirements. A liquid volume of about 8 μL was used for the incubation of fish eggs and the cultivation of fish embryos. A high reproducibility in the segment handling was achieved by using a three-phase system consisting of an alkane liquid carrier phase, the aqueous phase involving the fish eggs and an air bubble which supported a reliable transport of the larger aqueous segments. It was possible to evaluate the response of the developing eggs and embryos on increasing concentrations of the surfactant sodium dodecylsulfate (SDS) and heavy metal ions (copper (II)—chloride) on single eggs. Hence, an absolute minimization of consumed target organisms was realized. Beside the dose-dependent survival of developing embryos, different special development features such as gastrulation, the development of eyes, the detachment of tail, blood circulation, heat frequency, pigmentation, hatch and the frequency of deformations have been evaluated in dependence on the concentrations of applied noxious substances (Fig. 8.20).

8.9 Potential of the Segmented Flow Technique for Toxicology and Further Challenges

The technique of micro segmented flow opens the gate to detailed studies of complex effects of environmental conditions on the development and the physiological activity of small organisms, small populations and small biocenotic communities. These investigations become possible without excessive consumption of chemicals and biological material and at reasonable costs through the application of single test volumes in the sub-μL range. One further requirement is the down-scaling of the test volumes in order to achieve even higher test series and to have a further reduction of

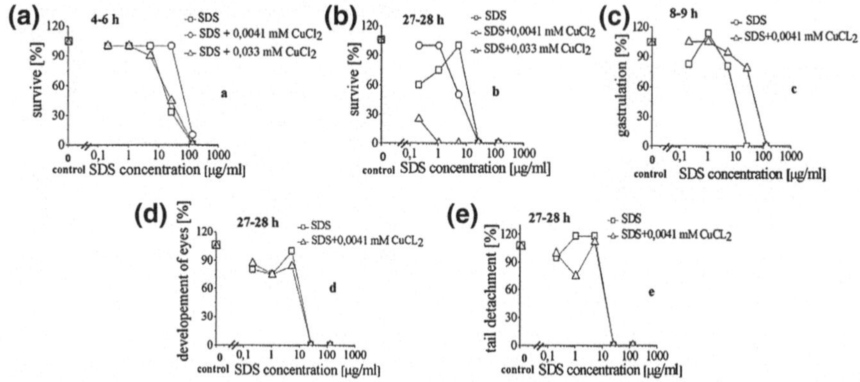

Fig. 8.20 Dose-dependent response of developing zebrafish embryos inside micro fluid segments against SDS and in combination with $CuCl_2$: **a** Survival rate of the zebrafish embryos at an age of 4–6 h. **b** Survival rate at an age of ca. 27 h. **c** Observed gastrulation among 8–9 h old embryos. **d** Eye development at an embryo age of 27 h. **e** Tail detachment at an embryo age of 27 h. Reprinted with permission from ref [40]. Copyright 2007 Royal Society of Chemistry

the required amounts of test substances. On the one site this further miniaturization depends on the minimum of the actual required volume for the given biological target and on the other site on future enhancements of the miniaturization of liquid handling and measurement techniques. The current state of the art on droplet-based microfluidics allows screening runs with single test volumes in the mid-pL range (10…100 pL). These volumes are comparable with the volumes of eukaryotic cells and are therefore too small for certain screenings. But in contrast, these volumes are large in comparison with many prokaryotic microorganisms with typical cell volumes in the lower fL-range. It is expected that these organisms can be handled up to population sizes in the order of about 10^3 when the mid-pL range is used for screenings. These small droplet volumes allow generating millions of well-separated individual fluid compartments with a total fluid volume of less than one milliliter. And there are fluidic devices which are able to generate these tiny droplets with frequencies in the kHz range. Hence, the time need for the generation of these large sample series can be below one hour. However, for guiding these segments through micro channels, these compartments need channel dimensions with a corresponding diameter. A real segmented flow is probably not possible when millions of segments have to be transported through channels with a diameter of about 30–80 µm since the resulting fluid resistance and pressure drop will be very high and the mechanical response of fluid segments on the liquid actuation is difficult to be controlled efficiently. As alternative, such small droplets can be handled as an emulsion in larger channels, what demands for interface-stabilizing surfactants and a labelling of droplets.

Another challenge resolves from the investigation of interorganismic relations [78]. The small fluid segments are ideally for studying minimal biocenotic systems (Fig. 8.21). Thus, they allow evaluating ecological interactions between different organisms under systematic variation of different start parameters such as substrate

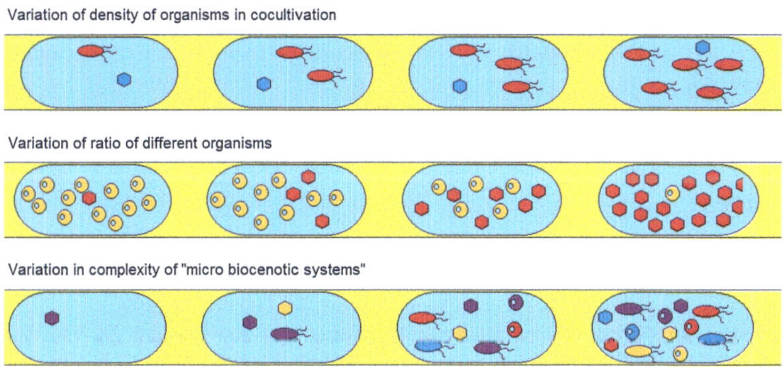

Fig. 8.21 Schema of minimal biocenotic systems

type, concentration, pH, salt content and other chemical factors as well as in dependence on the cell density or on the starting ratio of cell numbers of different types.

More challenging than the application of different organisms is: the read-out of densities of cells of different types, the read out of physiological activity or cell state and the determination of the type and concentration of cell-released substances effecting the intercellular relations. These challenges demand for the development of rapid processing imaging techniques and techniques for fast and specific read-out of molecular information from fluid segments.

The use of micro sensor beads is one step in this direction. Recently, the chemical information such as changes in pH or oxygen content are less specific and allow only a rough estimation of chemical changes in the segment liquid. More details about the state of cells can be gained when it would become possible to introduce sensor beads in to the cytoplasm of target cells so that these can report about changes in the inner cellular medium. This principle could work by using endocytosis active target cells.

Even more important would be the development of miniaturized sensing principles for detecting intercellular communication factors at low concentrations and the read-out of highly specific molecular information from the segment liquid as well as from the cells themself. Therefore, techniques have to be adapted to micro fluid segments which address vibrational finger prints of molecules like IR or Raman spectroscopy or allowing a selective read-out of information by bio molecular methods such as antigen/antibody interactions or selective molecular amplification such as in the case of PCR for nucleic acid characterization.

Finally, the technique of micro segmented flow would not solely be of interest for analytical or screening procedures, but could also come into a key role in biotechnology. It is expected that the parallelized handling of cell cultures will also allow implementing cell biological techniques and molecular biological techniques for larger sequences of micro fluid segments. Overall, this may even lead to improvements on the development of automated techniques for large experimental series in advanced biotechnology.

Acknowledgments This research was financially supported by the German Federal Environmental Foundation (DBU) and by the BMBF (OPTIMI project KFZ-16SV3701 and the BactoCat project KFZ-031A161A).

References

1. K. Drescher, W. Boedeker, Biometrics **51**, 716–730 (1995)
2. W.H. Konemann, M.N. Pieters, Food Chem. Toxicol. **34**, 1025–1031 (1996)
3. F.R. Cassee, J.P. Groten, P.J. van Bladeren, V.J. Feron, Crit. Rev. Toxicol. **28**, 73–101 (1998)
4. G.J.A. Speijers, M.H.M. Speijers, Toxicol. Lett. **153**, 91–98 (2004)
5. M.M. Mumtaz, D.B. Tully, H.A. El-Masri, C.T. De Rosa, Environ. Health Perspect. **110**, 947–956 (2002)
6. R.C. Rocha-e-Silva, L.A.V. Cordeiro, B. Soto-Blanco, Comp. Biochem. Physiol. C Toxicol. Pharmacol. **151**, 294–297 (2010)
7. J.P. Groten, V.J. Feron, J. Suhnel, Trends Pharmacol. Sci. **22**, 316–322 (2001)
8. M. Aufderheide, Exp. Toxicol. Pathol. **60**, 163–180 (2008)
9. D. Krewski, D. Acosta, M. Andersen et al., J. Toxicol. Environ. Health B Crit. Rev. **13**, 51–138 (2010)
10. G. Ulrich-Merzenich, D. Panek, H. Zeitler, H. Vetter, H. Wagner, Indian J. Exp. Biol. **48**, 208–219 (2010)
11. K. Martin, T. Henkel, V. Baier, A. Grodrian, T. Schon, M. Roth, J.M. Köhler, J. Metze, Lab Chip **3**, 202–207 (2003)
12. A. Grodrian, J. Metze, T. Henkel, K. Martin, M. Roth, J.M. Köhler, Biosens. Bioelectron. **19**, 1421–1428 (2004)
13. A. Aharoni, G. Amitai, K. Bernath, S. Magdassi, D.S. Tawfik, Chem. Biol. **12**, 1281–1289 (2005)
14. J.T. Schumacher, A. Grodrian, K. Lemke, R. Romer, J. Metze, Eng. Life Sci. **8**, 49–55 (2008)
15. J.U. Shim, L.F. Olguin, G. Whyte, D. Scott, A. Babtie, C. Abell, W.T.S. Huck, F. Hollfelder, J. Am. Chem. Soc. **131**, 15251–15256 (2009)
16. M.S. Kim, J.H. Yeon, J.K. Park, Biomed. Microdevices **9**, 25–34 (2007)
17. P.M. van Midwoud, G.M.M. Groothuis, M.T. Merema, E. Verpoorte, Biotechnol. Bioeng. **105**, 184–194 (2010)
18. A. Schober, U. Fernekorn, B. Lubbers, J. Hampl, F. Weise, G. Schlingloff, M. Gebinoga, M. Worgull, M. Schneider, C. Augspurger, C. Hildmann, M. Kittler, M. Donahue, Materialwiss. Werkstofftech. **42**, 579–579 (2011)
19. U. Fernekorn, J. Hampl, F. Weise, C. Augspurger, C. Hildmann, M. Klett, A. Laffert, M. Gebinoga, K.F. Weibezahn, G. Schlingloff, M. Worgull, M. Schneider, A. Schober, Eng. Life Sci. **11**, 133–139 (2011)
20. A. Tirella, M. Marano, F. Vozzi, A. Ahluwalia, Toxicol. In Vitro **22**, 1957–1964 (2008)
21. I. Barbulovic-Nad, H. Yang, P.S. Park, A.R. Wheeler, Lab Chip **8**, 519–526 (2008)
22. L. Frenz, K. Blank, E. Brouzes, A.D. Griffiths, Lab Chip **9**, 1344–1348 (2009)
23. C.H.J. Schmitz, A.C. Rowat, S. Koster, D.A. Weitz, Lab Chip **9**, 44–49 (2009)
24. W.W. Shi, H. Wen, Y. Lu, Y. Shi, B.C. Lin, J.H. Qin, Lab Chip **10**, 2855–2863 (2010)
25. J.H. Sung, M.L. Shuler, Bioprocess Biosyst. Eng. **33**, 5–19 (2010)
26. H. Hufnagel, A. Huebner, C. Gulch, K. Guse, C. Abell, F. Hollfelder, Lab Chip **9**, 1576–1582, (2009)
27. J. Clausell-Tormos, C.A. Merten, Front. Biosci. **4**, 1768–79 (2012)
28. O.J. Miller, A. El Harrak, T. Mangeat, J.C. Baret, L. Frenz, B. El Debs, E. Mayot, M.L. Samuels, E.K. Rooney, P. Dieu, M. Galvan, D.R. Link, A.D. Griffiths, Proc. Nat. Acad. Sci. U.S.A. **109**, 378–383 (2012)
29. Y. Zhu, N. Wu, C.J. East, Front. Biosci. **5**, 284–304 (2013)

30. J.M. Köhler, T. Henkel, A. Grodrian, T. Kirner, M. Roth, K. Martin, J. Metze, Chem. Eng. J. **101**, 201–216 (2004)
31. B. Zheng, J.D. Tice, R.F. Ismagilov, Anal. Chem. **76**, 4977–4982 (2004)
32. J.M. Köhler, T. Kirner, Sens. Actuators A Phys. **119**, 19–27 (2005)
33. B. Zheng, R.F. Ismagilov, Angew. Chem. Int. Ed. **44**, 2520–2523 (2005)
34. J.Q. Boedicker, L. Li, T.R. Kline, R.F. Ismagilov, Lab Chip **8**, 1265–1272 (2008)
35. J. Clausell-Tormos, A.D. Griffiths, C.A. Merten, Lab Chip **10**, 1302–1307 (2010)
36. V. Trivedi, A. Doshi, G.K. Kurup, E. Ereifej, P.J. Vandevord, A.S. Basu, Lab Chip **10**, 2433–2442 (2010)
37. T.S. Kaminski, S. Jakiela, M.A. Czekalska, W. Postek, P. Garstecki, Lab Chip **12**, 3995–4002 (2012)
38. I. Schneegass, J.M. Köhler, J. Biotechnol. **82**, 101–121 (2001)
39. I. Schneegass, R. Brautigam, J.M. Köhler, Lab Chip **1**, 42–49 (2001)
40. A. Funfak, A. Brosing, M. Brand, J.M. Köhler, Lab Chip **7**, 1132–1138 (2007)
41. F. Yang, Z.G. Chen, J.B. Pan, X.C. Li, J. Feng, H. Yang, Biomicrofluidics **5**, 024115 (2011)
42. J. Clausell-Tormos, D. Lieber, J.C. Baret, A. El-Harrak, O.J. Miller, L. Frenz, J. Blouwolff, K.J. Humphry, S. Koster, H. Duan, C. Holtze, D.A. Weitz, A.D. Griffiths, C.A. Merten, Chemi. Biol. **15**, 875–875 (2008)
43. W.W. Shi, J.H. Qin, N.N. Ye, B.C. Lin, Lab Chip **8**, 1432–1435 (2008)
44. M.M. Crane, K. Chung, J. Stirman, H. Lu, Lab Chip **10**, 1509–1517 (2010)
45. N. Chronis, Lab Chip **10**, 432–437 (2010)
46. R.F. Ismagilov, Angew. Chem. Int. Ed. **42**, 4130–4132 (2003)
47. A. Huebner, S. Sharma, M. Srisa-Art, F. Hollfelder, J.B. Edel, A.J. Demello, Lab Chip **8**, 1244–1254 (2008)
48. K. Martin, K. Lemke, T. Henkel, BioSpektrum **12**, 743–745 (2006)
49. M. Roth, K. Martin, T. Henkel, in *Proceedings of 14th Heiligenstädter Kolloquium*, 177–180, (2008)
50. A.B. Theberge, F. Courtois, Y. Schaerli, M. Fischlechner, C. Abell, F. Hollfelder, W.T.S. Huck, Angew. Chem. Int. Ed. **49**, 5846–5868 (2010)
51. M.E. Vincent, W.S. Liu, E.B. Haney, R.F. Ismagilov, Chem. Soc. Rev. **39**, 974–984 (2010)
52. W.B. Du, M. Sun, S.Q. Gu, Y. Zhu, Q. Fang, Anal. Chem. **82**, 9941–9947 (2010)
53. W. Tanthapanichakoon, N. Aoki, K. Matsuyama, K. Mae, Chem. Eng. Sci. **61**, 4220–4232 (2006)
54. T. Henkel, T. Bermig, M. Kielpinski, A. Grodrian, J. Metze, J.M. Köhler, Chem. Eng. J. **101**, 439–445 (2004)
55. D. Malsch, N. Gleichmann, M. Kielpinski, G. Mayer, T. Henkel, D. Mueller, V. van Steijn, C. R. Kleijn and M. T. Kreutzer, Microfluid. Nanofluid. **8**, 497–507, (2010)
56. W. Wang, C. Yang, Y.S. Liu, C.M. Li, Lab Chip **10**, 559–562 (2010)
57. J.M. Köhler, A. Groß, Microfluid. Nanofluid. **3**, 653–663 (2007)
58. A. Funfak, M. Fischlechner, E. Donath, J.M. Köhler, AICHE Spring National Meeting, Houston/Texas, Conf. Proc., oral paper 121c (2007)
59. K.R. Strehle, D. Cialla, P. Rosch, T. Henkel, J.M. Köhler, J. Popp, Anal. Chem. **79**, 1542–1547 (2007)
60. A. Marz, T. Henkel, D. Cialla, M. Schmitt, J. Popp, Lab Chip **11**, 3584–3592 (2011)
61. M. Budden, S. Schneider, G.A. Groß, J.M. Köhler, Sens. Actuators A Phys. **189**, 288–297 (2013)
62. P.M. Gunther, S. Schneider, G.A. Groß, J.M. Köhler, in *Proceedings of Microsystem Technology Congress Darmstadt*, vol. 10, pp. 945–947, (2011)
63. T. Bergan, I.B. Carlsen, J. Antimicrob. Chemother. **15**, 147–152 (1985)
64. S.R. Ford, R.L. Switzer, Antimicrob. Agents Chemother. **7**, 555–563 (1975)
65. A. Funfak, R. Hartung, J.L. Cao, K. Martin, K.H. Wiesmuller, O.S. Wolfbeis, J.M. Köhler, Sens. Actuators B Chem. **142**, 66–72 (2009)
66. D. Kuersten, J. Cao, A. Funfak, P. Mueller, J.M. Köhler, Eng. Life Sci. **11**, 580–587 (2011)

67. J.L. Cao, D. Kursten, S. Schneider, A. Knauer, P.M. Gunther, J.M. Köhler, Lab Chip **12**, 474–484 (2012)
68. A. Funfak, J.L. Cao, O. Wolfbeis, K. Martin, J.M. Köhler, Microchim. Acta. **164**, 279–286 (2009)
69. R. Di Corato, P. Piacenza, M. Musaro, R. Buonsanti, P.D. Cozzoli, M. Zambianchi, G. Barbarella, R. Cingolani, L. Manna, T. Pellegrino, Macromol. Biosci. **9**, 952–958 (2009)
70. S. Nagl, O.S. Wolfbeis, Analyst **132**, 507–511 (2007)
71. M.I.J. Stich, L.H. Fischer, O.S. Wolfbeis, Chem. Soc. Rev. **39**, 3102–3114 (2010)
72. S.K. Lee, I. Okura, Anal. Commun. **34**, 185–188 (1997)
73. C. Baleizao, S. Nagl, M. Schaferling, M.N. Berberan-Santos, O.S. Wolfbeis, Anal. Chem. **80**, 6449–6457 (2008)
74. J.L. Cao, D. Kursten, S. Schneider, J.M. Köhler, J. Biomed. Nanotechnol. **8**, 770–778 (2012)
75. A. Funfak, J.L. Cao, A. Knauer, K. Martin, J.M. Köhler, J. Environ. Monit. **13**, 410–415 (2011)
76. A.M. Fayaz, K. Balaji, M. Girilal, R. Yadav, P.T. Kalaichelvan, R. Venketesan, Nanomed. Nanotechnol. Biol. Med. **6**, 103–109 (2010)
77. J.M. Köhler, T. Henkel, Appl. Microbiol. Biotechnol. **69**, 113–125 (2005)
78. J. Park, A. Kerner, M.A. Burns, X.X.N. Lin, Plos One **6**, e17019 (2011)

Chapter 9
Screening for Antibiotic Activity by Miniaturized Cultivation in Micro-Segmented Flow

Emerson Zang, Miguel Tovar, Karin Martin and Martin Roth

Abstract Despite arduous dedication of the scientific community during the last decades, most screening efforts failed to reveal new antibiotic substances. Droplet-based microfluidics provide a powerful platform to effectively exploit natural metabolic diversity and revitalize the search for new antimicrobials. In this chapter, we explore main requirements to develop and apply droplet-based microfluidic screening strategies for the discovery of novel antibiotics from natural products.

9.1 Introduction: Antibiotics and Antimicrobial Resistance

Short History of Antibiotics and Antimicrobial Resistance: With the discovery of first antibiotics in the 1930s (penicillin, sulfonamides), the eradication of infectious diseases was thought to be just a matter of time. However, even before the introduction of penicillin in 1942, it was discovered that *E. coli* is capable of producing penicillin-hydrolysing β-lactamases, hampering its efficiency as a therapeutic agent [1]. Other molecules representing different compound classes introduced during the "golden era" of antibiotic discovery (1950s) shared a similar fate. Almost simultaneously with their introduction to the market, at least one compound-specific resistance mechanism was unveiled for chloramphenicol, erythromycin or streptomycin, thus challenging the effectiveness of these alleged "silver bullets" [2, 3]. Nonetheless, resistant pathogens were yet not ubiquitous and antibiotics helped save millions of lives threatened by infectious diseases. Since the early 1960s, however, only four new classes of antibiotics were brought to market (Fig. 9.1) [4], while resistant pathogens are becoming increasingly omnipresent. Although time has been gained by chemical tailoring of existing scaffolds, misusage, cross-resistances and a limited number of

E. Zang (✉) · M. Tovar · K. Martin · M. Roth
Leibniz Institute for Natural Product Research and Infection Biology e.V. Hans-Knöll-Institute (HKI), Beutenbergstraße 11a, 07745 Jena, Germany
e-mail: emerson.zang@hki-jena.de

J. M. Köhler and B. P. Cahill (eds.), *Micro-Segmented Flow*,
Biological and Medical Physics, Biomedical Engineering,
DOI: 10.1007/978-3-642-38780-7_9,

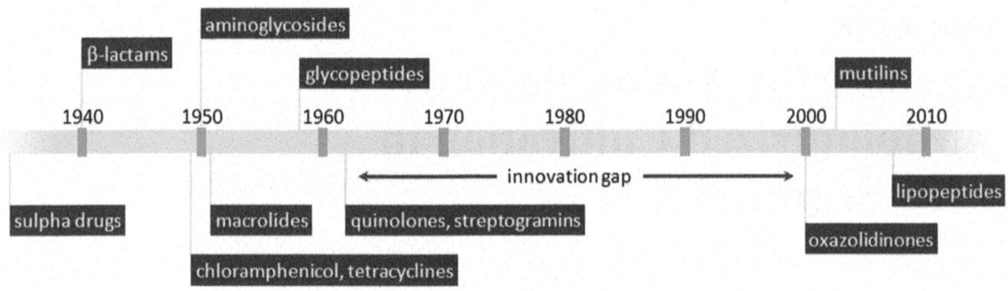

Fig. 9.1 The innovation gap for new classes of antibiotics between 1962 and 2000 (after Fischbach and Walsh [4])

useful derivates constantly diminish this competitive edge. The progressing emergence of new multidrug-resistant (MDR) strains further exacerbates the critical situation, rendering the requirement for new substance classes increasingly urgent [5].

9.2 Current State of Screening for New Antimicrobial Products

Failure of Target-Oriented Screening: After the golden era of antimicrobial compound discovery, classic empirical screening approaches frequently failed to reveal new substances—mainly due to permanent rediscovery of already known compounds. Aside from the carbapenems in 1985, all antibiotics approved for clinical use between the early 1960s and 2000 were synthetic derivates of existing scaffolds [6]. As an attempt to overcome drainage of the lead pipeline, a paradigm shift in screening technology took place: with the dawn of the 1980s, emphasis was put on target-oriented screening approaches, where pathogen-specific putative target-proteins were tested against libraries of natural products as well as libraries of semisynthetic or fully synthetic substances [7]. Despite enormous efforts of pharmaceutical companies dedicated to these strategies, the outcome was disappointing. Although several promising compounds were found, most of them proved to have insufficient membrane permeability [8].

Actinobacteria-Still a Rich Source of New Scaffolds: There is no doubt that the class *Actinobacteria* is still a rich source of new compounds: the theoretical number of substances to be potentially assembled by non-ribosomal peptide synthetases (NRPS) and polyketide synthetases (PKS) is almost infinite. Moreover, there is evidence of a huge pool of so called orphan pathways—gene clusters encoding secondary metabolites not expressed under common cultivation conditions—that remain to be investigated in already known strains [9]. Yet, many of the secondary metabolite producing actinomycetes are still unculturable—often due to inappropriate cultivation methods or low assertiveness of the majority of strains compared to more abundant, fast-growing species like *Streptomyces*. Baltz [10] estimates, that ~99 % of actinomycetes are yet to be discovered. As a consequence, experts propose the return to discovery of antimicrobial compounds from actinomycetes [4, 11], combined with the achievements of high-throughput-screening (HTS).

High-Throughput Screening of Actinomycetes: Since 1536-well microtiter plates already reach the limit for reliable liquid handling, other technical approaches must be established to maximize throughput [12]. An interesting strategy is pursued by Cubist Pharmaceuticals (Lexington, USA): in a first step, microbial spores from soil samples are encapsulated in 2 mm calcium alginate macrodroplet beads together with culture medium containing antimicrobials to prevent growth of microorganisms other than actinomycetes. After germination and growth, the presence of antibiotics is tested with an *E. coli* strain bearing resistances against the most common broadband antibiotics. Thereby, Cubist Pharmaceuticals is capable of screening 10^7 actinomycetes for antimicrobial activity *per annum*, a magnitude in which the discovery of yet unknown antibiotic producers becomes likely [13].

An alternative approach for high-throughput cultivation of rare species was presented by Akselband et al. in 2006 [117]. Marine bacterial cells were encapsulated in agarose gel-microdroplets (GMDs) of 30-50 μm by bulk emulsification. After incubation, confined microcultures were fluorescently labeled and sorted in a FACS device. Despite enormous HTS capabilites, this approach does not allow for complex assay regimes, since GMDs and confined microcultures are inaccessible for further assays steps, e.g. addition of a reporter strain.

To overcome these limitations, we discuss biological and technical aspects of bacterial growth in aqueous sub-microliter droplets to establish the detection of antimicrobial activity within microfluidic chip devices. By this, we aim to contribute to a complete whole-cell high-throughput screening for novel antibiotic substance classes from actinobacteria.

9.3 Microbial Assays in Droplet-Based Microfluidic Systems and in Micro-Segmented Flow

9.3.1 General Considerations for Microbial Assays in Droplet-Based Systems

Droplet-based microfluidics provides excellent means to further increase throughput in antimicrobial screening. Figure 9.2 shows a general workflow for detection of antimicrobial activity from a microbial spore suspension. A remarkable feature of this assay is the generation of millions of pure cultures that allow for independent germination and microcolony formation—undisturbed by other highly abundant, fast-growing bacteria [14]. By using the pathogen of interest as a reporter organism, it is ensured that drugs with pathogen-specific activity are selected. This inhibition is then detected as a reduced fluorescent signal from a genetically engineered reporter (e.g. with GFP or derivates). Different approaches are imaginable, mainly determined by the nature of the applied promoter for expression of the respective gene.

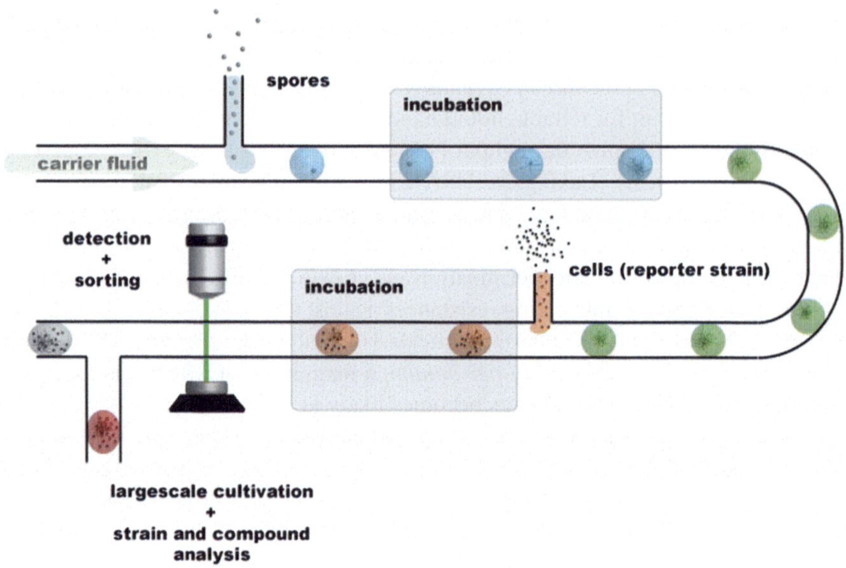

Fig. 9.2 General scheme of screening for antibiotic-producing *Actinobacteria*

Performing whole-cell-assays on a droplet-based microfluidic platform requires a fine tuning of the growth medium regarding optical (background fluorescence, turbidity), rheologic (viscosity, surface tension) [15] and physico-chemical (wettability, critical micellar concentration) demands. In many cases, this translates into suboptimal conditions for bacterial growth with respect to pH-value, mass transfer, and homogenity of substrates and metabolites. Although media droplets can be considered as microscale-bioreactors, they lack common means of process monitoring and regulation, which complicates the optimization of media and strains for an ideal result. In this section, the main requirements to the growth media and reporter strain for cultivation and reliable fluorescence-based detection of antimicrobial activity in droplets are elucidated.

9.3.2 Culture Media for Droplet-Based Screening

In standard laboratory applications, minimal media are frequently used for cultivation of bacteria. Their advantage is the exact nutrient composition, which allows on one hand for high cultivation reproducibility and on the other precise process balancing, which is of great importance in any application where sound understanding of the cell's metabolism is necessary. With regard to actinomycetes, minimal media are traditionally used for basic genetic studies [16, 17]. Moreover, they were employed in studies on RNA polymerases in *Streptomyces coelicolor* [18, 19], in physiological investigations of *Streptomyces viridochromogenes* [20, 21] and for the assessment of production of antibiotics in actinomycetes [22]. The most prominent minimal medium M9, first referenced by Sambrock et al. [23], contains only mineral salts and 5 g/L glucose.

In contrast to synthetic media, complex broths generally provide better growth conditions and also promote the secondary metabolism of actinomycetes. Hence, with the purpose of finding a suitable culture medium for microfluidic screening of antibiotics-producing actinomycetes, five media with reportedly good promotion of actinomycete growth were selected and optimized towards low phosphate concentration, since it is known that secondary metabolites are often produced at phosphate depletion [24, 25]. An important criterion for exclusion of a broth was the presence of insoluble ingredients, such as oatmeal, soybean meal, corn steep or calcium carbonate, which causes turbidity and hence impedes optical read-out methods to be applied in the final microfluidic assay. 14 different representative actinobacteria were cultivated for 4 days in the chosen media and tested for biomass formation and antimicrobial activity against *E. coli*, *B. subtilis* and *S. cerevisae* in agar diffusion tests (Table 9.1).

Growth was observed in all strain/medium-combinations, and each medium triggered in at least four species production of compounds which inhibited growth of one or more of the reporter strains. Even synthetic M9 medium proved being capable of promoting growth of actinobacteria, but compared to complex media tested in this scope, about 80 % less biomass was produced and inhibition zones were smaller on average, indicating inferior production of antimicrobials.

Modified malt medium ("MMM"), containing malt-, yeast- and beef-extract, appeared to be the most potent. It triggered production of antibiotics in seven out of 14 tested strains, and its supernatants generated the highest number of inhibition zones, i.e. 12 out of 42 possible combinations of actinobacterium and reporter organism. For Boehringer-8 medium similar results were obtained, but the zones of inhibition were in a few cases less distinct compared to the zones of inhibition generated by supernatants of MMM. Due to these findings, MMM was chosen for further experimentation in droplets.

Since active pH-control is not feasible in microfluidic droplets, buffering is the only available strategy to avoid undesired pH-shifts. In the case of MMM (pH 7.2), buffering might have been beneficial for some strains where the pH-value shifted remarkably after 4 days of cultivation. Interestingly, those cultures in MMM where the final pH value was above 7.7 (*K. azatica, A. mediterranei and A. lactamdurans*) or below 5.97 (*N. pusilla* and *S. sibiricum*), the supernatant did not show any antimicrobial activity at all. In case buffering is required for antibiotic screening, it must be desisted from phosphate buffers due to the aforementioned reasons.

Depending on the concentration of complex organic contents (extracts, hydrolysates), the culture media may exhibit particularly high background fluorescence, especially at short wavelength (Fig. 9.3). If fluorescence-based read-out techniques are applied for detection of reporter cell growth or inhibition, strong background fluorescence reduces the sensitivity of the assay. This can be circumvented by dilution of the medium, which might as well be beneficial for the achieved diversity of cultivated strains: it is assumed that less nutritive media, especially with respect to

Table 9.1 Produced biomass (BM [mg]) and inhibition zone diameters (IZ [mm]) for different strain/medium combinations (cultivation: 10 ml in shake flask, 4 days, 28 °C)

Medium		LB			B8			MMM			M9			Glu-So		
Strain		*E*	*S*	*B*	*E*	*S*	*B*	*E*	*S*	*B*	*E*	*S*	*B*	*E*	*S*	*B*
Actinomadura kijaniata IMET 9741	BM	79.4			73.55			45.11			10.19			54.1		
	IZ	0	0	0	0	0	0	0	0	17	0	0	20	0	0	21
Actinoplanes sp. HKI 715	BM	37.22			65.79			22.57			22.91			52.26		
	IZ	0	0	0	0	15	0	0	0	0	0	0	0	0	0	0
Amycolatopsis mediterranei IMET 7651	BM	22.54			37.52			31.77			10.22			80.94		
	IZ	0	0	0	0	0	0	0	0	0	0	0	0	0	0	0
Amycolatopsis lactamdurans HKI 323	BM	29.96			60.37			37.45			5.19			49.77		
	IZ	0	0	0	15	0	0	0	0	0	0	0	0	0	0	0
Kitasatospora azatica HKI216	BM	36.5			23.44			32.2			10.9			30.87		
	IZ	0	0	0	0	0	0	0	0	0	0	0	0	0	0	0
Kutzneria viridogrisea HKI 207	BM	41.49			49.78			55.91			11.49			78.2		
	IZ	0	0	0	17	30	17	12	18	2	0	0	10	20	30	22
Nonomureae pusilla IMET 9586	BM	12.05			10.21			22.88			7.9			18.53		
	IZ	0	0	0	0	0	0	0	0	0	0	0	0	0	0	0
Saccharopolyspora erythrea HKI 184	BM	77.37			73.32			64.43			8.09			54.09		
	IZ	0	0	25	0	0	26	0	0	24	0	0	0	0	0	25
Saccharothrix mutabilis subsp. capreolus HKI 264	BM	50.62			50.81			86.42			16.85			67.39		
	IZ	0	0	0	0	0	0	0	0	15	0	0	0	0	0	0
Streptomyces noursei JA 3890	BM	108.41			59.08			78.92			21.62			54.79		
	IZ	0	30	0	0	60	18	0	50	18	0	33	0	0	11	15
Streptomyces rapamycinicus IMET 43975	BM	66.5			41.9			48.15			8.73			45.63		
	IZ	0	15	15	0	27	14	0	27	15	0	0	15	0	25	11
Streptomyces sp. HKI 714	BM	30.72			30.77			41.88			6.81			48.56		
	IZ	0	25	26	0	23	26	0	26	30	0	0	0	0	25	28
Streptosporangium sibiricum HKI 30	BM	40.21			53.74			71.24			11.98			63.04		
	IZ	0	0	0	0	0	0	0	0	0	0	0	0	0	0	0
Streptosporangium sibiricum FH 8054	BM	77.37			73.32			64.43			8.09			54.09		
	IZ	0	0	0	0	0	0	0	0	0	0	0	0	0	0	0
triggered strains		4/14			7/14			7/14			4/14			6/14		
inhibition events		6/42			12/42			12/42			4/42			11/42		

Antimicrobial activity of supernatants was assessed by agar diffusion tests against *Escherichia coli* (*E*), *Saccharomyces cerevisiae* (*S*) and *Bacillus subtilis* (*B*)

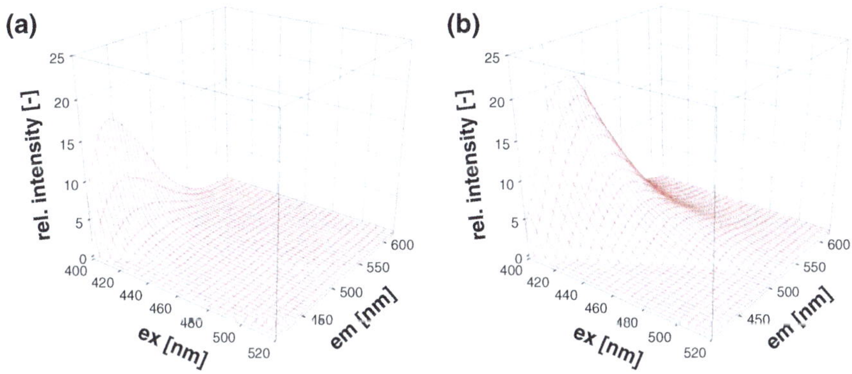

Fig. 9.3 Fluorescence spectrum of M9-medium (**a**) and LB-medium (**b**)

Table 9.2 Organisms and media applied in recent studies in the field of droplet-based microfluidics

Organism	Medium	Modification	Source
U937	RPMI (Invitrogen)	10 % fetal bovine serum	[46]
HEK293T	DMEM (Invitrogen)	10 % fetal bovine serum	[87]
Jurkat	RPMI (Invitrogen)	10 % fetal bovine serum	
S. cerevisae	SD-medium	-	[88]
P. fluorescens, *R. rhodochrous,* *E. coli*	LB-medium	-	[59]
E. coli	M9-medium	-	[100]
S. aureus	LB-medium	-	[45]

the carbon source, decelerate the propagation of fast-growing species ("lab weeds") and thus favour slow-growing, less abundant strains [14]. However, the incubation time might increase from days to several weeks.

As shown above, commonly used media often provide a reasonable starting point for tailoring a growth medium according to specific requirements of the cultivated organism and the detection strategy. Table 9.2 lists examples of media employed for the cultivation of cells and microorganisms in recent publications related to droplet-based microfluidics.

9.3.3 Detection Mechanisms for Droplet-Based Screening

The reporting mechanism is the linchpin of every assay. Depending on the application, droplet-based microfluidics provide various means for assessment of ongoing biological, chemical and biochemical processes in single droplets. To maintain droplet integrity and avoid cross-talk between individual droplets, most detection

Table 9.3 Non-invasive readout techniques for microfluidic droplets

	Read-out	Applicability for AB-screening	Reference
Optical	FRET	–	[101]
	Fluorescent proteins	++	[27, 101–103]
	Fluorescent dyes (life/dead staining)	–	
	Fluorogenic reactions	+	[104]
	Luminogenic reactions	(+)	[35, 36]
Non-optical	Amperometric	–	[105]
	Capacitive	–	[63, 106]
	Raman spectroscopy	(+)	[107]

techniques are non-invasive (Table 9.3). For sensing antibiotic-induced inhibition, optical read-outs are the method of choice, since reporter cell-generated signals can be easily tracked and filtered from complex biological and chemical background found in bacterial cultures. Monitoring synthesis of fluorescent proteins (FPs) like GFP, RFP and their derivatives [26] provides a simple means to verify cell fitness. Ideally, measured fluorescence correlates directly with the growth of the reporter strain or with the magnitude of protein synthesis, respectively [27]. Main benefits of FPs include the avoidance of a further assay step in which a fluorogenic substrate is added and their long-term stability. Nevertheless, care must be taken when choosing an FP, since they exhibit great differences in a variety of parameters such as brightness, maturation time and photo stability [28]. In general, derivatives of the wildtype-protein exhibit favourable properties, as for example the "m*Fruit*-series" developed by Shaner et al. [28], derived from the red-fluorescent protein of *Discosoma sp.* Yet, if mass transfer in the incubation device is poor (e. g. in glass chips), maturation and thus fluorescence of the FP can be unsatisfactory due to oxygen limitation [29, 30]. As a work-around, anaerobic fluorescent proteins (e.g. evoglow®, evocatal) can be employed as reporter [31]. However, susceptibility to photo bleaching of such FP should be taken into account, although it might be considered negligible in the case of short exposure times.

Enzymes with a fluorogenic substrate are also generally suited for monitoring cell viability. Baret et al. [32] implemented a β-galactosidase as the reporting enzyme in a highly sensitive fluorescence-activated droplet sorting (FADS) assay, catalyzing the turnover of fluorescein-di-β-D-galactopyranoside to its fluorescent product. Similarly, Agresti et al. [33] established a microfluidic assay to optimize horseradish peroxidase (HRP) itself. Such an approach should also be applicable for an anti-microbial screening assay, using HRP as a representative of entire protein synthesis. Enzymatic assays generally benefit from the conversion of many fluorogenic substrate molecules catalysed by a single enzyme molecule, resulting in leverage of the ratio between synthesized protein to fluorescence. This may be exploited for testing single-cell activity, as already demonstrated for yeast cells [33]. A remarkable drawback of enzyme-based assays is the necessity of rendering the enzyme accessible—either through an additional cell-breakup step (enzymatic lysis [34],

Table 9.4 Fluorescence-based read-out techniques in droplet-based microfluidics

Protein/Enzyme	Substrate	Reference
β-galactosidase	Fluorescein-di-β-D-galactopyranoside (FDG), fluorescein	[104, 108]
PTP1B (target for diabetes mellitus)	Fluorescein diphosphate (FDP)	[104]
HRP	Amplex Ultrared (AUR)	[33]
Arylsulfatase	Fluorescein disulfate, bis(methylphosphonyl)-fluorescein	[34]
YFP (constitutive)	–	[27]
mCherry	–	own unpublished data

electroporation, polymyxin) or by cell surface displaying. Baret et al. [32] relied on a minority of cells in each droplet experiencing autolysis for the above-mentioned FADS assay. In case of an antibiotic screening where co-cultivation of reporter cells and actinobacteria takes place, an increased risk of false negatives is given, since the substrate might not exclusively be hydrolysed by the reporter strain, but also by actinobacteria with yet unknown enzymatic activity. Similarly, live/dead staining, e.g. with resazurin, is also not recommended, for it must be assumed that droplets contain in most cases a living microcolony of actinobacteria, which would frequently cause false positives. Table 9.4 gives an overview of recently applied fluorescence-based assays in the field of droplet microfluidics.

Luminescence-based approaches, such as the firefly luciferase system (lucFF) or the bacterial luciferase derived from *Vibrio fischeri* (luxCDABE), are known to exhibit higher sensitivities and much shorter response times than FP-based assays [35]. Nevertheless, the read-out requires longer integration times, which usually exceeds the droplet's residence time in the measurement section.

Independent of the detection principle, the respective gene must be controlled by a suitable promoter. Three general types of expression control can be taken into consideration: constitutive, compound-inducible and stress-inducible. For several reasons, compound-inducible are preferable to constitutive approaches. First, reporter-cell associated background fluorescence is drastically reduced. This improves the read-out quality and allows for reduction of the droplet volume, which is an important prerequisite for enhancing the throughput of the assay. Accordingly, the inoculation density can be increased, since detection of antimicrobial activity does not merely rely on increment of biomass but also on the degree of protein synthesis. Secondly, taking this concept to its extreme by inoculating with very high cell densities, the time-to-response is reduced: expression and maturation of reporter genes is usually less time-consuming than a sufficient multiplication of biomass. However, if droplets are inoculated with high cell densities, screening is biased towards inhibition of protein synthesis rather than any other mode of antibiotic activity. If detection of the latter is desired, the increment of biomass from an inoculation with low cell densities must be screened. Nevertheless, compound-inducible reporter genes may also help to reduce the frequency of false negatives in such a case. Finally, addition of an inductor to the spore suspension triggers quasi-constitutive expression once the reporter strain enters the droplet. Yet, most inducers do not interfere or affect the microbiological culture.

Stress-inducible promoters represent an entirely different approach for monitoring antibiotic activity. Depending on their origin, stress promoters are activated upon DNA damage, metabolic changes and quorum sensing related events as well as protein-, DNA-, RNA-, and fatty acid synthesis interference [36]. On the basis of these promoters, sets of antibiotic biosensors have been generated with *B. subtilis* [36] and *E. coli* [37–40] as host organisms. The reporting agent is luciferase in both cases, producing temporary luminescence upon addition of luciferin—its luminogenic substrate. However, the application of stress promoters in HTS for antimicrobial activity is connected to several difficulties: Eltzov et al. reported varying response times for different antibiotics when using the same promoter [39], which results in temporally staggered peaks of reporter activity, and thus variability in the optimal timing for droplet analysis. A further stress-promoter related complication is caused by the missing proportionality of substance toxicity and promoter response. Compounds with strong antimicrobial activity do not obligatorily trigger a vigorous promoter response, while a substance without any pathogen-impeding function might produce a false positive due to promoter activation. In addition, every stress-promoter based assay possesses an upper detection limit set by the minimal inhibition concentration of each combination of antibiotic substance and reporter strain. If this limit is exceeded, the cells might be inhibited or killed before generating a response. Moreover, due to the tendency of stress promoters to leak, the difference between maximal response and background activity (response ratio) is relatively low, impeding reliable detection. In 96- and 384-well microtiter-based assays with *E. coli* biosensors, the highest achieved induction was 255-fold for a combination of a porin related antisense-RNA promoter (micF) and the sulfonamide antibiotic sulfamethoxazole, after 10 h of incubation [39]. The average maximal response ratio of all 14 promoters tested in this study only accounted to 45.3. Also measured in a 384-well microtiter plate, the *B. subtilis* biosensors merely reached a 13.8-fold enhanced luminescence after induction [36]. In contrast, >500-fold increase in total expression is achievable with common compound-induced promoters, e.g. the T7-polymerase system (NEB). However, efforts are being undertaken to enhance response ratio and intensity of stress promoters by sequence-elongation, random mutagenesis, site-directed mutagenesis and promoter duplication [41]. Yet, a striking argument in favor of a stress promoter-based assay is its "inverted" response compared to approaches detecting inhibition of bacterial growth or protein synthesis since screening for fluroescent droplets is technically less demanding than screening for non-fluorescent droplets.

9.3.4 Reporter Organisms for Droplet-Based Screening

Depending on the type of pathogen to be targeted with a potential novel antibiotic, differentiation between three general types of reporter organisms is reasonable: Gram-positive, Gram-negative and fungi. Each type has one or several "work horses" (e.g. *B. subtilis, E. coli* and *S. cerevisiae*) which can also well be utilized in the microfluidic system: they are widely studied, grow fast, produce large amounts of recombinant protein and are flexible with respect to growth conditions, especially regarding C, N and P sources and pH-value. Most importantly, genetic modifications and engineering can be easily performed, due to the manifold of plasmids, kits and protocols that have been tailored for their physiology.

Since filamentous fungi show a tendency to grow out of droplets (personal observation), the group of fungi can solely be represented by yeasts. However, many of today's pathogenic fungi are yeasts (e.g. *Candida* species) and hence worth targeting.

Upon designing a reporter strain, the antibiotic resistances used as selective marker should be carefully taken into account. Discovery of antibiotics belonging to the respective class is ruled out, which might be counterproductive for the actual screening. Yet, rediscovery of already known, abundant classes of antimicrobials is preferably avoided, so that the reporter strain's resistance against this class is beneficial. Cubist Pharmaceuticals took advantage of this screening concept by generating an *E. coli* reporter strain bearing multiple antibiotic resistance genes integrated into the chromosome. This strain was employed for an alginate-bead-based assay for the discovery of novel antibiotics with new modes of action [10].

9.3.5 Aspects of Co-cultivation of Different Microbial Species

After addition of the reporter cells to the droplets bearing actinobacteria microcultures, a co-cultivation takes place, which is related to specific difficulties. Essential nutrients might be depleted due to the growth of actinobacteria and thus not available to the reporter strain. Moreover, as already noted above, media promoting secondary metabolite production of actinobacteria mostly lack phosphate, which is needed for satisfactory growth of reporter strains. Finally, the reporter could be also affected by a shifted pH due to prior cultivation of actinobacteria. Nevertheless, the reporter itself is suspended and dispensed in growth medium, which allows for addition of extra nutrients and buffering agents to the existing droplet. To compensate depleted nutrients, the added medium may also be overconcentrated.

9.4 Detection of Antibiotic Activity in Droplets and Screening for Novel Antibiotics

9.4.1 Possibilities and Constraints of Antibiotic Screening in Droplets

More than 50 % of today's natural products with antibiotic activity derive from microbial secondary metabolites synthesized by actinomycetales [42]. Although approximately 99 % of soil actinobacteria are hitherto still undiscovered, classic, cultivation-based screening approaches fail to reveal new species, mainly due to suppression of less assertive strains and insufficient throughput [10]. According to Watve and Berdy, the decline in discovery of novel antibiotics is not to be ascribed to a decreasing amount of unknown substances rather than a reduction in screening efforts by industries and academia [42, 43]. The microfluidic assay presented here has the potential to revive the interest in antimicrobial screening, since it is cheap (negligible costs for consumables) and fast; thus potentially profitable. Droplet-based microfluidics allows for ultrahigh-throughput generation of pure actinobacteria cultures. Each droplet constitutes a defined nano-scale bioreactor that protects the enclosed bacterial culture against harmful environmental influences, allowing fragile and slow-growing species to develop. The small culture volume also helps to take advantage of intrinsic noise within a species, facilitating detection of antimicrobials that might usually be produced merely in subinhibitory concentrations [44]. Whole-cell-based screening, the most promising approach according to leading experts ([4, 13, 116]), is performed by addition of reporter cells for detection of antimicrobial activity. With this approach, not only target-specificity of a substance is ensured, but also its uptake by the pathogen. However, droplets do not have a solid shell as bioreactors of larger dimensions, which renders the droplet prone to physical and biological influences affecting its size and integrity. Further potential constraints of microbial cultivation in droplets are limited mass transfer and local metabolite accumulation due to missing agitation as well as restricted sensing and control (pO_2, pH, carbon source) (Table 9.9). Moreover, microfluidic channels are subjected to biofouling and clogging, particularly when filamentous organisms as actinomycetes are employed. The following paragraphs elucidate approaches and strategies to assess and possibly resolve all mentioned potential difficulties, and ultimately explore the feasibility of a complete, droplet-implemented screening for antimicrobial substances.

The field of droplet-based microfluidics can be divided into mostly surfactant-free, micro-segmented flow and emulsion-based systems, relying on surfactants. Both approaches are theoretically suited for antibiotic whole-cell-screening and were employed in recent studies related to the detection of antibiotic activity [27, 45–48]. The advantages and disadvantages of each approach will be discussed by means of representative case studies.

9.4.2 Screening for Novel Antibiotics in Micro-Segmented Flow

In general, micro-segmented flow comprises two components: a carrier oil ("continuous phase") and aqueous droplets ("dispersed phase"). As continuous phase, either pure, long-chained carbohydrates (tetradecane, hexadecane, etc.), silicone oils or mineral oils are chosen, depending on the chip material and the nature of its surface. The main criteria for selection of a carrier fluid are immiscibility with the aqueous phase and the prevention of surface-wetting by the dispersed phase, which would result in droplet cross-contamination. Within the experimental setup presented here, tetradecane served as carrier fluid, while the aqueous phase was MMM or LB-medium. Figure 9.4 shows the chip design integrating all required unit operations for antibiotic whole-cell-screening except for final droplet sorting. ~10 nL-droplets are generated from a spore suspension at a frequency of 20 droplets/second, using a T-junction with double-inlet. The dispensing channel for the aqueous phase disembogues in a nozzle shape to provoke interface-tension-dominated droplet tear-off and thus reduce polydispersity at low flow rates [49]. Droplets are stored in the microfluidic chip by stopping the flow after filling the first incubation loop with droplets, spaced by tetradecane. The chip is disconnected, sealed and incubated for several days at 28 °C to promote germination of actinobacteria spores. Subsequently, the chip is reconnected and after restarting the stream of carrier fluid, medium containing reporter cells is added at the second T-junction, simply by applying a continuous flow at a rate that dosage to existing droplets occurs, rather than generation of new droplets (Fig. 9.5). During a second incubation step, now at 37 °C, the reporter cell populations increase in biomass and express the reporting protein. The incubation time is dependent on the applied reporting system, and thus might vary between 2 and 48 h (see Sect. 9.2). The read-out either takes place at halted flow by means of fluorescence microscopy, or is performed in the flowing system with a photomultiplier-based sensor. Sorting of droplets is intended to be performed off-chip.

The double-inlet allows for generation of media gradients upon droplet formation, providing the opportunity to test different growth conditions in a single experiment. Since a serial flow regime is inherent to the presented system, the composition of each droplet can be easily deduced by tracking its position in the series. Consequently, determination of a suitable media composition for optimal promotion of secondary metabolite production in actinobacteria is feasible. This approach can also be applied for activation of orphan pathways unveiled by genome mining [9].

The presented chip displays optimal performance regarding droplet generation: droplets are monodisperse, equally distanced and a lubricant layer at the channel surface is constantly maintained. After 4 days incubation of droplets generated from a suspension containing *S. noursei* spores (10^7 spores/mL $\approx$ 1 spore/droplet), germination and development of typical micropellets could be observed in the majority of droplets. However, distances between the droplets exhibited a much higher variance than upon generation. In rare cases, the inter-droplet distance was reduced to a minimum, which finally led to merging of droplet pairs or even small droplet series. It is known that droplets stored in microchannels move due to fluid-fluid surface tension

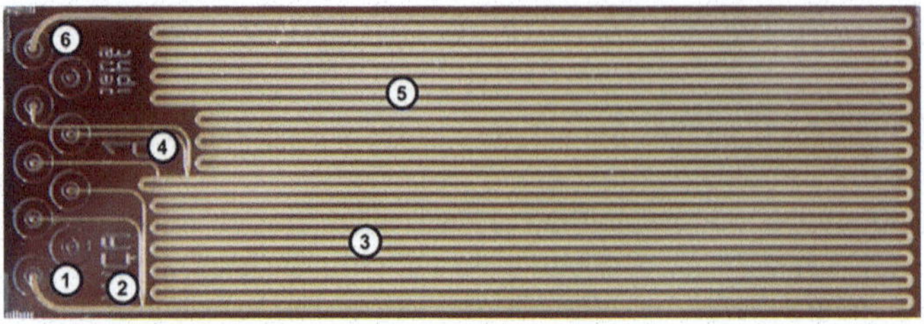

Fig. 9.4 Chip design integrating all structures required for antimicrobial whole-cell-screening. (*1*) Inlet carrier oil (*2*) 2-way droplet generator (*3*) 1st incubation loop (*4*) addition of reporter strain (*5*) 2nd incubation loop (*6*) outlet

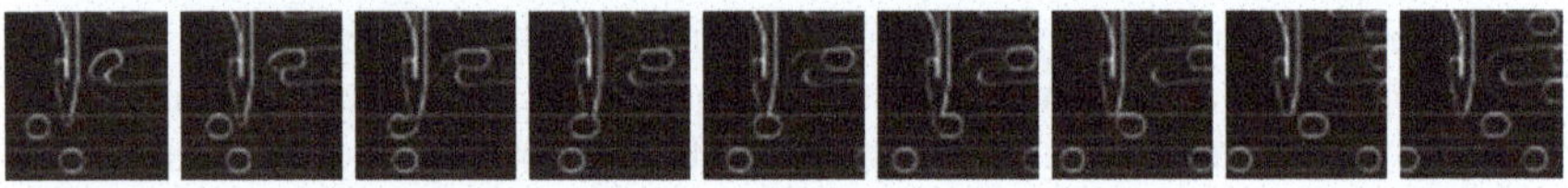

Fig. 9.5 Addition of reporter strain to passing droplet

or solid-liquid surface tension gradients [50]. The first is referred to as "Marangoni convection", caused by temperature (thermo-capillary convection) or chemical gradients along the interphase of two fluids, resulting in a mass flow towards lower interfacial tension. In the system presented here, the probability of chemical gradients leading to local differences in surface tension is high, since microcolonies are not equally distributed over the droplet and metabolites temporally accumulate due to the absence of agitation. Solid-liquid surface tension gradients can either be generated by temperature variations along the channel wall, or by chemical differences—e.g. uneven silanization of the channel surface leading to gradients in hydrophobicity. In both cases, droplets will flow towards higher channel wettability to release interfacial energy.

Upon restarting the flow of carrier oil for addition of reporter cells, maintaining equal distances between the incubated droplets proved to be nearly impossible: while the majority of droplets moved as expected with the stream of carrier oil, some appeared to be stuck in their position, which in consequence led to uncontrollable merging events. A possible explanation for this finding might be a layer of gas bubbles observed on the channel surface, which seemed to hinder droplet displacement. Other plausible reasons might be wetting effects occuring due to settling during long-term incubation or due to the production of biosurfactants [51] that destabilize the lubricant layer. Moreover, since not all droplets contain mycelium, they potentially differ in viscosity, which in turn affects each droplet's velocity to a varying extent [52].

Due to the complexity of the presented assay and the problems inherent to on-chip droplet incubation, the question whether the system is suitable for detection of reporter cell inhibition by an antibiotic has not been answered yet. Droplets generated from antibiotic solutions allow for simulation of in-droplet antibiotic production by actinobacteria, while omitting the most critical step of droplet incubation (including bacterial growth) and thus reducing assay complexity.

To investigate specific reporter cell inhibition in droplets, the whole-cell-assay chip was used for the generation of a droplet series with a gradient in the concentration of the antibiotic nourseothricin—a protein synthesis inhibitor. Without stopping the flow, *E. coli* reporter cells constitutively expressing GFPuv were injected at the intended position (pos. 4 in Fig. 9.4). Subsequently, the droplets were incubated on-chip for 48 h at room temperature (pos. 5 in Fig. 9.4). After the incubation period, image-based fluorescence analyses revealed droplets with regular growth and dark, non-fluorescent droplets where the inoculated cells did not propagate. The clear cut between both droplet species within the series indicated the minimal inhibition concentration of nourseothricin with regard to the employed reporter strain. A Z'-factor of 0.59 was determined from the positive- and negative controls, as a measure for the assay quality [53]. A value of 0.5–1 characterizes an excellent screening.

To go a step further, droplets were generated from culture supernatant of *Streptomyces noursei*, a producer of nourseothricin. This allowed for a more realistic simulation of the in-droplet production of antimicrobial substances by providing the chemical complexity inherent to media in the late phase of cultivation. The concentration of supernatant was altered in a sinusoidal manner along the droplet series. *E. coli* cells producing the red-fluorescent protein mCherry under control of the lac-promoter were injected into the droplets, which already contained IPTG as inducing agent. Again, image analysis displayed inhibition of the reporter by the culture supernatant, allowing for clear determination of a MIC-value. However, the Z'-factor indicated sub-optimal assay quality, probably due to unknown metabolites in the supernatant that affect the behaviour of the reporter cells. Interestingly, besides droplets with "normal" and without cell-derived fluorescence, droplets with strongly enhanced fluorescence were detected, in particular at the transitions between inhibitory and non-inhibitory conditions (Fig. 9.6). This might indicate the occurence of hormesis—i.e. enhanced metabolic activity of microorganisms in the prescence of subinhibitory concentrations of toxic compounds [54, 55]. This effect might be exploited in future screenings: instead of merely picking non-fluorescent droplets (= inhibition) for further investigation, detection and sorting of droplets with significantly enhanced fluorescence could be beneficial, since these potentially indicate the presence of an antibiotic in subinhibitory concentration.

Despite the non-optimal culture conditions to be found in droplets incubated on-chip, particularly with respect to oxygen supply, these results prove the detectability of reporter cell inhibition in the presented chip system.

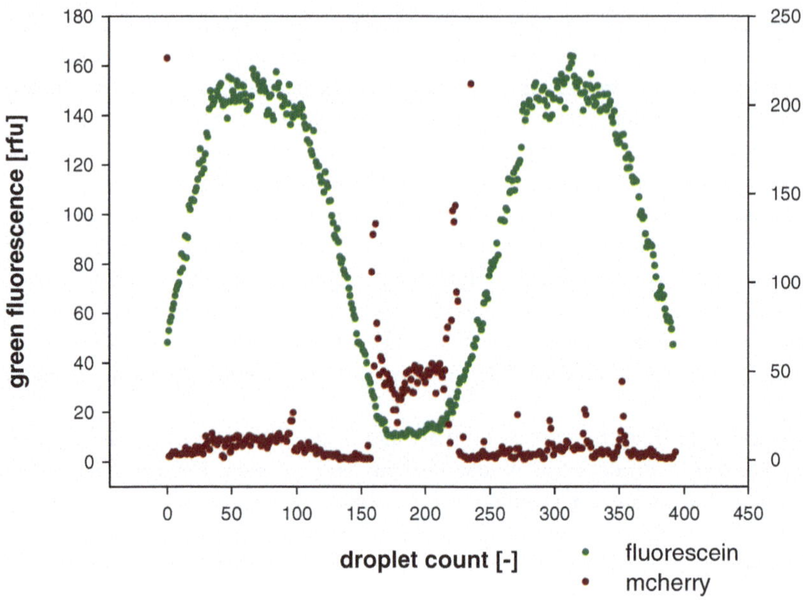

Fig. 9.6 Reporter strain (*red fluorescent*) growth inhibtion in droplets generated with culture supernatant of antibiotic-producing *Streptomyces noursei*. Fluorescein (*green fluorescence*) was used as a marker for supernatant concentration

9.4.3 Improving Robustness of Screening in Micro-Segmented Flow

As indicated above, a complete screening for novel antibiotics from actinobacteria spores with micro-segmented flow is afflicted with substantial problems. However, slight changes in this concept promise improved performance.

Introduction of a third immiscible phase: Instead of separating droplets with carrier fluid, a third immiscible phase may be introduced to the system to avoid contact of aqueous droplets and consequent merging. Zheng and Ismagilov [56] applied this concept in a similar context to test contents of nL-droplets against a large number of reagents (and vice versa). Aqueous plugs were separated by air bubbles, both dispersed in a fluorinated carrier fluid. However, the authors observed carry-over of droplet contents via the dispensing channel. To circumvent undesired cross-contamination, droplets of interest had to be separated by two droplets containing phosphate-buffered saline (Fig. 9.7). In another work by Chen et al. [57] different fluidic systems comprising three liquid phases were developed. The third liquid always served as a separation phase between two reagent plugs. A main challenge was to find combinations of three phases, where engulfment of one dispersed phase into droplets of the second dispersed phase did not occur. To satisfy that condition, the authors predicted that the magnitude of interfacial tension between the carrier fluid and one dispersed phase may not exceed the sum of interfacial tension between

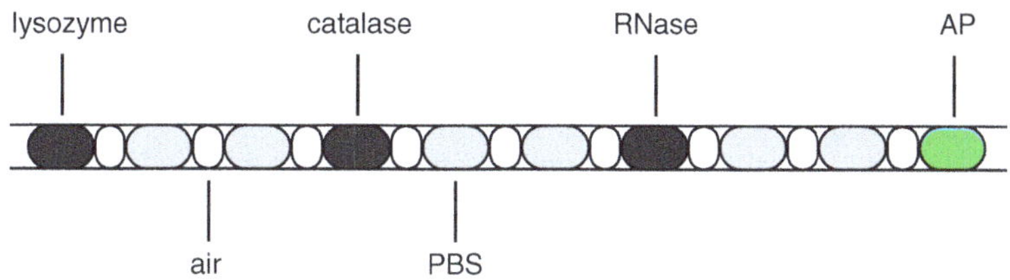

Fig. 9.7 Droplet train separated by air bubbles in a multiple enyzme assay against fluorescein diphosphate. Droplets with enzyme are spaced by droplets containing phosphate-buffer saline (*PBS*) to avoid cross-talk between droplets of interest via the substrate dispensing channel [56]

Table 9.5 Immiscible liquid system comprising three phases with one aqueous dispersed phase

Continuous phase	Dispersed phase (A)	Dispersed phase (B)
FC3283/PFO[a] (10:1 v/v)	1,3-diphenyl-1,1,3,3-tetramethyldisiloxane	water
FC3283/PFO[a] (10:1 v/v)	dimethyl tetrafluorosuccinate	water

[a] 1,1,2,2-tetrahydroperfluorooctanol

carrier fluid and the other dispersed phase plus the interfacial tension between both dispersed phases. These demands were met by two systems comprising aqueous reagent droplets, both with FC3283 as carrier fluid containing 10 % (v/v) PFO as surfactant (Table 9.5). A similar approach was established by Baraban et al. [27]: to provide robustness to a millifluidic droplet analyzer with Novec HFE-7500 (+0.006 % tri-block copolymer surfactant) as continuous phase, aqueous droplets were separated by plugs of mineral oil.

Incubation in tubing: All systems described above, comprising three liquid phases, need at least small amounts of surfactant to avoid wetting of the channel walls by the dispersed phases. Thus, simple dispensing of fluids into existing droplets, as shown in Fig. 9.5 is hampered or even impossible, requiring the integration of electrodes into the chip to destabilize the interphase between carrier oil and aqueous phase by application of an electric field [58]. Of course, the implementation of electrodes increases the demands on chip fabrication and technical periphery. Hence, instead of on-chip incubation of droplets, several authors postulated incubation in PTFE- or FEP tubing, even without use of surfactant or separating droplets of a third immiscible phase: to investigate survival rates of microorganisms in microfluidic droplets after up to 144 h, Martin et al. [59] guided 60-nL-plugs containing different model organisms (*Escherichia coli, Pseudomonas fluorescens, Rhodococcus rhodochrousand Saccharomyces cerevisae*) into teflon tubing of 0.5 mm inner diameter. For analysis, series of 5–30 droplets of the same type were first pooled and then plated out, so that equidistant droplet trains were not as crucial as in an antibiotic screening assay comprising incubation and dosing operations. In follow-up experiments by Grodrian et al. [60], actinobacteria spores were extracted from soil samples, singularized in droplets and incubated at 28 °C for seven days. Again, subsequent analysis was kept simple: to separate droplets for investigation, the capillary was cut into pieces and contents (droplets) unloaded into 100 µl of buffer. Occurrence of any merging events is not reported in this work, implying that it did not appear or only to a negligible extent.

Cao et al. [47] achieved long-term incubation (24 h) of up to 1,100 droplets in PTFE-tube coils for cultivation of *E. coli* in the presence of antibiotic agents. Occurrence of droplet fusion is conceded, but details with regard to frequency are omitted. Nevertheless, the fusion events do not remain unnoticed, since droplet size, -distance and -position are monitored.

Whether external incubation of droplets in tubing and without the help of surfactants leads to increased assay robustness compared to on-chip incubation remains to be elucidated, but in the cases discussed above the authors did not report contrary observations. Since PTFE-tubing does not require silanization prior to usage, a higher chemical homogenity with regard to hydrophobicity of the channel walls might account for minimized droplet movement during incubation.

9.5 Emulsion-Based Microfluidic Screenings: An Overview

9.5.1 Droplet Generation and Handling for Highly Parallelized Operations

The fluidic system for emulsion-based microfluidics generally comprises three components: a carrier fluid, a surfactant (dissolved in the carrier fluid) and the aqueous phase. Due to its amphiphilic nature, the surfactant molecule accumulates at the interphase of dispersed and continuous phase (Fig. 9.8), thus lowering the surface tension of aqueous droplets upon generation. In subsequent steps of the microfluidic assay, droplets may come into contact without the risk of merging due to steric repulsion of surfactant molecules on the droplets' surface. In contrast to surfactant-free systems, "parallel" incubation of droplets in (external) containers is feasible, allowing for immense increase in storage capacity and thus throughput. At the same time, the robustness is enhanced, since maintaining equidistant droplets is no longer necessary.

Various recent publications demonstrate the enormous potential of surfactant -stabilized, droplet-based microfluidics in applied biological sciences (Table 9.6).

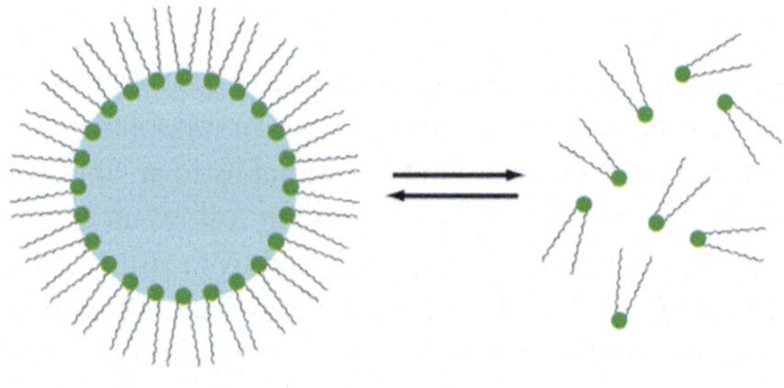

Fig. 9.8 Accumulation of surfactant molecules on the surface of aqueous droplets

Table 9.6 Publications on surfactant-stabilized, droplet-based microfluidics in the field of cell-biology and microbiology

Author	Assay	Year
Agresti et al. [33]	Directed evolution	2010
Kintses et al. [34]		2012
Fallah-Araghi et al. [92]		2012
Beer et al. [109] Kiss et al. [110]	Single-copy quantitative (RT)-PCR	2008
Brouzes et al. [46]	High-throughput single-cell cytotoxicity testing	2009
Mazutis et al. [76]	Isothermal DNA-amplification	2009
Tawfik and Griffiths [112]	In-vitro translation	1998
Dittrich et al. [113]		2005
Fallah-Araghi et al. [92]		2012
Mazutis et al. [76]		2009
Miller et al. [104]	High-resolution dose-response screening	2012
Pekin et al. [77]	Quantitative detection of rare mutants	2011
Granieri et al. [114]	Phage-display	2010
El Debs et al. [115]	Monoclonal antibody production and target binding studies	2012

All of the presented experimental setups rely on the same microfluidic unit operations, which are discussed below.

An important prerequisite for unit operations in emulsion-based microfluidics is a suitable carrier fluid/surfactant combination. Different combinations are employed, depending on the type of application—but only a few can be considered as prevalent: Mineral oil or hexadecane with Span80 as surfactant provide a simple and easily obtainable carrier fluid system, which can be applied in most assays, particularly those that do not include long droplet incubation periods [48, 61–65]. If higher droplet integrity is required, perfluorinated carrier fluids as FC-40, FC-77 or HFE-7500 (3M) are employed [66]. These oils exhibit very high chemical inertia and minimal water solubility, hence are less prone to droplet cross-talk. A further feature of perfluorinated carrier oils is their very high gas solubility, which can be of advantage if oxygen supply or gas formation of encapsulated microorganisms is critical. Biocompatible fluorosurfactants are often triblock copolymers composed of two perfluorinated polyether tails (PFPE) and a hydrophilic head, e.g. dimorpholino phosphate (DMP-PFPE) or polyethylene glycol (PEG-PFPE) [67–70]. They provide long-term droplet stability for many cell-based applications, but were until recently not commercially available, or only with rigorous restrictions (EA-surfactant, RainDance). PicoSurf^{TM} by Dolomite (Royston, England) fills that gap now.

Droplet generation is accomplished in most cases at cross-junctions by "flow-focussing" or at T-junctions, where the incoming aqueous stream is pinched off by the carrier fluid into monodisperse droplets [15, 71–73]. Due to hydrophobic channel surface modification (e.g. Repelsilan, Ombrello) and the presence of surfactant in the continuous phase, wetting of the channel walls by the emerging droplets is avoided. Even in cases where the droplet occupies the majority of the channel's cross section (“slug flow”), a lubricating layer of carrier fluid is maintained between the droplet and the channel wall.

Addition of cells or reactants to existing droplets can be accomplished by either pairing and merging targeted droplets with droplets delivering the contents in demand [74–77] or by dispensing the reagent or cell-containing fluid into targeted droplets passing a T-junction (“picoinjection“) [78]. Independent of the merging technique applied, the targeted droplets have to be spaced with carrier fluid after reinjection into the chip, which can also be achieved at cross- or T-junctions, possibly arranged in a sharp angle to reduce shear stress. Mazutis et al. [76] used FC-40 perfluorinated oil with reduced surfactant concentration (0.27–0.55 % w/w EA-surfactant, Raindance) for droplet spacing, which leads to depletion of surfactant molecules on the droplet's surface upon reinjection. Droplets being generated on-chip using the same surfactant concentration are paired with the reinjected ones by flow-rate synchronization and droplet pairs are finally merged in zig-zag-shaped channels inducing local surfactant depletion and thus facilitating fusion of droplet interfaces. To diminish subsequent undesired merging of droplets in the collection chamber, carrier oil with high surfactant concentration (2.8 % w/w) was injected upstream of the zig-zag-structure. Since diffusion of the surfactant micelles to the droplet surface is a function of surfactant concentration and time [32], desired droplet coalescence in the zig-zag-channels is merely affected. The one-to-one droplet fusion efficiency achieved with this technique is ~99 %. To avoid such complicated and cumbersome flow arrangements, droplet merging or addition of fluids to existing droplets can be induced with electric fields generated by on-chip electrodes located near the point of droplet pairing or fluid dispensing, respectively [79–81]. By applying a pulsed electric potential of less than 3 V, Priest et al. [82] demonstrated targeted destabilization of inter-droplet lamella resulting in pairwise droplet merging in a hexadecane/saline system with 2 % Span80 as emulgator. In the same work it was shown that a 10 μm insulating layer between the electrodes and the channel does not impede electrocoalescence, since capacitive coupling carries the voltage to the rupture area when AC electric fields are applied. This way, undesired electrochemical side effects detrimental to living cells are also circumvented. In a series of experiments published by Ahn et al. [83], two droplet populations with different average size (50 and 25 μm) were generated on-chip and merged pairwise. Interestingly, the authors exploited the rheologic effect of smaller droplets catching up to larger ones, to reliably reduce inter-droplet distance and facilitate electrocoalescence of the two droplet species. Using hexadecane as continuous phase with 5 % (w/w) Span80 as surfactant, 100 V AC at a frequency of 100 kHz was required for successful merging at a rate of 100 droplet pairs/s. Given an electrode distance of 200 μm, the applied voltage accounts to a field strength of 500 kV/m. This concept is also applicable for reinjected droplets which were incubated off-chip to replace one or both of the on-chip generated droplet species. However, the discussed methods of droplet pairing and merging require precise fine tuning of flow rates for synchronization of the two droplet species, which is in every case prone to errors and occasionally yields two-to-one fusion events or unfused droplets.

In contrast, “picoinjection” provides an approach where only one population of droplets must be handled, since fluids containing cells or agents are directly dispensed into the reinjected droplets from a perpendicular and pressurized channel at a T-junction with adjacent electrodes generating an electric field [78]. As an additional benefit of this technique, injection can be performed at kHz-frequencies and in a selective manner, e.g. triggered upon detection of fluorescence. Moreover, serial and combinatorial injections are feasible by implementing additional picoinjection structures along the droplet-conducting channel. However, the authors do not describe how the pressure in the feeding channel is controlled, which is relevant for such complex dispensing patterns. The downside of this and similar techniques is the risk of cross-contaminations and carry-over-effects through the dispensing channel, but both can be considered as not important in case of the aimed screening for antibiotics: cross-contamination of filamentously growing, monolithic micropellets of Actinomycetes is unlikely and carry-over of potentially produced antibiotics would be subject to strong dilution and thus negligible. For completeness it should be mentioned that Sivasamy et al. [64] developed a passive merging structure for dosage of reagents that represents an intermediate between a droplet merger and a dispensing unit: aqueous droplets evolve from a feeding channel directly into a square-shaped chamber in downstream proximity, forming an uncurved interface with the carrier fluid (mineral oil + 2 % Span80), which exhibits a relatively low Laplace pressure jump. According to the authors, the interface can thus be easily disturbed by approaching droplets, resulting in a high probability of merging. However, we were not able to reproduce these findings when testing similar structures with FC40+PEG-PFPE surfactant as continuous phase.

Apart from incubation and detection—which is not subjected in this section—droplet sorting is the last required unit operation for a complete, surfactant-stabilized, microfluidic screening for novel antibiotic substances. Applied structures generally consist of a bifurcated microfluidic channel and adjacent eletrodes generating a pulsed, non-uniform electric field upon triggering [84–86]. Without pulse, droplets flow into the output channel with lower fluidic resistance, which is determined by downstream width and length of the channels. When the electrodes are energized, the respective droplet is polarized and dragged towards higher field strength, which is located near the output channel with higher fluidic resistance. By this, droplets of interest—e.g. selected on the basis of fluorescence detection—can be separated from the droplet population. The required voltages depend on the droplet velocity and volume, the viscosity of the continuous phase, the shape and position of the electrodes and the difference in fluidic resistance of the output channels. For sufficient displacement of 25 μm diameter aqueous droplets in hexadecane, Ahn et al. [84] applied a 500 μs-pulse of 10 kHz and 700 V across the electrodes. This allowed for reliable droplet sorting at frequencies of 1,600 Hz.

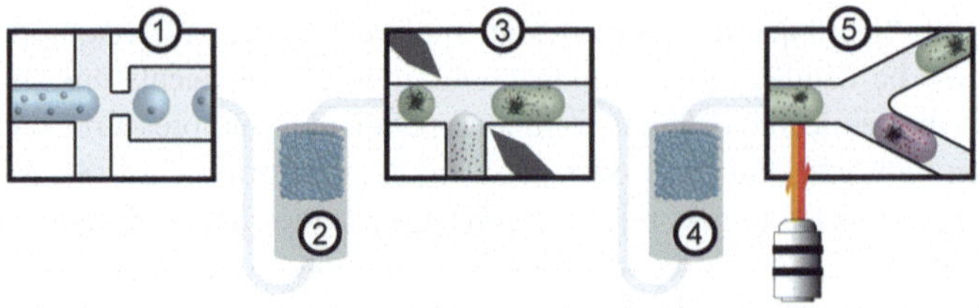

Fig. 9.9 Screening for antibiotic activity from soil-derived actinobacteria spores singularized and germinated in pL-droplets: (*1*) droplet generation (*2*) incubation of spores (*3*) addition of reporter cells aided by electrodes (*4*) incubation of reporter cells (*5*) fluorescence detection and droplet sorting

Fig. 9.10 Microfluidic fused silica chip

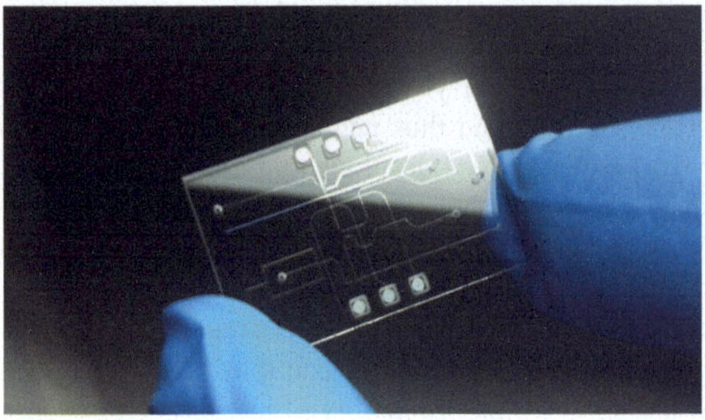

9.5.2 Screening for Novel Antibiotics with an Emulsion-Based Microfludic Approach

The general workflow for the emulsion-based search for novel antibiotics on the basis of the unit operations discussed above is depicted in Fig. 9.9. In contrast to screening in a micro-segmented flow, droplets are incubated in bulk and addition of reporter strains is assisted by electric fields.

To test the applicability of the surfactant-stabilized microfluidic concept for antibiotic screening, chips were designed for droplet generation, addition of reagents (dosage) and droplet sorting, according to the references discussed above (Fig. 9.10). Chips were fabricated by LioniX BV (Enschede, Netherlands) from fused silica wafers: the top-plate (0.5 mm) bears 50 μm deep channels generated by deep reactive ion etching using a 3 μm silicon layer as masking material. Vias for fluidic and electric contacting were created by powderblasting. Metallic electrodes were applied to the bottom-plate (1.0 mm) using a lift-off process on a sputtered stack of platinum (180 nm) on top of a titanium adhesion layer (15 nm). Prior to bonding of both halves, a layer of silicon oxynitride was deposited for levelling and insulation, using plasma-enhanced chemical vapor deposition. Before use, chips were rendered hydrophobic by rinsing with dimethyldichlorosilane (Repelsilan, Amersham Biosciences). Additionally, different devices were tested for droplet incubation: besides

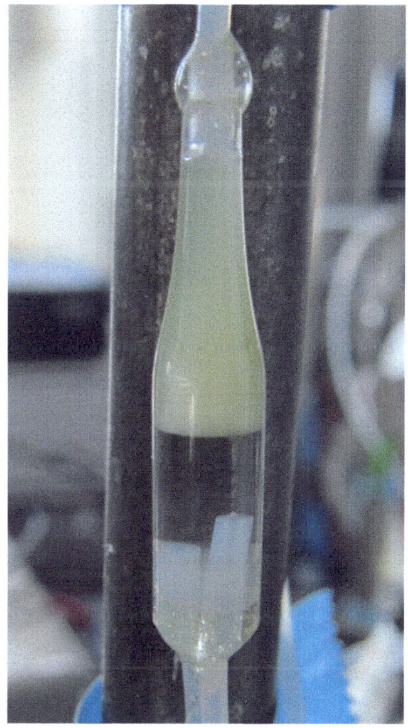

Fig. 9.11 Incubation vial made from a clear-glass micro-insert for HPLC-vials. Creamed droplets can be observed in the upper part

simply guiding droplets into PTFE-tubing (0.5 mm ID), droplets were transfered into 1.5 ml microreaction tubes or silanized pasteur pipettes followed by overlaying with pure and sterile culture medium to minimize droplet evaporation. For another home-made incubation device (in the following referred to as "incubation vial"), the tip of a clear-glass micro-insert for HPLC-vials (15 mm, 0.1 ml, VWR) was removed with a glass cutter and connected to PTFE-tubings after sealing the bottom with a silicone plug and epoxy resin adhesive (Fig. 9.11). Subsequently, the device was rendered water-repellent with Ombrello (moton auto-motive, Germany). Within this setup, no overlaying with aqueous medium was required.

The aforementioned microfluidic unit operations were evaluated with regard to their suitability for antibiotic screening from actinobacteria spores. Emphasis was given to droplet generation, incubation and reinjection, since these steps are prone to various stresses of physical and biological nature (Table 9.7) that influence droplet populations to an unknown extent. Hence, conditions for all subsequent operations are set during these initial steps, particularly by the degree of monodispersity of the resulting droplet population: high polydispersity will hamper reliable addition of reporter cells and is disadvantageous to the final step of droplet sorting. To assess the impact of spores and their germination on the monodispersity of a droplet population, we tested droplet generation from spore suspensions and subsequent incubation under diverse conditions.

Table 9.7 Physical and biological factors influencing droplet populations

Physical	Evaporation
	Satellite formation
	Ostwald ripening
	Merging (local surfactant depletion)
	Cross-contamination (through wetting)
	Cross-contamination (through dispensing channel)
	Diffusion, cross-talk
Biological	Gas formation due to metabolic activity
	Shrinkage due to metabolic activity
	Merging (increased surface tension through catabolites)
	Outgrowth of hyphae

Table 9.8 Composition of a model actinomycetes spore suspension used for microfluidic experiments

Strain	Taxon	Produced antibiotics
IMET40285	*Streptomyces gelaticus*	Elaiomycin
IMET40235	*Streptomyces griseus*	Cycloheximid
HKI0323	*Amycolatopsis lactamdurans*	Cephamycin C
HKI0040	*Streptomyces canus*	Amphomycin
HKI0016	*Streptomyces hygroscopicus*	Staurosporin
JA02640	*Streptomyces californicus*	Viomycin
IMET40177	*Streptomyces griseorubiginosus*	Cinerubin
HKI0216	*Kitasatospora azatica*	Alazopeptin
HKI0423	*Nonomuraea recticatena*	
HKI0269	*Streptomyces celluloflavus*	Aureothricin
IMET41584	*Streptomyces albus*	Salinomycin

Without applying spores, it was found that a flow rate of 195 nl/s for the carrier oil (HFE 7500 + 1 % PicoSurf) and 94 nl/s for the aqueous phase (MMM) gives the best results with regard to polydispersity. Droplets were produced at frequencies above 500 Hz, exhibiting a polydispersity of 1.15 %, which is clearly below the accepted standard of 2–3 % according to literature [87]. When a model spore suspension (in MMM) (Table 9.8) was employed instead of water as aqueous phase, similar values for the polydispersity were achieved, indicating that the presence of spores does not affect droplet generation. However, larger droplets bearing an obtrusive piece of mycelium were observed with non-significant frequency. These irregularities are likely to be minimized with more rigid centrifugation and filtration during spore extraction, which reduces the proportion of loose filaments in the suspension. Although low polydispersity values were obtained within each experiment, we observed high fluctuations in average droplet size between different droplet generation experiments: at various combinations of aqueous phase (pure MMM, MMM with spores and MMM with reporter cells), incubation device (Pasteur

Table 9.9 Restraints of droplet-based microfluidics with regard to cell-based assays

limited mass transfer
no agitation
limited sensing (pH, pO_2, etc.)
pH-shift
depletion of growth media
biofouling of surfaces
clogging

pipette, microreaction-tube and incubation vial) and surfactant concentration (1 and 1.25 %), the apparent mean droplet size varied between ~115 and ~139 pL, although the same flow rates were applied. Even with identical setups, the measured average droplet sizes revealed recognizable inter-experimental variations. While these can only derive from image analysis-related issues (variations in light intensity, focus plane, detection algorithm), greater variations at unequal experimental setups could also be attributed to changes in backpressure (due to varying incubation device and length of output tubing) and viscosity of the aqueous phase. The variations in the mean volume of different droplet populations complicate comparing inter-experimental changes in polydispersity, but they do not impede intra-experimental comparisons of a single droplet population after generation and incubation.

To evaluate the physical impact of droplet incubation and reinjection on the polydispersity of a population, droplets generated from pure MMM were incubated for 0, 24 and 48 h before partial reinjection (Fig. 9.12). To our surprise, droplets that were immediately (0 h) reinjected showed the highest divergence in average volume (110.8 pL) compared to the measurement upon droplet generation (115.61 pL). With increasing incubation time, the mean droplet volume approached the original value. These findings cannot be ascribed to a certain physical effect rather than variations in imaging conditions. More importantly, polydisperity did not increase substantially with prolonged incubation time: upon droplet generation, a value of 1.24 % was measured, while after 48 h the polydispersity was 1.30 %.

These results indicate that Ostwald ripening occurred only to a negigible extent within this incubation period. Nevertheless a marginally increased number of satellite droplets was observed, which probably emerge due to shear upon reinjection and slight over-concentration of surfactant in the carrier oil. Additionally, a small number of larger droplets was detected during each reinjection round. Since their volume was not a harmonic multiple of the average droplet size, they did not originate from droplet merging events. Their occurrence remains to be elucidated, although they did not impede further unit operations. In the next step, the impact of actinomycetes on the droplet population was investigated: instead of using pure medium as aqueous phase, droplets were generated from a spore suspension (in MMM). Subsequently, the droplets were transferred to an incubation vial allowing for spore germination (at room temperature). As expected, droplets are nearly as monodisperse as upon generation when reinjected immediately, since germination has not occured yet.

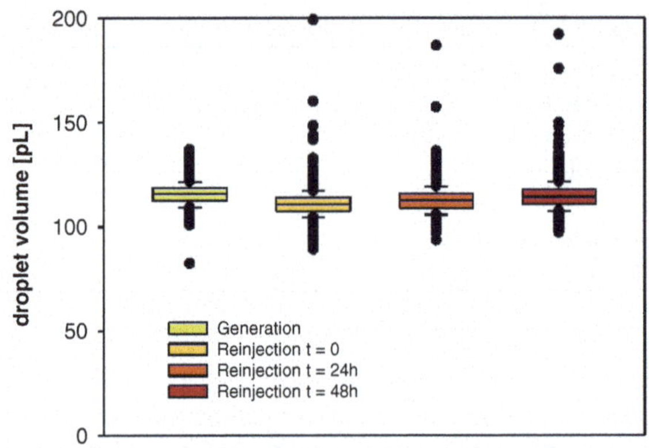

Fig. 9.12 Droplet volumes upon generation and after reinjection. Aqueous phase: pure MMM

Even after 24 h of incubation the polydispersity increased only from 1.12 % (upon generation) to 2.07 %, and first droplets bearing mycelium were observed (Fig. 9.13). After 48 h incubation, the majority of droplets showed growing actinomycetes. Interestingly, droplets with microcolonies shrank while empty droplets experienced an increase in volume (Figs. 9.14 + 9.15). This phenomenon, which is numerically reflected in an elevated polydispersity of 9.27 %, can be ascribed to the metabolic activity of growing bacteria: carbon sources are partly catabolized to highly diffusive products (mainly CO_2) that easily escape the droplet. As a result, the osmolarity decreases, leading to an efflux of water towards unoccupied droplets, since they exhibit a relatively higher osmolarity [88, 89]. In extreme cases, the mycelium occupies the whole volume of a droplet, which is thus deformed in its surface. However, undesired outgrowth of hyphae is only observed in very rare cases after extensive incubation periods. It should be pointed out that hardly any fused droplets were observed, indicating that the employed model actinomycetes did not produce any metabolites that significantly affect droplet integrity. Whether this finding also applies to larger model libraries or even to the immense diversity of actinomycetes found in soil samples remains to be investigated, but it provides first evidence that maintaining microcultures of unknown actinobacteria in surfactant-stabilized droplets is feasible.

A potential caveat to both micro-segmented flow and the emulsion-based approach is also related to incubation: secondary metabolites with antimicrobial activity emerging from the droplet of origin to other droplets in close proximity might cause false negative or positive hits, depending on the antimicrobial potency as well as the mobility of the respective molecule. This phenomenon, also termed "cross-talk", was recently investigated by various researchers. Bai et al. [66] tested the retention of fluorescein in trapped droplets dispersed either in mineral oil/Span80 or FC77/EA-surfactant, respectively. As already indicated above, the author found that the perfluorinated carrier oil (FC77) exhibited much lower transport of fluorescein molecules, due to lower fluidity at the water/oil interface. Nevertheless, Courtois et al. [90] and Skhiri et al. [91] found that surfactant concentrations widely exceeding the critical micellar contration (CMC) can enhance cross-talk by formation of

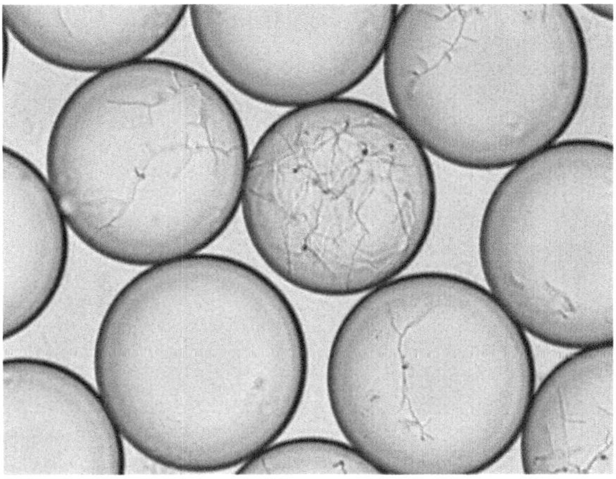

Fig. 9.13 MMM-droplets with mycelium or micropellets (after 24 h)

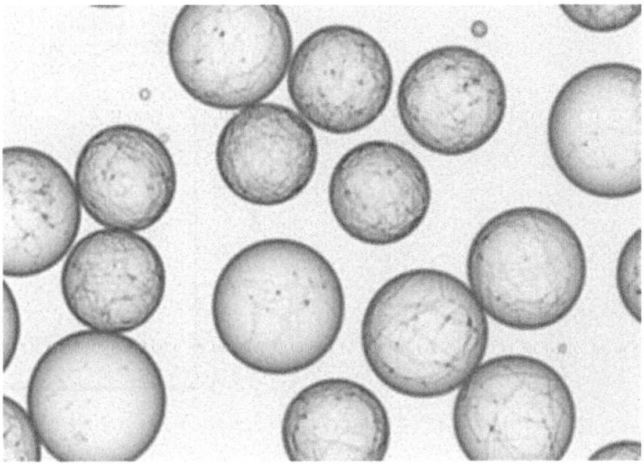

Fig. 9.14 MMM-droplets with mycelium or micropellets (after 48 h)

reverse micelles entrapping and transporting molecules that are dissolved in the droplets. Together with diffusion, this mechanism is claimed to play the key role in inter-droplet exchange of molecules. Both authors also present a counteracting measure to molecule transport between droplets: by addition of BSA to the aqueous phase, the retention of resorufin was enhanced 18-fold [91], and transport of fluorescein could also be reduced signifcantly. Since this finding is ascribed to the general property of BSA to increase solublity of other molecules, it can be assumed that this also applies for unknown antimicrobial substances entrapped in droplets. Although molecule mobility is related to its charge and is thus unpredictable for the bulk of unknown molecules produced by actinobacteria, cross-talk can probably be reduced to acceptable dimensions by employment of BSA or similar additives.

In a further step, addition of reporter cells to droplets containing germinated spores with a picoinjector was tested. As mentioned above, high polydispersity of the incoming droplet population was assumed to be detrimental to reliable picoinjection. Different droplet sizes lead to varying inter-droplet distances after spacing which

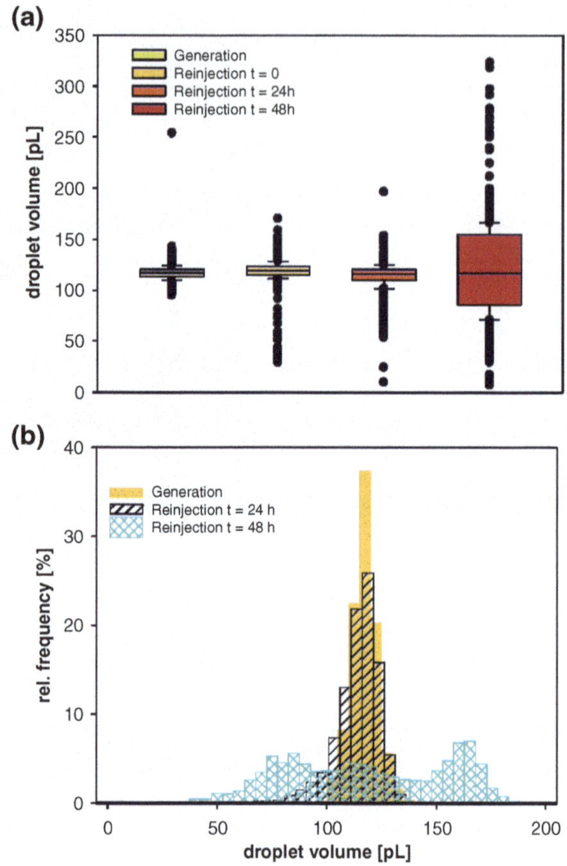

Fig. 9.15 Droplet volumes upon generation and after reinjection (0, 24 and 48 h). Aqueous phase: MMM + spores. B: Distribution of droplet volumes upon generation and after incubation

in turn results in the addition of unequal fluid volumes. If the diameter is smaller than the channel cross-section, the droplet might even not come into contact with the meniscus of the dispensing channel and thus not be subjected to reporter cell addition. However, upon reinjecting 3-day old droplets generated from the model spore suspension with 140 picolitre average volume, all droplets were still large enough to occupy the entire cross-section of the incoming channel upstream of the picoinjection structure (size 50 μm). The outgoing, plug-shaped droplets appeared equal in volume. Yet, addition of variable amounts to the droplets would not be detrimental to the overall assay performance, since cell densities will equalize with growth time in droplets without antibiotic activity, diminishing the risk of false positives. However, in case droplet polydispersity becomes an obstacle in future assays—e.g. by limiting throughput—droplets can be sorted according to their size prior to each reinjection round as counteracting measure. The functionality of passive, size-dependent sorting structures was demonstrated recently [92–94].

The last remaining unit operation to be tested is sorting of droplets representing a hit. Again, it must be assumed that high polydispersity of incoming droplets is challenging with regard to droplet sorting. However, as in the case of reporter cell picoinjection, passive droplet-sorting according to size prior to fluorescence guided sorting might be helpful. The maximum frequency of reliable droplet separation from the main population has to be assessed and is surely dependent on the exact shape of applied channel and electrode structures as well as the droplet volume and velocity. Difficulties might arise from the contradiction of reporting principle and detection mechanism: since reporter cells are inhibited, droplets containing an antibiotic will be non-fluorescent, while the evaluation of droplets is fluorescence-based.

However, this issue might be easily solved by detection and sorting of the fluorescent droplet species, which also leads to pooling of non-fluorescent droplets at the other end. Nevertheless, this approach would result in an extremely high frequency of sorting events, which inflicts higher demands on the sorting periphery. The feasibility of such an ultra-high-frequency sorting must be tested. Alternative ways to tackle this problem might include general addition of a fluorophore that emits at a different wavelength or detection of all droplets by capacitive sensing prior to fluorescence analysis [63].

Once droplets of interest are sorted into the "value" outlet, they have to be separately extractable so that they can serve as inoculum in an upscaling chain—starting from a well of a microtiter plate, for example. A suitable chip-to-world interface remains to be developed, which is also a requirement for other droplet-based microfluidic assays and thus only a matter of time.

9.6 Summary and Outlook on Antimicrobial Screenings in Micro-Segmented Flow and Emulsion-Based Systems

The constant emergence of new, life-threatening pathogens and resistance mechanisms requires the development of novel antibiotic substances and substance classes. Target-oriented approaches, as they were postulated at the beginning of the 1980s, failed to reveal new antimicrobials. Despite the discovery of suitable candidates for target-inhibition, most substances were not able to penetrate the bacterial cell wall. Hence, leading experts proposed the return to whole-cell-based screening of soil-derived actinobacteria—an approach that delivered the majority of all discovered antibiotics so far. However, classic screening of actinomycetes is cumbersome and suffers from low throughput, which limits its success rate. By providing millions of microscaled reaction compartments, droplet-based microfluidics allows for high-throughput cultivation of Actinobateria and promises subsequent whole-cell testing for production of antimicrobial substances. Here, micro-segmented flow, implemented on a monolithic chip device, and an emulsion-based approach were tested for antibiotic screening capabilities by separately investigating all contributing unit operations.

Droplet generation and incubation led in both discussed systems to germination of encapsulated spores and formation of micropellets. Penetration of the droplet/carrier oil interphase by growing hyphae was only observed in rare cases after very long incubation periods of several weeks. However, subsequent addition of reporter cells could not be achieved with micro-segmented flow, since inter-droplet distances were not uniform and undesired droplet fusions occurred. Several methods to resolve this conflict were proposed, including the introduction of a third immiscible phase to separate droplets and external incubation in PTFE tubing. To test the system for general detectability of antibiotic activity, droplet series with a concentration gradient of the protein-synthesis-inhibitor nourseothricin and culture supernatant of nourseothricin-producing *S. noursei*, respectively, were generated. After addition of *E. coli* reporter cells and subsequent incubation, droplets were analyzed for fluorescence. For both, the pure nourseothricin and the culture supernatant, the minimal inhibitory concentration could be clearly determined.

For the modular, surfactant-stabilized system it was shown that growth of actinomycetes leads to an increase in droplet polydispersity. Nevertheless, by applying moderate incubation times, the polydispersity can be kept in an acceptable range, still allowing for reliable addition of reporter cells by picoinjection. In case droplets with higher polydispersity must be handled—due to prolonged incubation times for example—antecedent size-dependant droplet-sorting with passive sorting structures is proposed. Fluorescence-dependent sorting of droplets with pulsed electric fields was already demonstrated in several studies, but applicability for polydisperse droplet populations bearing mycelia of actinomycetes still has to be confirmed. To get hold of droplets sorted on-chip, an interface allowing for selective guidance of single droplets, e.g. in the wells of a microtiter plate, needs to be developed.

Issues and questions concerning the magnitude of undesired droplet-cross-talk, quality of the read-out provided by different reporting principles and achievable throughput need to be resolved. Moreover, accessible diversity of actinobacteria and frequency of unknown species found in droplets has to be investigated, e.g. by pyrosequencing of 16S-rRNA fragments. If required, pooling of soil samples, variation of media composition and employment of streptomycete-specific phages [95] might further increase the probability of discovering yet unknown species. Expression of orphan pathways might be stimulated through addition of aqueous soil extracts [96, 111]. As a promising variation of the presented screening approach from actinobacteria spores, high-throughput-investigation of soil-derived metagenomes in prokaryotic hosts might be taken into consideration [97, 98], although probabilities of generating hits must be thoroughly evaluated [99].

The general feasibility of a droplet-based microfluidic screening of soil-derived actinobacteria spores for novel antimicrobial substances was confirmed. Although further development and fine-tuning is required, the presented case-studies raise the prospect of a highly efficient assaying system as a source of new active substances against emerging pathogens.

References

1. E.P. Abraham, E. Chain, An Enzyme from Bacteria able to Destroy Penicillin. Nature **146**, 837–837 (1940)
2. M. Demerec, Origin of Bacterial Resistance to Antibiotics. J Bacteriol **56**, 63–74 (1948)
3. T.H. Haight, M. Finland, Resistance of bacteria to erythromycin. Proc Soc Exp Biol Med **81**, 183–188 (1952)
4. M.A. Fischbach, C.T. Walsh, Antibiotics for emerging pathogens. Science **325**, 1089–1093 (2009)
5. M.A. Cooper, D. Shlaes, Fix the antibiotics pipeline. Nature **472**, 32–32 (2011).
6. D.J. Newman, G.M. Cragg, Natural products as sources of new drugs over the last 25 years. J. Natl. Prod. **70**, 461–477 (2007)
7. S.S. Cohen, A strategy for the chemotherapy of infectious disease. Science **197**, 431–432 (1977)
8. K.M. Overbye, J.F. Barrett, Antibiotics: where did we go wrong? Drug Discov. Today **10**, 45–52 (2005)
9. M. Nett, H. Ikeda, B.S. Moore, Genomic basis for natural product biosynthetic diversity in the actinomycetes. Nat. Prod. Rep. **26**, 1362 (2009)
10. R.H. Baltz, Renaissance in antibacterial discovery from actinomycetes. Curr. Opin. Pharmacol. **8**, 557–563 (2008)
11. D.D. Baker, M. Chu, U. Oza, V. Rajgarhia, The value of natural products to future pharmaceutical discovery. Nat. prod. Rep. **24**, 1225–1244 (2007)
12. L.M. Mayr, P. Fuerst, The future of high-throughput screening. J. Biomol. Screening **13**, 443–448 (2008)
13. R.H. Baltz, Marcel Faber Roundtable: is our antibiotic pipeline unproductive because of starvation, constipation or lack of inspiration? J. Ind. Microbiol. Biotechnol. **33**, 507–513 (2006)
14. K. Zengler, G. Toledo, M. Rappe et al., Cultivating the uncultured. Proc. Nat. Acad. Sci. U.S.A. **99**, 15681–15686 (2002)
15. C.N. Baroud, F. Gallaire, R. Dangla, Dynamics of microfluidic droplets. Lab Chip **10**, 2032–2045 (2010)
16. D.A. Hopwood, M.J. Bibb, K.F. Chater, *Genetic manipulation of Streptomyces?: a laboratory manual* (John Innes Foundation, Norwich, 1985)
17. D.A. Hopwood, Genetic analysis and genome structure in streptomyces coelicolor. Bacteriol Rev. **31**, 373–403 (1967)
18. M.J. Buttner, K.F. Chater, M.J. Bibb, Cloning, disruption, and transcriptional analysis of three RNA polymerase sigma factor genes of Streptomyces coelicolor A3 (2). J. Bacteriol. **172**, 3367–3378 (1990)
19. M.J. Buttner, RNA polymerase heterogeneity in Streptomyces coelicolor A3 (2). Mol. Microbiol. **3**, 1653–1659 (1989)
20. R.H. Coleman, J.C. Ensign, Regulation of formation of aerial mycelia and spores of Streptomyces viridochromogenes. J. Bacteriol. **149**, 1102–1111 (1982)
21. D.A. Hodgson, Glucose repression of carbon source uptake and metabolism in Streptomyces coelicolor A3 (2) and its perturbation in mutants resistant to 2-deoxyglucose. J. Gen. Microbiol. **128**, 2417–2430 (1982)
22. J.F. Martín, A.L. Demain, Control of antibiotic biosynthesis. Microbiol. Rev. **44**, 230–251 (1980)
23. J. Sambrook, E. Fritsch, T. Maniatis, *Molecular Cloning*, vol. 2 (Cold Spring Harbor Laboratory Press, A Laboratory Manual, USA, 1989)
24. M.A. Farid, El-Enshasy HA, El-Diwany AI, El-Sayed SA, Optimization of the cultivation medium for natamycin production by Streptomyces natalensis. J. Basic Microbiol. **40**, 157–166 (2000)

25. J.F. Martín, Phosphate control of the biosynthesis of antibiotics and other secondary metabolites is mediated by the PhoR-PhoP system: an unfinished story. J. Bacteriol. **186**, 5197–5201 (2004)
26. A. Huebner, M. Srisa-Art, D. Holt, et al., Quantitative detection of protein expression in single cells using droplet microfluidics. Chem. Commun. (Camb) 1218–1220 (2007)
27. L. Baraban, F. Bertholle, M.L. Salverda et al., Millifluidic droplet analyser for microbiology. Lab Chip **11**, 4057–4062 (2011)
28. N.C. Shaner, R.E. Campbell, P.A. Steinbach et al., Improved monomeric red, orange and yellow fluorescent proteins derived from Discosoma sp. red fluorescent protein. Nat. Biotechnol. **22**, 1567–1572 (2004)
29. R. Heim, D.C. Prasher, R.Y. Tsien, Wavelength mutations and posttranslational autoxidation of green fluorescent protein. Proc. Natl. Acad. Sci. U.S.A. **91**, 12501–12504 (1994)
30. N.C. Shaner, P.A. Steinbach, R.Y. Tsien, A guide to choosing fluorescent proteins. Nat. Methods **2**, 905–909 (2005)
31. T. Drepper, R. Huber, A. Heck et al., Flavin mononucleotide-based fluorescent reporter proteins outperform green fluorescent protein-like proteins as quantitative in vivo real-time reporters. Appl. Environ. Microbiol. **76**, 5990–5994 (2010)
32. J.-C. Baret, F. Kleinschmidt, A. El Harrak, A.D. Griffiths, Kinetic aspects of emulsion stabilization by surfactants: a microfluidic analysis. Langmuir **25**, 6088–6093 (2009a)
33. J.J. Agresti, E. Antipov, A.R. Abate et al., Ultrahigh-throughput screening in drop-based microfluidics for directed evolution. Proc. Natl. Acad. Sci. U.S.A. **107**, 4004–4009 (2010)
34. B. Kintses, C. Hein, M.F. Mohamed et al., Picoliter cell lysate assays in microfluidic droplet compartments for directed enzyme evolution. Chem. Biol. **19**, 1001–1009 (2012)
35. K. Hakkila, M. Maksimow, M. Karp, M. Virta, Reporter genes lucFF, luxCDABE, gfp, and dsred have different characteristics in whole-cell bacterial sensors. Anal. Biochem. **301**, 235–242 (2002)
36. A. Urban, S. Eckermann, B. Fast et al., Novel whole-cell antibiotic biosensors for compound discovery. Appl. Environ. Microbiol. **73**, 6436–6443 (2007)
37. M.I. Bahl, L.H. Hansen, S.J. Sorensen, Construction of an extended range whole-cell tetracycline biosensor by use of the tet(M) resistance gene. FEMS Microbiol. Lett. **253**, 201–205 (2005)
38. A.A. Bianchi, F. Baneyx, Stress responses as a tool to detect and characterize the mode of action of antibacterial agents. Appl. Environ. Microbiol. **65**, 5023–5027 (1999)
39. E. Eltzov, D.Z. Ben-Yosef, A. Kushmaro, R. Marks, Detection of sub-inhibitory antibiotic concentrations via luminescent sensing bacteria and prediction of their mode of action. Sens. Actuators **129**, 685–692 (2008)
40. A.C. Vollmer, S. Belkin, D.R. Smulski et al., Detection of DNA damage by use of Escherichia coli carrying recA':lux, uvrA':lux, or alkA':lux reporter plasmids. Appl. Environ. Microbiol. **63**, 2566–2571 (1997)
41. S. Yagur-Kroll, S. Belkin, Upgrading bioluminescent bacterial bioreporter performance by splitting the lux operon. Anal. Bioanal. Chem. **400**, 1071–1082 (2011)
42. J. Bérdy, Bioactive microbial metabolites. J. Antibiot. **58**, 1–26 (2005)
43. M. Watve, R. Tickoo, M. Jog, B. Bhole, How many antibiotics are produced by the genus Streptomyces? Arch. Microbiol. **176**, 386–390 (2001)
44. M.B. Elowitz, A.J. Levine, E.D. Siggia, P.S. Swain, Stochastic gene expression in a single cell. Science **297**, 1183–1186 (2002)
45. J.Q. Boedicker, L. Li, T.R. Kline, R.F. Ismagilov, Detecting bacteria and determining their susceptibility to antibiotics by stochastic confinement in nanoliter droplets using plug-based microfluidics. Lab Chip **8**, 1265–1272 (2008)
46. E. Brouzes, M. Medkova, N. Savenelli et al., Droplet microfluidic technology for single-cell high-throughput screening. Proc. Natl. Acad. Sci. U.S.A. **106**, 14195–14200 (2009)
47. J. Cao, D. Kürsten, S. Schneider et al., Uncovering toxicological complexity by multidimensional screenings in microsegmented flow: modulation of antibiotic interference by nanoparticles. Lab Chip **12**, 474 (2012)

48. K. Churski, T.S. Kaminski, S. Jakiela et al., Rapid screening of antibiotic toxicity in an automated microdroplet system. Lab Chip **12**, 1629–1637 (2012)
49. T. Henkel, T. Bermig, M. Kielpinski et al., Chip modules for generation and manipulation of fluid segments for micro serial flow processes. Chem. Eng. J. **101**, 439–445 (2004)
50. T. Squires, S. Quake, Microfluidics: Fluid physics at the nanoliter scale. Rev. Mod. Phys. **77**, 977–1026 (2005).
51. I. Thampayak, N. Cheeptham, W. Pathom-Aree et al., Isolation and identification of biosurfactant producing actinomycetes from soil. Res. J. MicrobioL. **3**, 499–507 (2008)
52. B.J. Jin, Y.W. Kim, Y. Lee, J.Y. Yoo, Droplet merging in a straight microchannel using droplet size or viscosity difference. J. Micromech. Microeng. **20**, 035003-035003-10 (2010)
53. J.H. Zhang, T.D. Chung, K.R. Oldenburg, A simple statistical parameter for use in evaluation and validation of high throughput screening assays. J. Biomol. Screening **4**, 67–73 (1999)
54. J. Davies, G.B. Spiegelman, G. Yim, The world of subinhibitory antibiotic concentrations. Curr. Opin. Microbiol. **9**, 445–453 (2006)
55. D. Hill, B. Rose, A. Pajkos et al., Antibiotic susceptabilities of Pseudomonas aeruginosa isolates derived from patients with cystic fibrosis under aerobic, anaerobic, and biofilm conditions. J. Clin. Microbiol. **43**, 5085–5090 (2005)
56. B. Zheng, R.F. Ismagilov, A microfluidic approach for screening submicroliter volumes against multiple reagents by using preformed arrays of nanoliter plugs in a three-phase liquid/liquid/gas flow. Angewandte Chemie. (International ed) **44**, 2520–2523 (2005)
57. D.L.L. Chen, L. Li, S. Reyes et al., Using three-phase flow of immiscible liquids to prevent coalescence of droplets in microfluidic channels: Criteria to identify the third liquid and validation with protein crystallization. Langmuir **23**, 2255–2260 (2007)
58. S. Herminghaus, Dynamical instability of thin liquid films between conducting media. Phys. Rev. Lett. **83**,2359–2361 (1999)
59. K. Martin, T. Henkel, V. Baier et al., Generation of larger numbers of separated microbial populations by cultivation in segmented-flow microdevices. Lab Chip **3**, 202–207 (2003)
60. A. Grodrian, J. Metze, T. Henkel et al., Segmented flow generation by chip reactors for highly parallelized cell cultivation. Biosens. Bioelectron. **19**, 1421–1428 (2004)
61. N. Bremond, A. Thiam, J. Bibette, Decompressing emulsion droplets favors coalescence. Phys. Rev. Lett. (2008).
62. H. Hufnagel, A. Huebner, C. Gulch et al., An integrated cell culture lab on a chip: modular microdevices for cultivation of mammalian cells and delivery into microfluidic microdroplets. Lab Chip **9**, 1576–1582 (2009)
63. E.V. Moiseeva, A.A. Fletcher, C.K. Harnett, Thin-film electrode based droplet detection for microfluidic systems. Sensor Actuat. B- Chem. **155**, 408–414, (2010)
64. J. Sivasamy, Y.C. Chim, T.N. Wong et al., Reliable addition of reagents into microfluidic droplets. Microfluid. Nanofluid. **8**, 409–416 (2010)
65. H. Yang, X. Qiao, M.K. Bhattacharyya, L. Dong, Microfluidic droplet encapsulation of highly motile single zoospores for phenotypic screening of an antioomycete chemical. Biomicrofluidics **5**, 44103–4410311 (2011)
66. Y. Bai, X. He, D. Liu et al., A double droplet trap system for studying mass transport across a droplet-droplet interface. Lab Chip **10**, 1281–1285 (2010)
67. J.-C. Baret, Surfactants in droplet-based microfluidics. Lab Chip **12**, 422–433 (2012)
68. D.J. Holt, R.J. Payne, C. Abell, Synthesis of novel fluorous surfactants for microdroplet stabilisation in fluorous oil streams. J. Fluorine Chem. **131**, 398–407 (2010a)
69. D.J. Holt, R.J. Payne, W.Y. Chow, C. Abell, Fluorosurfactants for microdroplets: interfacial tension analysis. J. Colloid Interface Sci. **350**, 205–211 (2010b)
70. C. Holtze, A.C. Rowat, J.J. Agresti et al., Biocompatible surfactants for water-in-fluorocarbon emulsions. Lab Chip **8**, 1632–1639 (2008)
71. S.L. Anna, N. Bontoux, H.A. Stone, Formation of dispersions using "flow focusing" in microchannels. Appl. Phys. Lett. **82**, 364 (2003)
72. G.F. Christopher, S.L. Anna, Microfluidic methods for generating continuous droplet streams. J. Phys. D-Appl. Phys. **40**, R319–R336 (2007)

73. T. Thorsen, R.W. Roberts, F.H. Arnold, S.R. Quake, Dynamic pattern formation in a vesicle-generating microfluidic device. Phys. Rev. Lett. **86**, 4163–4166 (2001)
74. K. Liu, H.J. Ding, Y. Chen, X.Z. Zhao, Droplet-based synthetic method using microflow focusing and droplet fusion. Microfluid. Nanofluid. **3**, 239–243 (2007)
75. L. Mazutis, A.F. Araghi, O.J. Miller, et al. (2009a) Droplet-based microfluidic systems for high-throughput single DNA molecule isothermal amplification and analysis. Anal. Chem. **81**, 4813–4821
76. L. Mazutis, J.-C. Baret, A.D. Griffiths, A fast and efficient microfluidic system for highly selective one-to-one droplet fusion. Lab Chip **9**, 2665–2672 (2009b)
77. D. Pekin, Y. Skhiri, J.-C. Baret et al., Quantitative and sensitive detection of rare mutations using droplet-based microfluidics. Lab Chip **11**, 2156–2166 (2011)
78. A.R. Abate, T. Hung, P. Mary et al., High-throughput injection with microfluidics using picoinjectors. Proc. Natl. Acad. Sci. U.S.A **107**, 19163–19166 (2010)
79. J.S. Eow, M. Ghadiri, A.O. Sharif, T.J. Williams, Electrostatic enhancement of coalescence of water droplets in oil: a review of the current understanding. Chem. Eng. J. **84**, 173–192 (2001)
80. X. Niu, F. Gielen, A.J. deMello, J.B. Edel, Electro-coalescence of digitally controlled droplets. Anal. Chem. **81**, 7321–7325 (2009)
81. M. Zagnoni, J.M. Cooper, On-chip electrocoalescence of microdroplets as a function of voltage, frequency and droplet size. Lab Chip **9**, 2652–2658 (2009)
82. C. Priest, S. Herminghaus, R. Seemann, Controlled electrocoalescence in microfluidics: Targeting a single lamella. Appl. Phys. Lett. **89**, 134101-134101-3 (2006)
83. K. Ahn, J. Agresti, H. Chong, et al., Electrocoalescence of drops synchronized by size-dependent flow in microfluidic channels. Appl. Phys. Lett. **88**, 264105-264105-3 (2006a)
84. K. Ahn, C. Kerbage, T.P. Hunt, et al., Dielectrophoretic manipulation of drops for high-speed microfluidic sorting devices. Appl. Phys. Lett. **88**, 024104–024104 (2006b)
85. F. Guo, X.H. Ji, K. Liu et al., Droplet electric separator microfluidic device for cell sorting. Appl. Phys. Lett. **96**, 193701 (2010)
86. D.R. Link, E. Grasland-Mongrain, A. Duri et al., Electric control of droplets in microfluidic devices. Angewandte Chemie-Int. Ed. **45**, 2556–2560 (2006)
87. J. Clausell-Tormos, D. Lieber, J.C. Baret et al., Droplet-based microfluidic platforms for the encapsulation and screening of Mammalian cells and multicellular organisms. Chem. Biol. **15**, 427–437 (2008)
88. L. Boitard, D. Cottinet, C. Kleinschmitt et al., Monitoring single-cell bioenergetics via the coarsening of emulsion droplets. Proc. Natl. Acad. Sci. U.S.A. **109**, 7181–7186 (2012)
89. T.W. Hofmann, S. Hanselmann, J.W. Janiesch et al., Applying microdroplets as sensors for label-free detection of chemical reactions. Lab Chip **12**, 916–922 (2012)
90. F. Courtois, L.F. Olguin, G. Whyte et al., Controlling the retention of small molecules in emulsion microdroplets for use in cell-based assays. Anal. Chem. **81**, 3008–3016 (2009)
91. Y. Skhiri, P. Gruner, B. Semin et al., Dynamics of molecular transport by surfactants in emulsions. Soft Matter **8**, 10618 (2012)
92. A. Fallah-Araghi, J.-C. Baret, M. Ryckelynck, A.D. Griffiths, A completely in vitro ultrahigh-throughput droplet-based microfluidic screening system for protein engineering and directed evolution. Lab Chip **12**, 882–891 (2012)
93. H.N. Joensson, M. Uhlén, H.A. Svahn, Droplet size based separation by deterministic lateral displacement–separating droplets by cell-induced shrinking. Lab Chip **11**, 1305 (2011)
94. Y.-C. Tan, Y. Ho, A. Lee, Microfluidic sorting of droplets by size. Microfluid. Nanofluid. **4**, 343–348 (2008)
95. I. Groth, C. Rodríguez, B. Schütze et al., Five novel Kitasatospora species from soil: Kitasatospora arboriphila sp. nov., K. gansuensis sp. nov., K. nipponensis sp. nov., K. paranensis sp. nov. and K. terrestris sp. nov. Int. J. Syst. Evolut. Microbiol. **54**, 2121–2129 (2004). doi:10.1099/ijs.0.63070-0
96. A.L. Demain, Induction of microbial secondary metabolism. Int. Microbiol. **1**, 259–264 (1998)

97. I.A. MacNeil, C.L. Tiong, C. Minor et al., Expression and isolation of antimicrobial small molecules from soil DNA libraries. J. mol. microbiol. Biotechnol. **3**, 301–308 (2001)
98. M.D. McMahon, C. Guan, J. Handelsman, M.G. Thomas, Metagenomic analysis of Streptomyces lividans reveals host-dependent functional expression. Appl. Environ. Microbiol. **78**, 3622–3629 (2012)
99. R.H. Baltz, Antimicrobials from actinomycetes: back to the future. Microbe-Am. Soc. Microbiol. **2**, 125 (2007)
100. J. Park, A. Kerner, M.A. Burns, X.N. Lin, Microdroplet-enabled highly parallel co-cultivation of microbial communities. PLoS One **6**, e17019 (2011)
101. M. Srisa-Art, E.C. Dyson, A.J. deMello, J.B. Edel, Monitoring of real-time streptavidin-biotin binding kinetics using droplet microfluidics. Anal. Chem. **80**, 7063–7067 (2008)
102. A.C. Hatch, J.S. Fisher, A.R. Tovar, et al., 1-Million droplet array with wide-field fluorescence imaging for digital PCR. Lab Chip. **11**, 3838–3845 (2011).
103. B.K. McKenna, J.G. Evans, M.C. Cheung, D.J. Ehrlich, A parallel microfluidic flow cytometer for high-content screening. Nat. Methods **8**, 401–403 (2011)
104. O.J. Miller, A. El Harrak, T. Mangeat et al., High-resolution dose-response screening using droplet-based microfluidics. Proc. Natl. Acad. Sci. U.S.A. **109**, 378–383 (2012)
105. Z. Han, W. Li, Y. Huang, B. Zheng, Measuring rapid enzymatic kinetics by electrochemical method in droplet-based microfluidic devices with pneumatic valves. Anal. Chem. **81**, 5840–5845 (2009)
106. B.P. Cahill, R. Land, T. Nacke et al., Contactless sensing of the conductivity of aqueous droplets in segmented flow. Sens. Actuators B: Chem. **159**, 286–293 (2011)
107. K.R. Strehle, D. Cialla, P. Rösch et al., A reproducible surface-enhanced Raman spectroscopy approach. Online SERS measurements in a segmented microfluidic system. Anal. Chem. **79**, 1542–1547 (2007)
108. J.-C. Baret, O.J. Miller, V. Taly et al., Fluorescence-activated droplet sorting (FADS): efficient microfluidic cell sorting based on enzymatic activity. Lab Chip **9**, 1850–1858 (2009b)
109. N.R. Beer, E.K. Wheeler, L. Lee-Houghton et al., On-chip single-copy real-time reverse-transcription PCR in isolated picoliter droplets. Anal. Chem. **80**, 1854–1858 (2008)
110. M.M. Kiss, L. Ortoleva-Donnelly, N.R. Beer et al., High-throughput quantitative polymerase chain reaction in picoliter droplets. Anal. Chem. **80**, 8975–8981 (2008)
111. T. Lincke, S. Behnken, K. Ishida et al., Closthioamide: An unprecedented polythioamide antibiotic from the strictly anaerobic bacterium clostridium cellulolyticum. Angewandte Chemie Int. Ed. **49**, 2011–2013 (2010)
112. D.S. Tawfik, A.D. Griffiths, Man-made cell-like compartments for molecular evolution. Nat. Biotechnol. **16**, 652–656 (1998)
113. P.S. Dittrich, M. Jahnz, P. Schwille, A new embedded process for compartmentalized cell-free protein expression and on-line detection in microfluidic devices. Chem. Bio. Chem. **6**, 811–814 (2005)
114. L. Granieri, J.-C. Baret, A.D. Griffiths, C.A. Merten, High-throughput screening of enzymes by retroviral display using droplet-based microfluidics. Chem. Biol. **17**, 229–235 (2010)
115. B. El Debs, R. Utharala, I.V. Balyasnikova et al., Functional single-cell hybridoma screening using droplet-based microfluidics. Proc. Natl. Acad. of Sci. **109**, 11570–11575 (2012)
116. J.A. DeVito, J.A. Mills, V.G. Liu et al., An array of target specific screening strains for antibaterial disvcovery. Nat.Biotechnol. **20**, 473–483, (2002)
117. Y. Akselband, C. Cabral, T. P. Castor, et al., Enrichment of slow-growing marine microorganismsfrom mixed cultures using gel microdrop (GMD) growth assay and fluorescence-activated cellsorting. J. Exp. Mar. Biol. Ecol. **329**, 196–205 (2006)

Index

J. M. Köhler and B. P. Cahill (eds.), *Micro-Segmented Flow*,
Biological and Medical Physics, Biomedical Engineering,
DOI: 10.1007/978-3-642-38780-7,

The manufacturer's authorised representative in the EU is Springer Nature Customer Service Centre GmbH, Europaplatz 3, 69115 Heidelberg, Germany. If you have any concerns regarding our products, please contact ProductSafety@springernature.com

Printed and bound by CPI Group (UK) Ltd, Croydon, CR0 4YY
12/08/2026
02194522-0002